Constraints and Solutions for Energy and Electricity Development

Zhixuan Wang
Editor

# Constraints and Solutions for Energy and Electricity Development

中国电力出版社
CHINA ELECTRIC POWER PRESS

Springer

*Editor*
Zhixuan Wang
China Electricity Council
Beijing
China

ISBN 978-981-10-8302-0 ISBN 978-981-10-8304-4 (eBook)
https://doi.org/10.1007/978-981-10-8304-4

Jointly published with China Electric Power Press, Beijing, China

The print edition is not for sale in China Mainland. Customers from China Mainland please order the print book from: China Electric Power Press.

Library of Congress Control Number: 2018938355



This Springer imprint is published by the registered company Springer Nature Singapore Pte Ltd.
The registered company address is: 152 Beach Road, #21-01/04 Gateway East, Singapore 189721, Singapore

# Preface

The sustainable development of energy and electric power is the basis of sustainable development of national economy and energy, and electric power is related to economic development, social stability, and national security. At present, China is in the middle of industrialization, in the third step of the three-step development strategy of modernization, and in the rapid development stage of industrialization and urbanization which is characterized by a sharp rise in energy and electricity consumption demand, putting forward higher requirements for energy production and supply. In this book, energy development is defined as energy production and supply; from the perspective of meeting China's energy demands to analyze the major constraints on energy development and seek possible solutions. As electric power is the core of energy development based on coal in China, analysis of constraints on electric power and coal development as well as of the solutions is essential to energy development. This book focuses on the research on electric power and coal.

With the scientific concept of development as guidance, this book adheres to safe, clean, and conservation development of energy and electric power as its fundamental concept and provides scientific analysis of the various factors related to energy and electricity production and consumption. Through discriminant analysis, we identified resource endowment, environmental protection, water resources, and climate change adaptations are the four major factors constraining energy and electric power development. The constraint of natural resources mainly refers to the limitation of sustainable development of electric power due to incompatibility between the growing demands of energy and electric power and the supply of resources. The constraint of environmental protection mainly means energy development is limited due to incompatibility between the demands for social and economic development, and environmental pollution and ecological damage caused by unsustainable development of energy, especially unscientific development and utilization of coal. The constraint of water resources mainly means energy exploitation is restricted by the quantity of water resources and destruction of water resources as a result of energy development. The constraint of climate change adaptations mainly refers to energy development and related greenhouse gas emissions, energy structure adjustment and energy conservation, and emission reduction measures, which are subject to the political decisions and administrative requirements made by the Chinese government.

Based on the analysis of the medium- and long-term energy and electric power demand, this book also analyzes the present situation of the above-mentioned four constraint factors, the relationship between the constraint factors and the impacted targets, the changing trend of the constraint factors, correlations between constraint factors, and the resolutions based on technologies and economic policies. Through comprehensive analysis and system optimization, this book proposed programs and policy recommendations to overcome the constraints. To resolve the constraint of energy and resources, countermeasures must be a positive change in energy development model. Based on domestic resource endowments, following the objective law of energy and economic development and understanding globalization and market of the world with the trend of low-carbon energy, we should adhere to a comprehensive development strategy focusing on coal as the main body of energy, electricity as the center of development, and oil and gas along with continuous development of renewable energy. In order to solve the constraint of environmental protection, coal industry should vigorously reform the industrial structure and continue to take various effective measures to actively promote coal washing and processing and extend the industrial chain; the coal-fired power industry should effectively control air pollution at the regional level, by adjusting the layout and structure of power plants, strengthening management, and using advanced pollution control technologies and devices; the hydropower industry should focus on constraints of ecological and social environment in different development stages. For the constraint of water resources, this book suggests a comprehensive consideration of the layout of the thermal power to match regional water resource conditions during the "12th Five-Year Plan" and "13th Five-Year Plan" periods and in order to improve spatial adaptation of thermal power and water resources. In terms of the constraint of the climate change, it proposes that as the major source of China's carbon dioxide emissions, the electric power industry should accelerate the reformation of the electric power structure, vigorously develop non-fossil renewable energy, constantly improve coal power generation efficiency and clean utilization level, and accelerate the construction of legal and regulatory system for the market mechanism in favor of reducing greenhouse gas emissions.

From long-term and macroscopic perspective, this book will help clarify several problems which must be solved in China's sustainable energy development. First, China's sustained energy demand and growth will put great pressure on energy supply. Second, relative shortage of resources restricts the development of the energy industry. Third, the coal-dominated energy structure and poor energy efficiency have negative impacts on environmental protection. Fourth, the relatively outdated energy technologies limit the improvement of the energy supply capacity. Fifth, the changes in the international energy market pose a major threat to nation's energy security. To resolve the above-mentioned problems, energy structure reformation must be further accelerated. In 2012, coal, petroleum, natural gas, and renewable energy including hydropower and nuclear power accounted for 66.0, 18.4, 5.8, and 9.8% of total energy consumption, respectively. Compared with the rapid growth of total energy consumption, the reformation of energy consumption structure is quite slow. To meet the target of non-fossil energy consumption at 15% of total energy consumption by 2020, it is necessary to perform further studies on change in of the energy development model and improve the proportion of hydropower, wind power, solar power, and other renewable energy sources.

This book includes research methods of data analysis, regulatory policy review, analysis of constraint factors, and the prediction of major issues related to the nation's "12th Five-Year Plan." In the course of the study, first of all, we performed targeted data collection. On the one hand, we summarized and analyzed existing data collected by the author's team in industrial statistics and long-term archived database as well as the research results of the research group members in recent years; on the other hand, we gathered national or industrial authoritative reports as well as relevant research results in China and other foreign countries. Secondly, we conducted numbers of surveys in the form of establishing expert panels, organizing special surveys, and held relevant workshops. Finally, we used the method of comparative study to analyze China's energy development patterns. In addition, we learnt lessons from foreign experience, focused on the correlation of qualitative and quantitative analysis and the interactive coordination between this research and other related studies, and made this research comprehensively and innovatively.

Zhixuan Wang is in charge of the framework, general ideas, technical route design, and overall compilation of the book; Li Pan is responsible for overall organization and coordination; Jingjie Zhang takes charge of the organization of all chapters and system integration. The authors of all chapters involved in discussions throughout the book. Chapter 1 is written by Zhixuan Wang and Jingjie Zhang; Chap. 2 by Li Pan and Jingjie Zhang; Chap. 3 by Fan Yang, Jie Tan, and Xin Sui; Chap. 4 by Dongsheng Wang and Jing Yin; Chap. 5 by Fan Yang and Jie Tan; and Chap. 6 by Zhixuan Wang, Li Pan, Jingjie Zhang, and Peitang Guo. Wang Chenglong did the data checking. Xin Qiu made overall edit, quality control, and assurance for Chinese to English translation. We would like to express my heartfelt thanks to the authors of the relevant literature cited in this book.

Due to the limitation in time and capability, omissions and improprieties are inevitable. We sincerely invite readers to help us criticize and correct them.

Beijing, China
November 2015

Zhixuan Wang

# Contents

# Abstract

With the thriving of the energy revolution characterized by using renewable energy resources to generate green electric power and gradually replace traditional fossil fuels, this book studies and puts forward the method for analyzing the factors constraining energy development. It revolves around China's energy production and electric power conversion, focuses on four major constraints including resource endowment, environmental protection, water resources, and climate change, analyzes the energy development and medium- to long-term energy demand in the world and China, and constructs the index system of sustainable energy development and the target values of the planning in terms of developing energy and electric power in four dimensions: being safe, clean, resource-saving, and economical, and in two aspects: productivity and production relations. Based on the four major constraint factors, it studies and analyzes the relationship between constraint factors and the targets impacted by them, the changing trend of the constraint factors, correlation between constraint factors and technical and economic policies that can solve these constraints. It puts forward that we must actively change energy development model, rely on China's natural resources, follow the objective law of energy and economic development, grasp the trend of the world energy globalization, marketization, and low carbonization, and insist the electric power-focused strategy of developing oil, gas, and new energy resources in a comprehensive way. In addition, the authors suggest that the coal industry should vigorously adjust the industrial structure and continue to take various effective measures to actively promote coal washing and processing and extend the industrial chain; the coal-fired power generation industry should further improve power generation efficiency and reduce carbon emissions intensity, use the joint prevention and control policy to control air pollution in key areas, adjust power supply layout and structure, give full consideration to improving regional water resource conditions and spatial adaptation of power supply, and use advanced pollution control technologies to effectively control pollutants; the hydropower development should provide solutions to overcome constraints of ecological environment and constraints of social environment in different stages.

This book can be used as a reference for who are engaged in energy research, planning and environmental protection as well as a reference for college and university's undergraduate and graduate students.

# Chapter 1
# Energy and Development Constraint Factor Analysis

**Zhixuan Wang and Jingjie Zhang**

## 1.1 Proposed Research on Energy Development Constraint Factors

### *1.1.1 Systematic Energy Research Is the Requirement in an Era of Energy Transformation*

Studies show that energy utilization is the driving force and symbol of human civilization. From primitive civilization, agricultural civilization, and industrial civilization to the present ecological civilization, the change in energy utilization always marks the beginning of a new form of civilization. The use of firewood ushered in the primitive civilization and enabled mankind to enter the agricultural civilization under the guidance of the fire. The invention of the steam engine in the eighteenth century triggered the industrial revolution and productivity developed rapidly due to the extensive use of fossil fuels. In the nineteenth century, the use of electric energy greatly promoted the development of social economy and human civilization. Human society entered a new development stage marked by globalization which created a new era of industrial civilization symbolized by electricity. In the twenty-first century, mankind has begun to enter an era of knowledge economy. On the basis of the inheritance and innovation of agricultural civilization and industrial civilization, the rise of energy revolution characterized by rapid development of power generation with renewable energy resources marks the advent of ecological civilization.

Human beings have never stopped focusing on the study of energy-related issues. In the vast sea of the literature, there are many different perspectives on the issues of energy, based on the consensus of all mankind (such as the view of

Z. Wang (✉) · J. Zhang
China Electricity Council, Beijing, People's Republic of China
e-mail: wangzhixuan@cec.org.cn

 
Z. Wang (ed.), *Constraints and Solutions for Energy and Electricity Development*,
https://doi.org/10.1007/978-981-10-8304-4_1

sustainable development), and also based on differences in political, economic, cultural, technological, developmental, and information levels. Particularly, two masterpieces written in 2012 which have had a great impact on the energy field and even on the economic and political fields are worth reading.

One is *The Quest*, written by the prominent American scholar Daniel Yergin who specializes in global energy economy. The book revolves around three basic questions: will there be sufficient energy for the world developing rapidly? How to ensure the safety of the world's energy systems? What kind of impact will the concerns about the environment, especially, about the climate change, have on the future energy development? It mainly focuses on the impact on energy due to geopolitics, technological innovation, and the change in human daily life. The book describes the history and future trends of oil, natural gas, coal, and electric power development, describes the positions of different countries on the climate change, and elaborates on topics such as the rebirth of renewable energy sources, the reasons that clean energy source and efficiency of electric vehicles have become the top priority. It is mentioned in the book that, "It will be a challenge to meet the growing world energy demand. In the middle of the last decade, the 'Peak Oil Theory,' the popular viewpoint that the world oil will be exhausted soon, was widespread. However, with adding new oil supply by increase in investment and innovations, this fear faded away. The era of 'cheap oil' might be over, but the 'oil era' is not coming to an end." "Due to the huge scale and complexity of the energy supply system, it might take a long time to brew revolution. However, if the era of energy transformation really comes, the $6 trillion world energy market will stage a competition. That is to say, now the main body of the current energy supply, oil, natural gas, and coal will compete with the wind energy, solar energy, and biomass energy that have newly entered the world energy market and will try to gain a larger market share. If such a large-scale transformation really occurs, it will undoubtedly have a significant impact on emissions, the overall economy, geopolitics, and the status of the countries." As a result, the author is certain that the "oil era" is not coming to an end and reserves his viewpoint of large-scale energy transformation. However, what the author is certain of is "no matter what the future energy structure will be like, energy and energy challenges will be the theme of the future."

The another book is *The Third Industrial Revolution* written by the famous American economist Jeremy Rifkin, and its Chinese version was published in 2012 in China. It is pointed out in the book that, "By the 1980s, more and more signs showed that the industrial revolution driven by fossil fuels reached its peak and the human induced climate change is resulting in a huge global crisis." However, "in the USA, few people are concerned about the predictions of peak oil and ignore the warning of the consequences of the climate change and many signs show that our economy is not healthy." "No matter how hard it is, Europeans, at least, are trying to figure out how to deal with the fact that fossil fuels will soon be depleted. Unfortunately, most Americans still do not accept the fact that the fossil fuel era is nearly over. They are unwilling to admit the fact that the economic model we benefited from is hardly suitable for now." He points out that we are in the final stage of the second industrial revolution and the oil century. This is an unacceptable reality, as this reality will force humans to quickly establish a new energy system and industrial model. He suggests that the third industrial revolution will change the world and the Internet technology and renewable energy will be combined to create a strong foundation for the third industrial revolution. The transformation, distributed production, energy storage (in the form of hydrogen), distribution, zero emission transportation, and Internet distribution constitute the five pillars of the new economic model. In the new era, hundreds of millions of people will produce their own green energy at home, in offices, and at factories and share with others on the "energy Internet" in the same way we publish and share information on the Internet. He believes that we already have science and technology, planning and other conditions for the third industrial revolution. The 40 years of infrastructure construction in the third industrial revolution will create numerous new business opportunities and jobs. The completion of this project will mark the beginning of the new era characterized by cooperation, social network and industry experts and technical labor force.

It is clear that both authors who are concerned differently about China's energy issues have carried out relevant research and put forward suggestions. By creation of *The Quest*, Daniel Yergin particularly points out that China is the only country which occupies two chapters in this book. Jeremy Rifkin also pins great hopes on China. He points out, "The Chinese people need to be concerned about China's position in future 20 years. Will China be trapped in the second industrial revolution and continue to rely on fossil fuels and technologies, or actively get involved in the third industrial revolution and vigorously develop renewable energy technologies?" "If China chooses the third industrial revolution, China is likely to become the leading country in Asia and lead Asia into the next great economic era."

However, there is no doubt that China's energy development is one of the most important issues in China's economic and social development. China is facing and will face more severe energy challenges than any other big countries. China's energy problem is much more severe and complicated than what were mentioned by the two authors. When all other countries in the world argue how long oil will last, China is concerned about how long coal will last. When all other countries in the world discuss how to deal with climate change and step into the low-carbon era, China is still concerned about when smog and haze surrounding us will dissipate. Indeed, China is the largest wind turbine producer and the largest solar panel producer in the world. China boasts the richest solar energy resources, the most abundant wind resources and hydropower resources in the world. We also have considerable amount of biomass energy and geothermal energy resources. China's shale gas reserves may also be ranked the first in the world. Despite these advantages, it is difficult to answer the question of whether the rapidly developing China will have sufficient energy. If the answer is yes, what about the cost and what kind of technologies will be required? Similarly, it is difficult to answer the question of how to ensure the safety of China's energy system, how to protect the environment, how to deal with climate change, and what kind of impact the climate change adaptations will have on energy development. It is because we have done little in energy research. For example, we have not yet released the Energy Law; some basic concepts of energy development are inconsistent with relevant regulations, plans, and policies; environmental protection policies and energy policies conflict with each other; the development of different energy resources is imbalanced; energy development plans are separated from each other; the electric power planning has not been issued for years; the coal–electricity contradiction has existed for a long time. In addition, there are many problems such as uncoordinated development of power plants and power grids, imbalanced development of coal-fired power and hydropower, coexistence of lack of power supply and waste of redundant power supply, wind energy and hydroenergy curtailment, uncoordinated electricity pricing mechanisms, and a great variety of problems of nuclear power, hydropower, and renewable energy development. China must find out solutions for energy development based on China's national conditions so that China's energy development will follow the trend of world energy change and become the mainstay of China's economic and social development. Therefore, we must strengthen the systematic research on China's energy issues to meet the requirements of the new era of energy revolution.

### *1.1.2 Balancing of Energy, Economics, and Environment Is Still the Objective of Energy System Research*

Energy, economics, and environment have become evermore interrelated. Neither one exits without other two. Development in society, technology, and the natural environment, along with increase in material and spiritual demands, perceptions, and understanding of energy, economics, and environment have also undergone a persistent change. For example, the five pillars of the third industrial revolution itself embody an expanded understanding of energy. From an economic perspective, changes in regional economies will increasingly influence in a broader geographical regime, even influencing global economics. It impacts on economy quantitatively and qualitatively, including the improvement of quality such as security and stability. From an environmental perspective, pollution produced by traditional emission practices has raised environmental concerns, which have led to expanded biological protections, climate change adaptations, and pathways to make the life more comfortable. These changes make complicated relationships between energy, economics, and environment, and related research is constantly unable to keep up with the rapid shifts in trends and the need for complex solutions.

In regard to China's energy development issues, we must first use a scientific concept of development to establish a guide for our thinking. For instance, should the development of economics and environment be coordinated together, or should there be an environment-first approach? Should economic development be constrained by energy development restrictions, or should we seek to spur energy development in order to drive economic development? Should energy be developed with particular minimum limit for environmental protection, or should the environment be protected to the greatest extent possible? These questions may appear easy to answer at first glance, but by compiling past and current legislations and regulations and policies; by looking into emergent conflicts and incidents related to energy, economics, and environment; and by observing rapidly changing events in daily life, we are confused trying to answer these conceptual questions.

Systematic research on energy, economics, and environment has only existed for several decades. In energy and economics, a series of demand models utilized to study energy planning and measure energy needs appeared in the mid-1970s. In energy and environment, models emerged that could be used to measure the qualitative impact of pollution and emissions on air quality. These models primarily utilized two-dimensional systems for applications. In the 1980s, the emergence of three-dimensional systems appeared as a research framework for energy, economics, and environment; its optimal target progressed from a simple cost-minimization target toward multi-target decisions. In the 1990s, as the global movement to respond to climate change, climate models were improved at an accelerated pace; on April 12, 2014, the 39th meeting of the United Nation's Intergovernmental Panel on Climate Change (IPCC) reviewed and passed the Fifth Assessment Report and Policymakers Summary. Currently, modeling related to elements such as economics, energy, environment, resources, population, security, and energy efficiency is continually improved. Some of the models applied broadly include the MARKAL model, multi-target planning model, CGE model, and the input–output model. See Table 1.1: categorized table of several primary energy models and Table 1.2: comparative table of several primary energy models.

According to Table 1.2 and a large quantity of literatures, modeling is an effective tool commonly adopted to study the relations between energy, economy, and environment, which covers a full range of aspects related to energy. Modeling helps quantify influence factors and predict the trend of development. A good model features superiority in analyzing sensitivities of influence factors. However, there are certain restrictions on the use of models. Taking the top-down model as example, it analyzes how economic development affects other sectors from the macroeconomic perspective, i.e., how changes in macroeconomy affect the relation between energy supply and demand. But it describes technological application, resource production, and consumption in such an abstract manner that result in poorly identified resource consumption and underestimated the potential of technological advancement. On the other hand, the bottom-up model is built on energy engineering technology which facilitates accurate prediction of cost, efficacy, and trend. Though more reliable, it omits the feedback of macroeconomy and non-technology market. As for the combined model, which merges the advantages while offsetting the defects of top-down and bottom-up models, it makes model development and implementation more difficult. It is more difficult for China to adopt the combined model, as China is still in the initial stage of modeling research, and has not thoroughly understood the relationships between individual systems, which infiltrate and restrict each other.

Limitations or drawbacks are also found in the application of models. First of all, there has not been, or will never be, a universally applicable model that equalizes

**Table 1.1** Categorized table of several primary energy models

| Categorization method | Differentiated category | Typical representative | Main research questions | Time span |
|---|---|---|---|---|
| Based on research content | Energy–economics models | MACRO | Energy economics | Long term |
| | Energy–environmental models | AIM | Energy expenditure and energy environment | Long term |
| | Energy–economics–environmental models | 3Es model | Energy economics, environment, and policy | Long term |
| | Comprehensive models | IIASA-WEC E3 | Energy technology, economics, environment | Long term |
| Based on research methodology | Energy simulation models | POLES | Energy economics | Long term |
| | Energy optimization models | MESSAGE | Energy technology, economics, policy | Long term |
| | Energy equilibrium models | CGE | Energy economics, environment | Mid-term |
| | Energy input–output models | HERMES | Energy economics | Mid-term |
| Based on model capabilities | Energy supply models | PRIMES | Energy economics, environment, technology | Long term |
| | Energy demand models | MEDEE | Energy technology, economics | Long term |
| | Energy technology models | ERIS | Energy technology and energy generation | – |
| Based on research scope | Global energy models | IIASA-WECE3 | Energy technology, economics, environment | Long term |
| | Regional energy models | GEM-E3 | Energy economics, environment | Long term |
| | National energy models | NEMS | Energy economics, environment, policy | Mid-term |
| | Departmental energy models | LEAP | Energy economics, environment | Long term |
| Based on modeling methodology | Top-down models | CGE | Energy economics, environment | Mid-term |
| | Bottom-up models | MARKAL | Energy technology, environment | Long term |
| | Mixed energy models | NEMS | Energy economics, environment, policy | Mid-term |

*Data Source* Wei, Yiming, and others. "Progress in Modeling for Energy-Economy-Environment Complex System and Its Applications." *Chinese Journal of Management* 2005 (2): 160

**Table 1.2** Comparison of several major energy models

| Model | Theoretical basis | Indication | Pros and cons | Typical example | Developer |
|---|---|---|---|---|---|
| Top-down model | Econometrics, general equilibrium theory, and linear programming theory | Analyze energy from macroeconomic perspective, and develop energy policy and plans | Built upon method of economics, conducive to economic analysis; but cannot describe technology in detail; can reflect mainly technologies accepted by the market; can take a large amount of data for prediction; underestimate the potential of technological advancement; fail to control the influence of technology advancement on economy; determine energy demand by economic indicator but lay more emphasis on change of energy supply | CGE | Norway |
| | | | | 3Es-Model | NUT/Japan |
| | | | | MACRO | IIASA |
| | | | | GEM-E3 | NTUA/EU |
| Bottom-up model | Linear programming theory; non-linear programming theory; multi-objective programming; system dynamics and input–output method | Strategic research of energy technology selection; analysis of influence of energy technology on environment; energy supply and demand prediction; cost analysis of energy technology; and analysis of energy policies | Built upon method of engineering, not particularly applicable to economic analysis; describe technology in detail; properly reflect the potential of technology; overestimate the potential of technology advancement; describe delivery technology in detail with sporadic data, but emphasize changing energy consumption; calculate by averaging the cost of technology choice; assume the relations | MARKAL | ETSAP/IEA |
| | | | | MESSAGE | IIASA |
| | | | | EFOM | EU |
| | | | | MEDEE | IEPE/ France |
| | | | | ERIS | PIS, NTUA and IIASA |
| | | | | LFAP | SEI/Sweden |
| | | | | AIM | NIES/Japan |

(continued)

**Table 1.2** (continued)

| Model | Theoretical basis | Indication | Pros and cons | Typical example | Developer |
|---|---|---|---|---|---|
| | | | between energy and other authorities are neglectable | | |
| Mixed model | Linear programming theory; non-linear programming theory; mixed integer programming; and econometrics | Predict energy supply and demand; analyze energy policies; analyze energy-related environmental measures; and analyze how energy technology may evolve and its cost | Represent advantages of both models above; consider cost involved in technology selection and the role of price elasticity; serve as a simulation of the whole energy system for the sake of analysis; facilitate more in-depth analysis of energy economics; study global, regional, or national issue; boast full-range features and complex structure<br>A type of comprehensive and sophisticated system that represents and simulates the existing energy system | NEMS | EIA/DOE of America |
| | | | | IIASA-WEC E3 | IIASA and WEC |
| | | | | PRIMES | JOULE/EU |
| | | | | POLES | JOULE/EU |
| | | | | MIDAS | JOULE/EU |

values to different attributes existed in energy, economy, and environment. For example, there is no standard to evaluate the index among energy independence, cleanness, and economy. Secondly, limited by the researcher's capability and conditions of models, most of the studies focus on the relations between two particular systems but neglect the third one. As such, it is difficult to address various demands from different regions, stages, and levels. Thirdly, large amount of input data is selected by the user, varying from person to person. Taking the widely used MARKAL model as an example, which is built on mono-objective linear programming for energy system analysis and comprised by 4000–6000 variables and formulas. The output from this model depends largely on the subjective judgment of the user, and the user is allowed to obtain a desirable result by adjusting parameters and data selection within certain range, when the result is not consistent with user's own judgment.

In summary, as different models lead to different results from the same question, models appear more like elegant furnishings or are used by researchers as so-called scientific basis to support the conclusions. Though results predicted by model are reviewed, such review is superficial where the model is treated as a whole unbreakable "black box," due to constraints of time, condition, and professionalism of the reviewers. Such review rarely verifies the structure, parameters, and calculating process of the model. What's more, the fact that model is validated by researchers themselves makes credibility questionable. Therefore, models often serve as a reference or assistance in most of the scientific decision-making. But they may play a greater role in the future, when modeling and calculating techniques are improved, experience in data is accumulated, and various supporting studies are expanded and systemized. When China is facing complicated energy issues, and the dilemma in energy development across different stages, the arguments from principles to planning and measures, model application and analysis is not the key to open the solution door for China's energy issues. Furthermore, prerequisites of model application are to understand the crux of issue in China's energy sector, which is far from complete.

In conclusion, since the design and use of models are not enough for the research on energy, environment, and economy in China, it is not reasonable to totally rely on models in our research. What matters more is to thoroughly study the cruxes in energy, economy, and environment, guided by properly established principles. As such, the resolution of crucial issues can lead to resolve systematic issues.

### *1.1.3 The Emergence of "Obligatory Indicators" Opens a Door for the Systematic Research on Energy Issues*

In March 2006, *Outline of the 11th Five-Year Plan for National Economic and Social Development of the People's Republic of China (2006–2010)* was released. In the Outline, 23 major indicators of economic and social development in the 11th

Five-Year Plan period were identified and the attributes of indicators were defined for the first time; i.e., indicators were categorized into anticipated and obligatory. The eight obligatory indicators mainly focused on the two main categories of population, resources and environment, and public services and quality of life. The indicators mainly include control of total population, reduction in energy consumption per unit of GDP, reduction in water consumption per unit of industrial incremental value, total cultivated land, reduction in total major pollutant emissions, forest coverage, population ratio of basic pension in urban areas, and coverage of the New Rural Cooperative Healthcare System in rural area. In the Outline, the attribute of the two kinds of indicator was clearly explained. It was made clear that anticipated indicators were the expected national development goals which would mainly rely on the independent behavior of market players for realization. The government should create a favorable macroeconomic environment, institutional environment, and market environment; adjust the direction and strength of macroeconomic regulation and control; comprehensively use various policies to guide the allocation of social resources; and strive to achieve these goals. Obligatory indicators, based on the anticipated indicators, further identify and strengthen the government's responsibilities, and they are the requirements made by the Chinese central government for local governments as well as some central government departments in public services and other fields involving in public interests. The government should properly allocate public resources and effectively use administrative authority to ensure the realization of these goals. Once the "obligatory indicators" were defined in the Outline, they have become an important administrative measure in China's energy and environment management, well-known across the country.

Although the obligatory indicators have a very authoritative source and certain connotation, research on themselves is far from adequate, and there is a lack of systematic and theoretical research on economic, energy, and environment issues. In fact, obligatory indicators mainly serve as a tool to evaluate the performance of the government and officials, and people have no interest of why the obligatory indicators are selected, how to identify the values of the obligatory indicators, how to break them down into sub-indicators or what kind of impact on the balance of energy, economy, and environment once the indicator goals are achieved. Realistically, after the obligatory indicators are accomplished, problems needed to be solved still remain. For example, in the 11th Five-Year Plan period, the obligatory indicator of reduction of sulfur dioxide emission was achieved, but the smog and haze problem became more serious.

Nonetheless, the introduction of the obligatory indicators opens a door for us to focus on the "obligatory" aspects of energy development and economic development and to systematically resolve energy issues.

## 1.2 Theoretical Exploration and Application of Research on Constraint Factors of Energy Development

### *1.2.1 Constraints and the Theory of Constraints*

In some Chinese dictionaries such as Ciyuan and Cihai, constraint is defined as bondage, restraint, and restriction. It is also defined as the limitation of the surrounding objects on the position and motion of an object to be studied. In addition, in some Chinese classics, for example, Wenzi Shangyi, constraint also means discipline, rule, and regulation. We trace the definition source of constraint so that we can get inspirations from the etymological and historical perspective and have a better understanding of our research subject. From the above analysis, we know that when "energy development" becomes the research subject, we need to identify the "position" and "motion" of "energy development," "surrounding objects" related to energy development, and the factors constraining energy development. "Surrounding objects" include but are not limited to the requirements for economic development and environmental protection while "constraints" include but are not limited to natural resource endowment, technologies, laws and regulations and policies.

In essence, research on the coordination of energy, economics, and environment is research on relevant constraints. Although a large number of researches on economic development under the energy and environmental constraints are documented, there are few reports on the research on the theories of constraints in this field. Therefore, the theories of constraints adopted in this chapter are mainly those used for business administration.

Founded by the Israeli physicist and business management guru Eliyahu Moshe Goldratt (1948–2011), the Theory of Constraints[①] (TOC) is the theory of identifying and eliminating the constraints on the development of an enterprise. Aimed at the constraints on achieving the goal in production and operation of manufacturing industry, this theory analyzes the inherent laws of production and operation under various conditions, proposes scientific logical thinking modes to identify and eliminate the constraints, and provides effective measure to systematically solve problems in order to promote sustainable development.

According to TOC, no matter how complicated the system is, it has an inherent simplicity. The effective production of each enterprise is controlled by limited number of factors. The inherent simplicity of the system can be used to enable the enterprise to make breakthroughs in a short period of time.

The basic elements of TOC are: ① The enterprise is treated as a system, and the system goal is to obtain more profit at present and in the future. ② All the factors that prevent the enterprises from achieving its overall goal are constraints, including material type of constraints (such as material, market, capital) and non-material type of constraints (such as rules and regulations, the quality assurance system, corporate culture, and management), but there are only a few constraints that have significant impacts on the system. ③ First to identify the "bottlenecks" that prevent the production system from achieving its goal, then to prioritize and allocate resources based on the bottlenecks, and finally to prioritize other excess resources based on the prioritization of the bottleneck resources. ④ Giving up the traditional concept of accounting cost and suggest adapting the concept of throughput accounting as the standard to measure whether the enterprise can make a profit (i.e., throughput, inventory and operational expense). It is believed that the improvement of an enterprise can only be evaluated on whole, not partial of the system. For example, the investment in inventory and operational cost can be reduced, but cannot be reduced below zero. Only effective throughput can be increased continuously.

The core concept of TOC is that any system can be imaged to consist of interconnected and interdependent components affecting each other, just like an interlocking chain. In the cost concept, cost is of capacity in nature and subject to the law of addition; i.e., the overall performance is the sum of the individual performance. However, in the throughput concept, as the throughput is of "strength" in nature, it is not subject to the law of addition. Instead, the overall performance is not equal to the sum of the individual performance, and it is decided by the weakest link in the chain. Each system has its respective goal. In order to improve the efficiency of the system or reach higher system goals, the constraints on the performance of each system should be studied. For example, cost reduction might lead to failure or short-term profit as reducing cost usually will bring very limited profit to an enterprise but largely weaken its competitiveness. In fact, long-term profit maximization can be achieved by constantly improving the performance of the enterprise in each link based on the three-step operations suggested by TOC.

The TOC methods and continuous improvement processes consist of five key steps: (1) Identify the system's constraints. (2) Investigate and utilize all resources related to the system's constraints. (3) Coordinate other resources with Step 2. (4) Resolve the system's constraints. (5) Go back to Step 1 for continuous improvement.

TOC mainly adopts the thinking process analysis as its technical tool. The thinking processes are one of the main working methods of TOC. Thinking processes include three aspects, identifying key problems (what to change?), providing solutions (what to change to?), and making action plans (how to cause the change?). The thinking processes mainly have the following technical methods:

(1) Reality Tree. Reality tree is a tool used to analyze realistic problems, and it is divided into current reality tree (CRT) and future reality tree (FRT). Starting from the "root," the reality tree analysis extends to "trunk" and "leaves." The root represents the fundamental cause, trunk and branches are the intermediate results, and leaves are final results. For the CRT, leaves are the phenomena of problems and the root is the cause or the core problem; for the FRT, the root is the plan for solving the problem, while leaves are the final results.

    The CRT is used to answer key question at present. It usually starts with examples, i.e., starting with existing problems on the surface, drawing the logic relation diagram connecting surface problems to find the crux of the problem which is located at the root of the reality tree in the diagram.

The evaporating cloud (EC) method is used to find solutions to some problems and formulate a singular logical diagram in a win-win situation. Problems in a company are like clouds, and it's difficult to find which is the main cause of problems. EC is to eliminate the false and retain the true by cleaning clouds and mists and find solutions to the problems.

The FRT is used to find the resolutions. The FRT is generated by inserting ECs into the diagram of the CRT, replotting the logical diagram, and then generating the FRT.

(2) Trimming the Negative Branches. The thinking process is a wide opening and highly anticipated activity. It requires comprehensive and adequate communication with stakeholders, especially the participation of those stakeholders who know the negative branches well in order to prevent the negative branches and failure of implementation.

(3) Prerequisite Tree. A logic diagram related to the action plans is needed for implementing the changes by department or personnel. Prerequisite tree is the key link in the thinking processes and the characteristic that makes TOC unique among all thinking processes for continuous improvement.

(4) Transition Tree. It is the process to gather together all the activities needed for successful implementation, to clarify relationships between among these activities, and to prioritize these activities. This process is known as "transition tree".

Through the above steps, innovators can identify and improve the constraints. In general, bottlenecks can be categorized into three types: resources, markets, and regulations. Starting from the simplicity nature, finding and managing the key constraint (known as bottleneck factor), the organizational system can be significantly improved in a short period of time. As a result, the performance of the system depends on the constraints of the system but the optimal local of department indicators are not the optimal overall system's indicators.

### *1.2.2 Energy Development Constraint Research Framework*

By summarizing the experience and lessons of energy and electric power research and inspired by the TOC, we can build the basic framework of energy development constraint research.

1. **Energy Development Is the Subject of the Research While Constraint Is the Focus of the Research**

Because "energy development" is the subject of the research, in order to further clarify the research content, we are going to define some important concepts.

(1) Energy. Energy, also known as energy resources, refers to all substances existing in nature that have been proved or assumed to be able to be converted into thermal energy, electric energy, light energy, mechanical energy, and other forms of energy. In other words, energy refers to a great variety of resources which can be directly used, processed, or converted to various forms of useful energy including primary energy sources such as coal, crude oil, natural gas, hydropower, nuclear power, wind power, solar power, biomass, geothermal, and ocean energy, and secondary energy sources such as electric power, coke, coal, gas and petroleum products, and other forms of new energy and renewable energy.

(2) Energy development. Development refers to the changes of people or matter from small to large, from simple to complex, and from low level to high level. There are micro-, meso-, and macroscales of development. In this book, we mainly focus on the macroscale issues of energy, economic, and environments, but as required in the research on major constraints, we have to conduct meso- and even microresearch. Macroscopically, energy development is aimed to meet the demands for national economic and social development. The basic content of energy development was summarized in the *12th Five-Year Plan for Energy Development* released by the State Council on January 1, 2013. It was pointed out in the *12th Five-Year Plan for Energy Development* that the energy development plan identified the major goals for energy development in 2015 based on the overall judgment of the economic and social development trend in the "12th Five-Year Plan" period, in accordance with the overall requirements of *12th Five-Year Plan for National Economic and Social Development of the People's Republic of China* and taking into consideration safety, resources, environment, technology, economy, and other factors comprehensively. The indicators corresponding to the goals are as follows:

(1) Total energy consumption and efficiency: including total energy consumption, total electricity consumption, energy consumption per unit of GDP, comprehensive energy efficiency, standard coal consumption of thermal power generation, and comprehensive energy consumption of petroleum refinery.

(2) Energy production and supply capacity: including level of safety assurance, emergency response plan, development of energy production and supply capacity and level of dependence on foreign oil.

(3) Energy structure: including proportion of non-fossil energy consumption, proportion of installed power generation capacity of non-fossil energy, proportion of natural gas in the primary energy consumption, and proportion of coal consumption.

(4) Construction of national comprehensive energy base: including national energy base's primary energy production capacity and transport capacity.

(5) Ecological environment protection: including carbon dioxide emissions per unit of GDP, sulfur dioxide emissions per kilowatt hour (kWh) of coal-fired power generation, nitrogen oxide emissions kWh, emission intensity of particulate

matter (PM2.5) by energy development and utilization and coal mine land reclamation rate.

(6) Reform of energy system and mechanism: including reform of energy price marketization, energy taxation mechanism, energy regulatory policies and standards, and industrial sector management system.

It is noticed that the scope of energy development should include energy production and consumption; the extension should include production, efficiency, environmental protection, pricing, supporting service, legal, managing system, and safety mechanism.

Therefore, the connotation of energy development is the change from bad to excellent of physical conditions, technical support, supporting service, safety and security during the processes of production and consumption of various energy resources change in order to meet the demand of economic and social development. The research on the constraint factors of energy development is to seek the restrictive factors in the change from bad to excellent and to provide solutions.

2. **Identify the Comprehensive and Systematic Goals for Energy Development**

Energy system is a very complex system. Energy transformation should follow the basic laws of thermodynamics and other natural science as well as the basic laws of economics. In other words, energy transformation must be subject to constraints of quality attributes (such as usefulness, cleanliness, convenience) and economic attribute (price) of energy resources. The basic cause of all energy problems is actually the struggle of these two contradictory constraints. Various combinations of energy resources lead to a variety of energy production and utilization models; however, the purpose of energy research aims at the optimal methods of energy utilization.

For example, in order to resolve the problem of smog and haze pollution, Beijing Municipal Government issued Beijing Clean Air Action Plan for 2013–2017. According to the Action Plan, in 2017 total coal consumption in Beijing will be controlled within 10 million tonnes (t), reducing by 13 million t from 2012 and the proportion of coal-fired power in energy consumption will drop below 10%. However, it is more important to consider how to reduce coal consumption, which sector's coal consumption to be reduced, and what is the replacement energy in order to optimize the proportion of coal in energy consumption, because different approaches of coal reduction may result in different outcomes. The main measure taken by Beijing Municipal Government was to replace coal-fired thermal power with gas-fired thermal power, but a careful analysis of this action plan suggests it will result in more harm than good. First, in 2012 coal-fired thermal power generation consumed a total of 9.27 million t of raw coal (6.43 million t used for power generation and 2.84 million t used for heating), accounting for 41.2% of total coal consumption in Beijing. In a breakdown, Gaojing Power Plant, Gaobeidian Power Plant, First Thermal Power Plant of Guohua Electric Power Company, and Beijing Jingneng Thermal Power Company (hereinafter referred to as four power plants) consumed 9.13 million t of coal, accounting for 98.5% of total coal consumption in Beijing. Beijing can meet 70% reduction task of the coal consumption by cutting down the coal consumption of the four power plants as required by the Action Plan. However, in 2012 the emissions of the three air pollutants of sulfur dioxide, nitrogen oxides, smoke dust in Beijing were 93,800, 177,500, and 66,800 t, respectively. As the four power plants took the world leading pollutant control measures, the proportion of the emissions of the above three pollutants of the four power plants was only 2.1, 3.2, and 0.78%, respectively, while the combined emissions of the three pollutants accounted for only 2.4% of total emissions in Beijing. It is clear that reducing 40% of coal consumption for coal-fired power plants only results in less than 2.4% reduction of total emissions. Considering to add the emissions from replacement gas-fired power plants, the emission reduction effect will be even worse. Second, replacing coal with gas can only slightly improve the emission reduction effect and has no impact on the emissions of carbon dioxide. It is because as long as natural gas is lacking in China, replacing coal with gas will have a limited impact on reducing the emissions of carbon dioxide. It has nothing to do by regions. Third, replacing scattered coal burning with natural gas should have a better emission control effect than replacing coal consumption of coal-fired power plants with natural gas power. The surplus natural gas in Beijing should be used to replace scattered coal burning in surrounding areas because it can effectively improve the regional ambient environmental quality. Fourth, the cost of building new gas-fired thermal power plants is very high. If the above-mentioned coal-fired power plants are replaced with gas-fired thermal power plants, it will require a static investment of about RMB 17.37 billion and the price of electricity supplied by gas-fired thermal power plants will be about RMB 0.7/kWh, about RMB 0.2/kWh higher than the price associated with coal-fired power plants. Meanwhile, shutting down coal-fired thermal power plants will result in a significant waste of resources. Take one of the four power plants as an example. It has about RMB 900 million worth fixed assets. Since 2004, about RMB 1.7 billion investment was in place of in transforming heating, desulfurization, denitration, cloth bag dust removal facilities, and coal shed. If the power plant is replaced with gas-fired thermal power plant, all the investment and assets will be wasted and the cost of reducing emissions of pollutants will also increase significantly. The power plant emits about 2153 t of air pollutants (smoke dust, sulfur dioxide, nitrogen oxides) annually and a gas-fired thermal power plant is expected to emit about 710 t of pollutants. Annually, it will reduce the emissions of pollutants by 1443 t and the annual operation cost will increase by RMB 1.02 billion. The cost of reducing the emissions is about RMB 706/kg, which are hundreds of times higher than the average pollution control cost in China. Fifth, in terms of efficiency, as the four power plants use the cascade utilization of energy and total energy efficiency is about 80%, gas-fired thermal power plants will have no such advantage. Sixth, compared to coal, gas source and supply is less reliable, which will endanger the security of energy. Especially for Beijing which is a densely populated city with important political position, gas power plants will increase the potential energy supply risk. Seventh, it increases the risk of the power plant investors, especially production and benefit risk resulted from unstable gas supply and the risk of profit loss due to high gas prices. Eighth, shutting down coal-fired power plants will seriously affect employment. The four power plants have thousands of employees, while the new gas turbine project can provide less than 1/10 of job positions. As a result, the majority of employees will face the issue of job transfer and resettlement.

Another example is that at the end of 2010 in order to complete the energy conservation and emission reduction task, a few regions restricted the normal industrial production, ignored reasonable residential electricity demand, and compulsorily stopped thermal power generating units, which greatly increased power generation by diesel generating units and aggravated severe environmental pollution. Learning from past experience, the contradiction between natural gas supply and demand and the problems resulted from gas instead of coal actually demonstrates that China has failed to optimize the energy structure, especially coal-fired power generation. If simply phasing out coal without optimizing the energy structure, problems such as serious imbalance of natural gas use will continue to occur at a huge economic cost. Furthermore, it will hardly improve environment and even lead to social instability.

Examples above suggest that we should not make a judgment based on single energy issue; instead, we should look into whole system and evaluate if this is the best. Therefore, energy planning must be a comprehensive plan. We should not make individual electric power plan, oil plan, and renewable energy plan outside the consolidated energy planning framework. We cannot simply add all individual plans to construct energy development plans. The plans for all energy types must be optimized under a consolidated framework, and the optimal results are based on overall energy development plan instead of individual energy sectors. The disordered development of China's renewable energy such as wind energy curtailment, hydroenergy curtailment, and disconnection between power supply and power grid is due to the fact that all regions fail to set consolidated goals.

Based on demands of energy development made by economic and social development and referred to domestic and foreign energy development process and experience, this book identifies the overall goal of energy development in four aspects, safety, clean, conservation, and economy.

(1) Safety. Energy safety is the foundation of national economic and social development. Its connotation should include independence, safe, reliable and stable energy supply. China takes peaceful development as one of its fundamental policies, and energy independence might become a compulsory constraint. Therefore, in the future development, China needs an energy system based on energy independence to extensively strengthen international energy cooperation and lay a solid foundation for national development.
(2) Clean. Environmental protection is one of China's fundamental policies. Pollution treatment, environment protection, alleviating the ecological pressure, and adapting to the global climate change under the new situation are the directions of clean energy development, which also include important aspects

such as level of clean energy production, pollutant emissions, greenhouse gas emissions, impact on ecological system.

(3) Conservation. Resource conservation is one of the China's fundamental policies. Energy development should adhere to resource conservation and development for a long run and resource conservation should be put the first priority. Improving the comprehensive energy utilization efficiency and reducing energy and resource consumption constitute the main content of energy conservation and development.
(4) Economy. Energy development should reduce cost from aspects of development, usage, production, and consumption. That means we should use advanced technologies and appropriate approaches to develop and produce energy by effectively managing and consuming energy.

3. **Simplify Complicated Issues**

Between the end of the nineteenth century and the beginning of the twentieth century, Italian economist Vilfredo Pareto found that a matter of a few ($\sim$20%) is vital and the most (80%) are trivial, which is known as Pareto's Principle or the 80/20 Rule. If we apply Pareto's Principle to the TOC, we can find that no matter how complicated a system is, it has its inherent simplicity within complicated problems; thus, the problems can be simplified. Simplification does not mean intensification or naturalization. In the process of simplification, under the guidance of the system engineering methodology, we should simplify the overall energy system and have an in-depth understanding of all components of the energy system. The core of simplification is to find out the essential factors that affect the development of matters and eliminate minor factors based on theories and practical experience. However, we should avoid dividing a complicated problem into simple problems independently from each other because it will destroy the integrity of the whole system and lead to the optimization of local parts but loss of the optimization of the whole advantages.

Energy development has many constraints. Based on the principle of comprehensive, systematic, and simplifying complicated problems, after conducting a large number of researches and consulting experts, the authors collect, summarize, and analyze the factors constraining energy development of coal, petroleum, natural gas, nuclear energy, hydroenergy, wind energy, solar energy, biomass energy, and electric power by classifying different constraining factors, as summarized in Table 1.3.

Table 1.3 indicates that although energy problems are very complicated, we can find out major constraint factors by analyzing the characteristics of production, transportation, and utilization of different energy forms and finally resolve the problems.

The key to simplify a complicated problem is not to focus too much on the surface and ignore the nature of the problem. For example, from the point of view of the atmospheric pollution dispersion theory, the impact from a point source on the air quality at a certain point on the ground not only depends on the emission rate

**Table 1.3** Energy development constraining factor set

| Major energy industry | Constraining category | Constraint factor |
|---|---|---|
| Coal | Safe production | Safety protection investment, construction of safety facilities, annual accident rate, mortality per million tons of coal |
| | Policies, laws, and regulations | Industrial policy, development planning, industry standards, exit mechanism, industry access, coal reserve system |
| | Ecological environment protection | Pulverized coal dust, coal gangue, fly ash, coal mud, mine gas, destruction of water resources, ecological environment bearing force, surface subsidence, destruction of vegetation, etc. |
| | Technical equipment | Exploration technology, mining technology, transport capacity, clean coal technology, coal chemical technology, resource recycling technology, resource comprehensive utilization technology, energy- and water-saving technology, coal seam gas coordinated development technology, ecological restoration, and compensation technology |
| | Mechanism system | Industrial centerization, industrial distribution, industrial structure, industrial organization |
| | Energy resources | Coal reserves, distribution of coal resources, mining environment, supply and demand balance, external dependence, water resource supply capacity, mine water pollution, groundwater level decline, groundwater depletion, drainage of groundwater, groundwater pollution |
| | Economic efficiency | Domestic and international coal price, coal trading market system |
| Petroleum | Energy resources | Petroleum reserves, distribution of oil resources, exploitation environment, supply and demand balance, external dependence |
| | Laws and regulations | Industrial policy, development planning, industry standards, petroleum reserve regulations |
| | Technical equipment | Exploration technology, development technology, transport capacity, oil refining technology, petroleum and chemical technology |
| | Ecological environment protection | Oil spill and leakage accident, carbon and oxygen compounds |
| | Economic efficiency | Domestic and international oil price, oil price formation mechanism |

(continued)

**Table 1.3** (continued)

| Major energy industry | Constraining category | Constraint factor |
|---|---|---|
| Natural gas (including shale gas, coalbed gas, tight gas) | Energy resources | Natural gas reserves, natural gas distribution, development environment, supply and demand situation, external dependence |
| | Technical equipment | Exploration technology, development technology, transport capacity, utilization efficiency, infrastructure construction |
| | Mechanism system | Management system, market mechanism |
| | Ecological environment protection | Carbon dioxide, nitrogen oxides, hydrogen sulfide, ammonia |
| | Economic efficiency | Domestic and international natural gas price |
| Nuclear energy | Nuclear safety | Nuclear power plant siting, nuclear waste disposal, nuclear safety supervision |
| | Energy resources | Nuclear resource reserves, distribution of nuclear resources |
| | Technical equipment | Technical personnel quality, equipment localization level |
| Hydroenergy | Energy resources | Water resource distribution, water abundance and shortage periods |
| | Resettlement of affected residents | Resettlement compensation |
| | Ecological environment protection | Submergence, river truncation, fish migration, biodiversity |
| | Technical equipment | Operation management of hydropower station, small hydropower technology, hydropower grid connection |
| | Regulatory policy | Hydropower development planning, river basin comprehensive planning, hydropower price |
| | Economic efficiency | Construction cost of large- and medium-sized hydropower projects, economic relationship between power stations and local governments, property rights of small hydropower stations |
| Wind energy | Energy resources | Wind energy resource distribution, wind energy continuity difference |
| | Technical equipment | Wind power grid connection issues, equipment localization level |
| | Policies, laws, and regulations | Capital policy, fiscal and taxation policy, wind power industry service system |

(continued)

**Table 1.3** (continued)

| Major energy industry | Constraining category | Constraint factor |
|---|---|---|
| Solar energy | Energy resources | Distribution of solar energy resources, poor stability of solar energy |
| | Technical equipment | Photoelectric power grid connection issues, equipment localization level |
| | Policies, laws, and regulations | Industrial policy, industrial planning, industry standards |
| | Economic efficiency | Construction cost, operating cost |
| Biomass energy | Energy resources | Dispersion and stability of biological resources |
| | Technical equipment | Low utilization rate, technical equipment level |
| | Regulatory policy | Industrial development plan, capital policy, fiscal and taxation policy |
| Thermal power | Fuel supply | Fuel price, coal and electricity utilization coordination |
| | Safe production | Safety in construction period, production and operation safety, accident emergency management |
| | Ecological environment protection | Sulfur dioxide, nitrogen oxides, carbon dioxide, smoke, heavy metals, fly ash, desulfurization gypsum |
| | Regulatory policy | Electric power development planning, electric power environmental protection planning, energy conservation and emission reduction policy, power structure adjustment, shutting down of small thermal power stations |
| | Technical equipment | Power generation technology, pollutant control technology, supporting power grid technology |
| | Economic efficiency | Electric power investment and financing, coal-fired power price coordination |
| | Energy resources | Regional water supply capacity, surface water pollution, groundwater pollution |

(amount of emissions per unit time), but also relates to transport, dispersion, and transformation of pollutants including wind direction, wind speed, height of stack, topography and land use and land cover, atmospheric stability, and temperature stratification (such as inversion layer). If the transformation from primary pollutants to secondary pollutants is taken into consideration, additional factors such as atmospheric physical and chemical reactions and precipitation (which must be taken into account for secondary PM 2.5 formation) need to be considered. Under normal circumstances, by emitting the same amount of air pollutants, higher stacks, windy conditions, organized emissions are always less environmental impact than lower stacks, lower wind speeds, and scattered emissions. However, under special weather conditions, the situation will become much more complicated. The impact of multiple emission sources on environment quality to a given place is even more complicated. Air quality must be predicted by inputting various certain conditions and using proper air quality models. Clearly, same amount of pollutant emissions may result in different impacts on environment quality depending on different emission source distributions and emission types. Therefore, environment quality can be improved by the adjustment of source structure and layout. To improve environment quality by reducing total amount of pollutant emissions can only be effective when the emission layout, structure, and type remain unchanged. For example, if a car in Beijing increases particulate matter emissions by 1 kg while a power plant in Inner Mongolia reduces particulate matter emissions by 2 kg, and total pollutant emissions nationwide are reduced by 1 kg. However, environment quality in populated areas (Beijing) becomes worse instead of improving because Inner Mongolia is too far away from Beijing. As a result, we cannot resolve environment quality problems simply by reducing total emissions across the country. In the "Ninth Five-Year Plan" period, Chinese government already released policies to control total air emissions, and correspondent government agencies began to demarcate the acid rain control zone and the sulfur dioxide pollution control zones (referred to briefly as the "two control zones") and carried out total sulfur dioxide pollution control in the two control zones, which was approved in the *Law on Prevention and Control of Atmospheric Pollution* revised in 2000 and implemented in the "Tenth Five-Year Plan" period (e.g., in 2002 the Approval of the State Council Concerning Acid Rain and Sulfur Dioxide Pollution Control in the two control zones during the "Tenth Five-Year Plan" Period was issued). In the "11th Five-Year Plan" period, the Chinese government released the obligatory goals for control of total emissions of sulfur dioxide and COD, and in the "12th Five-Year Plan" period, the obligatory goals were updated by adding control of total emissions of nitrogen oxides and ammonia nitrogen. In the past 20 years, China has adopted the pollution control strategy focusing on the control of total emissions of selected key pollutants. Although this strategy raised the whole society's awareness of environmental protection and greatly promoted the control of air pollutant emissions in the electric power industry, the worsening smog and haze indicates the strategy is not successful or at least it has a significant room for improvement.

The impacts on environment quality, especially the formation of smog and haze, are combined effects from multiple emission sources and multiple pollutants. Focusing only on the control of "major pollutants" might lead to the negligence of the control of other pollutants. For example, during the "11th Five-Year Plan" period only total emissions of sulfur dioxide were assessed, and majority labor of environmental protection administration was assigned to assess $SO_2$ control, left unmanaged emissions of nitrogen oxides and smoke dust.

It looks right on the surface to vigorously develop renewable energy, optimize energy-saving scheduling, phase out small thermal power plants, and gradually tighten the emission standards for thermal power plants. However, failing to plan overall air pollution control might lead to wind energy curtailment, hydroenergy curtailment, the locking out effect of thermal power units, largely loss of coal-fired generating units and coexistence of lack of power supply and waste of redundant power supply. All these problems have appeared repeatedly in the course of China's energy development.

Therefore, the simplification process is to grasp the essence of the problem instead of the phenomenon. It is very important to use Pareto's Principle to analyze problems in energy research.

4. **Further Analysis of Main Research Objects and Constraints of China's Energy Development**

If we treat all sectors and aspects in the energy system as a chain, each part of the system is one of the links of the chain. In order to achieve the desired goal, we must start to fix the problem at the weakest link, namely the bottleneck link of the chain, to significantly improve the whole system. If the bottleneck link decides the speed of realizing the energy goal, we must fix the bottleneck link in order to significantly improve the effective output (e.g., solving the problem of smog and haze) of the system in a short period of time. Based on the analysis above, the factors measuring the strength of the chain or the standards for evaluating the quality of the chain are "safety, clean, resource conservation and economy." However, factors affecting the strength of each chain link vary from each other. Therefore, we must analyze and find out the weak links in whole chain and the key constraint factors (bottleneck factors) of the weak links. In other words, based on the constraint factors of the development of various energy forms, we must further analyze the impact of various energy forms on China's energy "chain" and identify major energy research objects and constraint factors.

First of all, we should analyze the impact of the development of various energy forms on the whole energy and environment systems.

(1) Impact of oil and natural gas development. In recent years, the establishment and implementation of the US Energy Independence Strategy has made an important impact on global political and economic pattern. China must enhance the level of energy independence, especially to set up the political goal for the dependence on import of oil and natural gas. This issue to a large extent depends on political decisions, and China has made the energy independence strategy based on China's natural resources. In addition, the development of natural gas (including shale gas, tight gas, and coalbed gas) is impacted by reserves, mining conditions, and economic conditions, which we will not make an in-depth analysis in the book.
(2) Impact of coal development. Coal consumption has long accounted for about 70% of China's total energy consumption. Soot pollution has been one of the major causes of China's air pollution. As coal is high carbon energy, coal mining is closely related to ecological protection and water resources. Meanwhile, coal is the cornerstone of China's energy and one of the important

foundations of China's social and economic development. Therefore, coal and coal-fired power generation problems have been and will be the bottlenecks of China's energy development. In this book, we will focus on these problems. Clean utilization of coal is an inevitable choice. Through modern coal mining technology, ecological protection and recovery in mining areas, and clean coal technology, we should be able to keep ecological impact of coal mining and emissions of conventional pollutants produced in the coal utilization under control safely and economically. At the present stage, the emissions of carbon dioxide produced in the coal utilization process can be reduced by improving the coal utilization rate, and in the process of replacing coal and other fossil energy sources with renewable energy forms, CCUS and other technologies can be used to reduce total carbon emissions.

In the next few decades, coal power will remain one of the energy foundations with a wide range optimal configuration and it is the core support center of clean energy and circular economy. Without the strong support and adjustment on electrical safety and stability from coal power, it won't be possible to have a large application of clean energy power generation in China.

From the perspective of the whole energy chain, the coal problem is resulted by the sharp contradiction between greenhouse gas emission control demands and energy independence requirements. From the perspective of simplifying complicated problems, the coal development problem mainly includes coal-fired power generation problems and coal mining and supply problems, in essence while coal-fired power generation problems are the main problems. Technically and economically, the emissions of conventional pollutants from coal-fired power plants will not become the bottleneck constraints of coal power development.

(3) Impact of electric power industrial development. Energy electrification and terminal energy consumption electrification are the inevitable trends of historical and social development, and electric power is an important factor to resolve China's smog and haze problem. The development of the electric power industry plays a key role in China's energy development because resolving the bottleneck constraints of electric power development will fix China's energy development problems.

From the history, a breakthrough theory of electricity started in the eighteenth century. In 1799, Italian physicist A. Volta developed battery; in 1831, M. Faraday found the electromagnetic induction principle, laying a foundation for theory of electric motor; in 1866, W. Siemens invented the excitation motor; in 1875, the world's first coal-fired power plant was built in Paris North Railway Station, France; in 1876, an American named Bell invented the telephone; in 1879, T. Edison invented the light bulb; in 1882, Edison built the world's first formal thermal power plant–Pearl Street Light Bulb Factory in New York, completing the technical system of the electric power industry.

From the 1930s onwards, the application of electric power was rapidly expanded from lighting and telegrams to providing power for all industrial sectors and then penetrating into various fields of national economy and social life. Since the 1960s due to the development and broad applications of ultra-large-scale integrated circuit element, microprocessor, and optical fiber communication technology, the application of computer technology, and the development of communication technology, the electric power industry has undergone great changes. Thanks to the automation system, load monitoring and grid scheduling automation system of power plants and substations as well as the development, application and improvement of the management information system, electric power production has become more economical, efficient, secure, and reliable. Since the 1980s, the electric power industry has entered the new stage of automation and informatization and then further toward intelligent.

As the utilization of electric power not only improves the efficiency of energy usage, but also leads to the utilization of wind energy, solar energy, and other renewable energy sources that could not be used for social and economic development in the past. In addition, electric power has become the irreplaceable energy form for some emerging industries such as information system. The growth rate of electric power is higher than that of primary energy, which is fully reflected in the percentage of electric power usage and the proportion of electric power in terminal energy consumption.

In terms of the role of electric power, the electric power industry is the energy production and transformation industry, and the basic industry of the national economy as well as a resource-intensive industry. From the perspective of power generation mode, there are hydropower generation, nuclear power generation, coal-fired power generation, oil-fired power generation, natural gas power generation, and power generation of renewable energy sources such as wind energy, biomass energy, solar energy, geothermal energy, and tidal energy. Electric power is mainly supplied to households through transmission and distribution network. At present, global thermal power generation (coal-fired power generation, natural gas power generation, and oil-fired power generation) is still the major power generation mode and the main form of consuming fossil energy around the world. In addition, from the theories of sustainable development and a large number of application point of view, electric energy as the cleanest secondary energy form plays a key role in the balance among energy, economy, and environment. The proportion of primary energy converted into electric energy and the proportion of electric energy in the terminal energy consumption have become an important and comprehensive indicator to measure the social and economic development level, energy utilization efficiency, the economic efficiency, environmental protection level, the state of material, and spiritual civilization of the country. Therefore, the electric power industry plays a fundamental, critical, and long-term role in energy development.

In terms of the characteristics of electric power, electric power is a convenient, efficient, and high-quality energy form. Electric energy can be divided into infinite parts and conveniently converted into light, thermal, and mechanical energy. In addition, it can be controlled easily and used conveniently, randomly, timely, quantitatively, and at a fixed point. The utilization efficiency of electric energy in terminal use is significantly higher than other energy forms. Non-fossil energy forms such as hydroenergy, wind energy, and nuclear energy cannot be widely utilized until converted into electric energy. Therefore, it is an inevitable trend to convert new energy forms into electric energy for utilization. However, in the short term, it is impossible to use renewable energy to replace coal as the basic energy form in China.

In terms of the constraints of electric power development, the mandate of greenhouse gas emissions control constitutes the hard constraint on coal power development. When China's greenhouse gas emission reduction goal shifted from relative reduction to absolute and relative reductions coexisting, it created the hard constraint on coal power development. Meanwhile, energy independence will also become a hard constraint. China becomes the world's biggest energy and economic country following the principle of peaceful development. In the future development, China will need a flexible energy system based on energy independence and broad international energy cooperation as the solid foundation of national development. The progress of energy independence decides the coal development strategy. China's shale gas development is a key uncertain constraint factor of coal power development. Although China's shale gas development does not constitute the hard constraint of coal power development, it directly affects the scale, layout, function, and utilization rate of coal power. Shale gas development is mainly controlled by resource exploitation characteristics, technical and economic efficiency, environmental protection requirements, and water resource constraints.

At present, as people are generally aware of the existence of "smog and haze," smog and haze pollution imposes a great constraint on China's electric power development, which should be analyzed scientifically. The characteristics of smog and haze in China are: ① Smog and haze is worsening due to a combined effect of motor vehicle emissions, emissions from coal combustion, industrial pollution, urbanization, especially large-scale construction pollution of rapid expansion of large cities, agricultural production pollution, rural pollution, and urban living pollution. However from a macroscopic perspective, motor vehicle emissions and emissions from coal combustion are the main causes of smog and haze. ② In China about 50% of 4 billion tons of raw coal is consumed annually by electric power generation. Due to the emission reduction year after year from power generation coupled with the characteristics of power supply layout and its emission methods, the impact of coal power emissions has reduced year by year. The reality is that emissions from 800 million tons per year of scattered coal combustions, plus emissions from a large number coal-fired industrial combustions with low pollution control levels result in increasing environmental impacts. ③ Coal-fired cogeneration power plant is the best solution to improve environmental quality and ensure urban heating supply under the supply shortage of natural gas. In the last few decades, urban coal-fired cogeneration power was the hero of urban environmental improvement by replacing tens of thousands of small boilers with an increase in coal-fired cogeneration power ratio. Coal-fired cogeneration is not the cause of worsening smog and haze level. Blaming urban coal-fired cogeneration as the source of smog and haze is to put the boot on the wrong leg, which will result in wrong decisions and put off the pollution control strategy. ④ It is an inevitable trend and a strategic choice that clean and renewable energy sources such as wind energy and solar energy will replace traditional fossil energy forms, which, however, cannot be realized in the short term. So at present, it is impossible to resolve the smog and haze problem through the development of renewable energy. ⑤ To replace coal with natural gas will be an effective measure to control smog and haze, but natural gas (including coal gasification, tight gas, coalbed gas, and shale gas) will be short in supply in the next 10 years or a longer time, and the price of natural gas is much higher than coal. It remains uncertain whether the effect of the shale gas revolution on the energy structure and economy in the USA can be replicated in China. Even if the shale gas revolution in the USA was very successful and the proportion of coal power generating capacity was reduced, in 2013 the proportion of coal-fired power generating capacity in total power generating capacity was still high at 39%. ⑥ Adhering to the insist principle of coordinated development of energy, environment, and economics, we should achieve economic development and environmental protection at the lowest energy cost. Therefore, the key to resolve China's smog and haze problem is to increase the proportion of coal-to-electric power conversion and the weighting of electric power in the terminal energy consumption by converting scattered coal and oil into electric power.

Therefore, power development constraint, especially the constraint of converting scattered coal consumption into coal power generation, determines the constraints of energy development to a large extent.

(4) Impact of water resources and hydroenergy development. Water resources and hydroenergy resources are an important constraint factor of energy development. Water resources are closely related to oil, shale gas, and coal mining development. On the one hand, oil, shale gas, and coal mining have a great impact on or great damage to water resources; on the other hand, they require a lot of water resources. Thermal power generation is closely related to water resources too mainly because it needs a large amount of water in the power generation process. Modern thermal power plants greatly reduce water consumption per unit power generation through the water-saving technology and can further reduce water consumption through the air cooling technology (air cooling tower). As the air cooling technology will reduce power generating efficiency and increase fuel consumption, to some extent, using the air cooling technology is a trade-off by water resources. But generally speaking, water resources will not become the bottleneck of thermal power generation technology.

China boasts one of countries with the world's richest water resources. From the perspective of energy development, water resources are mainly used for hydropower generation and China has the largest hydropower generating capacity. Hydropower is a kind of clean energy, but the construction and operation of different hydropower projects will have different impacts on ecological environment and affected residents, which is also the main reason for different disputes about hydropower in different periods of time. In China, the coal-dominated energy structure is under a great pressure of climate change adaptation demands. Promoting hydropower and nuclear power is an important measure to increase the proportion of non-fossil energy. On the one hand, it is supported by regulations and policies; on the other hand, it is restricted by relevant environmental protection and ecological regulations. China has made great achievements in hydropower development. Although there is still a large room to improve, the development cost has become increasingly high. In addition, with the influence of natural factors, the economic contradiction caused by the complementarity between hydropower and coal power has become increasingly sharp. The constraints of water resources and hydroenergy resources are an important aspect of energy constraints.

(5) Impact of renewable energy, nuclear power, and new energy development. It is an inevitable trend that renewable energy and clean energy will replace traditional fossil-fueled energy. However, in spite of the rapid development of renewable energy and clean energy globally in recent years which seems irresistible, in reality there are many constraint factors and the development is difficult and tortuous. As the main purpose of this book is to explore the solutions to energy development based on the main contradiction in China's energy development, we will not focus on the research of renewable energy, nuclear power, and new energy development.

From another perspective, if all four of the problems mentioned above are resolved, the problem of renewable energy, nuclear power, and new energy development will also be resolved. Moreover, although nuclear power development is mainly determined by political factors or non-technical factors, renewable energy and new energy development does not have any restrictive factors in essence except technical and economic factors. As long as through technical innovations, the economic competitiveness of renewable energy and new energy should approach to the level of fossil energy by taking low-carbon development policies into account, renewable energy and new energy development will not be restricted by any bottleneck factor. Certainly, the constraints of developing renewable energy are the constraints of developing fossil energy.

Secondly, we should analyze the key constraint factors of energy objects.

Once the key energy research objects are identified, we should further look for the constraint factors of the identified research objects based on the principle of focusing on the energy system, simplification of energy problems, and key issues. According to the TOC, constraints include resources, market, and laws and regulations which can be categorized by different perspectives. Macroscopic energy issues can be categorized by characteristics of productivity and production relations, rigidity and elasticity, and short-term and long-term prospectives.

(1) Productivity constraints and production relation constraints. Productivity constraints mainly include resource constraint and technical constraint. ① As energy development is based the prerequisites that there are resources available for development and utilization, resources become a constraint of energy

development. ② Technology determines the level of energy development and utilization. Energy development depends on and is limited by technology, while feasible technology must be economically affordable. As a result, productivity constraints also include economic factors. Production relation constraints mainly refer to various energy relations and energy behaviors specified by rules and order and include the following two aspects: ① direct constraints that are expressly prohibited by all relevant laws, rules, regulations, documents, and standards; ② indirect constraints that are restrictive effect through the manipulation of operational actions (e.g., emissions charges).

(2) Rigid and elastic constraints. Rigid constraints mainly refer to constraints that cannot be eliminated by increasing reasonable cost. In terms of productivity, rigid constraints include scarce resources or poor technology; in terms of production relations, rigid constraints include prohibitive orders. Elastic constraints refer to constraints that can be eliminated by increasing cost. In a macroscopic research, energy development obligatory indicators such as reducing energy intensity by 20% and main pollutants by 10% during the "11th Five-Year Plan" period and reducing carbon intensity by 40–45% by 2020 can be treated as rigid constraints. Guiding indicators in energy development plans can be treated as elastic constraints. Rigid constraints and elastic constraints can be transformed into each other under certain conditions. For example, with the improvement of exploration and excavation technology, elastic constraints can be changed into rigid constraints by expanding prohibitive requirements through the amendment of laws. Meanwhile, with economic development and the improvement of economic affordability, rigid constraints can be changed into elastic constraints. For example, we can adopt the pollutant control technology at a higher, but affordable economic cost which can lead to less emissions.

(3) Short-term and long-term constraints. Short-term constraints refer to main factors that constrain energy development during a five-year plan period, for example, in the "12th Five-Year Plan" period. Long-term constraints refer to main factors that constrain energy development in future years such as 2020 and 2030.

See Table 1.4 for energy research object constraint factors and nature.

From Table 1.4, we can see that the main constraint objects of energy development include four aspects, namely resource endowment constraints, environmental constraints, water resource constraints, and climate change adaptation constraints. Resource endowment constraints mainly mean the sustained energy development is restricted by the shortage of resource supply due to growing energy demands. Environmental constraints mainly mean energy development is limited by environmental pollution and ecological damage caused by unclean energy development, which is inconsistent with the demands of social and economic development. The water resource constraints mainly mean shortage of water resources as a result of energy development. The climate change adaptation constraints are new constraints of energy development which China must face in striving for

**Table 1.4** Main energy research object constraint factors and nature

| Constraint object | Energy research object | Characteristics | Constraint nature | | | | Situation in 2012 |
|---|---|---|---|---|---|---|---|
| | | | Rigid | Elastic | Long term | Short term | |
| Energy resources | Total | Development stage<br>Resource endowment<br>Energy self-sufficiency rate | | √ | √ | | Energy consumption rapid growth stage; energy self-sufficiency rate being 91.4%; coal accounting for 76.6% of total energy production; coal accounting for 67.1% of total energy consumption |
| | Energy structure | | | √ | √ | | |
| Environmental protection | Electric power | Regulatory requirements technical constraints economic level | √ | | √ | √ | Basic load regulatory requirements; some environmental protection technologies reaching or approaching international advanced level |
| | Coal | | √ | | √ | √ | |
| | Hydroenergy | | √ | | √ | √ | |
| Water resources | Coal | Regional distribution water-saving technology distribution of water resources | √ | | √ | √ | Unbalanced regional distribution of water resources; great water-saving technology development potential; less scientific water resources allocation |
| | Coal power | | √ | | √ | √ | |
| Greenhouse gas | Electric power | International situation production and consumption pattern energy structure | √ | | √ | | Great achievements; huge challenges |

development equity, energy equity, and environmental equity under the background of globalization. Meanwhile as the above four aspects of constraints are in a mutual restraint relationship, the mutual restrictive relation is comprehensively expressed in economic aspects. Therefore, the core of the constraint research is the correlation and balance among resources (energy), environmental protection, and economic development (or cost constraint).

5. **Breaking Constraints**

Breaking constraints is to put forward resolutions to the constraint problems. The analysis methods and technical tools in TOC such as thinking processes and logic tree analysis provide the basic methods to crack down constraints. The thinking processes include three sections, namely identifying key problems (what to change), proposing solutions (what to change), and making action plans (how to change). Based on the constraint problems in energy development research, this book includes sections of how to identify, evaluate, analyze, and optimize these constraint factors and then suggests related policies to establish an interconnected system framework (see Fig. 1.1).

(1) Establishing evaluation indicators in constraint assessment. Establishing evaluation indicators in constraint assessment represents the main process of breaking the energy development constraints. In essence, breaking energy development constraints is the process of realizing the energy development goal, which is further to set up indicators under "safe, clean, resource-saving, and economical" conditions, and quantify these indicators in different development stages. In Fig. 1.1, the quantitative indicators throughout the whole process of breaking energy development constraints are the "scale" and spirit. The functions of these quantitative indicators are manifested in the following aspects. They can be used to identify and evaluate the strength level of constraints, to analyze the temporal and spatial variation of the constraint factors, to provide operational tools and basis of making of regulations, policies, and decisions, and to instruct and guide industries to conduct benchmarking activities and constantly improve management. This is the foundation of breaking constraints.

    The constraint indicators should be clearly defined, measurable, legitimate, operable (including when being used for evaluation and supervision), and variable in different stages. In addition, they should reflect the essential attributes of things.

    Energy constraint indicators can be categorized into four categories, namely comprehensive indicators, energy and resource conservation constraint indicators, environmental protection constraint indicators, and climate change adaptations constraint indicators. By nature, they can be categorized into mandatory indicators and guiding indicators. It should be particularly pointed out that there are a large number of indicators in the energy, economics, and environmental protection fields, but indicators must be set for the research object and reflect the essential attributes of the research object. For example, for

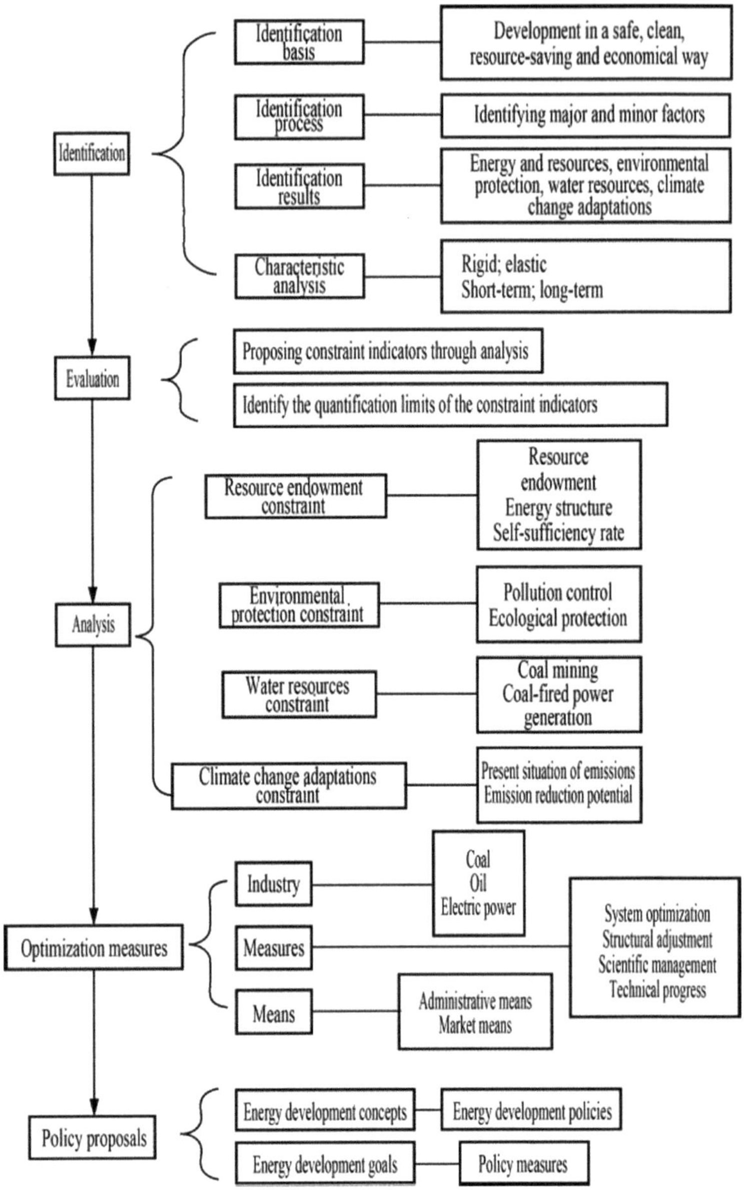


**Fig. 1.1** Concept road map of breaking energy development constraints

the research object of air pollution, the indicator "total coal consumption per unit land area" cannot reflect the essential attributes of air pollution because the amount of coal consumption cannot reflect the amount of emissions which is mainly dependent on the air pollution technical control level. Another example is that the indicator "carbon dioxide emission intensity per unit coal-fired generating capacity" reflects power generation efficiency and the effect of carbon emission reduction, but cannot reflect the adjustment of power supply structure. Instead the indicator "carbon dioxide emission intensity per unit generating capacity" can reflect the power supply structure adjustment because it also indicates the contribution of hydropower and renewable energy sources. As a result, in selecting indicators, the problem to be solved and the application scope of the indicators should be taken into consideration. The constraint indicators proposed in this book are mainly used for macroscopic research. To be specific, they can be used for making energy development strategies for a company group, a region or the nation of China, but not for an individual company.

(2) Establishing the constraint indicator system according to regulatory requirements and the role of the indicators in breaking constraints. Main constraint indicators can be divided into Level I indicators, Level II indicators, and reference indicators. Level I indicators are mainly made in accordance with national regulations or those that play a key role in the energy industry. They are mandatory indicators that can be clearly defined in national or industrial planning and used for the national assessment of the industry and enterprises. Level II indicators are guiding indicators that are not fully supported by national regulations or difficult to be regulated, but they can be used as important reference for government administration and play an important role in improving the management level of enterprises. Some Level II indicators are the same as Level I indicators in essence only that they measure the same fact from different perspectives. For example, the indicators "annual emissions of sulfur dioxide" and "emissions of sulfur dioxide per unit kilowatt hour" can be mutually converted under certain conditions. However, different indicators help enterprises better use or analyze and compare the constraint indicators. Reference indicators are mainly anticipated indicators and are not used for mandatory assessment.

    Through analysis we have put forward the energy development main constraint indicator system (see Table 1.5 for it).

The indicators in Table 1.5 are defined as follows:

(1) Energy intensity reduction rate. It is the ratio of the energy consumption per unit GDP in the statistical period to the energy consumption per unit GDP in the last statistical period. The calculation formula is as follows:

$$\text{Energy intensity reduction rate} = \frac{(\text{energy intensity in the statistical period} - \text{energy intensity in the last statistical period})}{\text{energy intensity in the last statistical period}} \times 100\%$$

(2) Coal consumption for power supply (thermal efficiency of electricity generation). Coal consumption for power supply, also known as standard coal

**Table 1.5** Energy development main constraint indicator system

| Category | Indicator | Attribute |
|---|---|---|
| Comprehensive | Energy intensity reduction rate (energy consumption reduction rate per unit GDP, %) | Level I indicator |
| Energy and resource conservation | Coal consumption for power supply (g/kWh) | Level I indicator |
| | Auxiliary power ratio (%) | Level II indicator |
| | Transmission line loss (%) | Level I indicator |
| | Power generation water consumption indicator (g/kWh) | Level I indicator |
| | Comprehensive utilization rate of solid waste (%) | Level II indicator |
| | The comprehensive utilization rate of desulfurized byproduct (%) | Reference indicator |
| | Mine water utilization rate (%) | Reference indicator |
| | Coal mine land reclamation rate (%) | Reference indicator |
| | Utilization rate of coal gangue (%) | Reference indicator |
| | Utilization rate of coalbed gas (%) | Reference indicator |
| | Utilization of mine gas (100 million $m^3$) | Reference indicator |
| | Utilization rate of mine gas (%) | Reference indicator |
| Environmental protection | Compliance rate of sulfur dioxide emissions from electric power generation (%) | Level I indicator |
| | Total annual sulfur dioxide emissions from electric power generation (10,000 t) | Level I indicator |
| | Sulfur dioxide emissions from electric power generation per unit thermal power generating capacity (g/kWh) | Level II indicator |
| | Proportion of flue gas desulfurization units (%) | Reference indicator |
| | Compliance rate of nitrogen oxide emissions from electric power generation (%) | Level I indicator |
| | Total annual nitrogen oxide emissions from electric power generation (10,000 t) | Level I indicator |
| | Proportion of flue gas denitration units (%) | Reference indicator |
| | Nitrogen oxide emissions per unit thermal power generating capacity (g/kWh) | Level II indicator |
| | Compliance rate of annual smoke and dust emissions from electric power generation (%) | Level I indicator |
| | Total annual smoke and dust emissions from electric power generation (10,000 t) | Level II indicator |
| | Smoke and dust emissions from electric power generation per unit thermal power generating capacity (g/kWh) | Level II indicator |
| | Mine water discharge compliance rate (%) | Reference indicator |

(continued)

**Table 1.5** (continued)

| Category | Indicator | Attribute |
|---|---|---|
| Climate change adaptations | Carbon dioxide emission intensity per unit generating capacity (g/kWh) | Level I indicator |
| | Proportion of non-fossil energy power generating capacity in total power generating capacity (%) | Level I indicator |

consumption for power supply, refers to the average standard coal consumption for per kWh electricity supplied by thermal power generation enterprises. The calculation formula is as follows:

$$\text{Coal consumption for power supply} = \frac{\text{total coal consumption for power generation}}{\text{total power supply of thermal power}}$$

Thermal efficiency of electricity generation refers to the proportion of thermal equivalent for per kilowatt hour of electricity generated by generating units in total heat consumption for power generation. In China usually the standard coal consumption for power supply is used while in foreign countries thermal efficiency of electricity generation is used. The two indicators can be mutually converted, i.e.,

$$\text{Thermal efficiency of electricity generation} = \frac{123}{\text{coal consumption for power supply}} \times 100\%$$

a. Notes: Standard coal, also known as coal equivalent, has unified calorific value standards. China's power industry converts the inferior calorific value of various energy carriers into standard coal (1 kg coal equivalent corresponds to the specified 29271 kJ). [In *General Rules for the Calculation of Comprehensive Energy Consumption* (GB/T 2589—2008), the specified value 1 kg coal equivalent corresponds to is 29307 kJ. It is mainly because there are two standards for the conversion between card and joule: at 20 °C 1 kcal = 4.1816 J while at 15 °C 1 kcal = 4.1868 J. As a result, there are two calorific values of standard coal, 29,271.2 and 29,307.6 kJ/kg].

The calculation formula is as follows:

$$B_b = \sum_{i=1}^{n} (B_i \times q_i/29{,}271)$$

where

$B_b$ total energy consumption of power generation enterprises;
$B_i$ the physical quantity of the $i$th kind of energy source consumed by power generation enterprises in industrial production;
$q_i$ the calorific value of the $i$th kind of energy source consumed by power generation enterprises in industrial production.

Total coal consumption for power generation refers to the consumption (converted into standard coal) of the fuel (including coal, oil and natural gas) used in industrial production in the statistical period without considering the change of coal bunker and sub-warehouse and excluding (fuel consumption should be deducted): fuel used for the drying off, boiling out, heating and no-load operation of new equipment or equipment after overhaul; fuel used for the pre-production load test running of new equipment; fuel used for planned overhaul and construction of infrastructure and changed projects; fuel used for phase operation of generators; fuel consumed by self-owned locomotives and ships for transportation outside the plant; fuel consumed by the step-up and step-down transformer (excluding auxiliary power transformer), variable wave machine and phase modification machine used for power transmission and distribution; fuel consumed by repair shop, garage, sideline production, comprehensive utilization, collective enterprises, external power supply and used for non-production purposes.

Total power supply of thermal power refers to total generating capacity of thermal power generation enterprises minus power consumption of power plants.

b. Indicator descriptions. Coal consumption for power supply (thermal efficiency of electricity generation) is the main economic and technical indicator of thermal power generation enterprises and reflects thermal power generation enterprises' equipment and system health level, repairment and maintenance technology level, operation and management optimization and regulation level and fuel comprehensive management level as well as the energy consumption per unit product of thermal power generation enterprises. Through this indicator, we can understand the energy consumption of different thermal power generation enterprises and promote these enterprises to achieve energy conservation and emission reduction by specifying energy-saving management, promoting energy-saving transformation, optimizing equipment operation, continuously tapping energy reduction potential and constantly improving coal consumption for power supply and thermal efficiency of electricity generation.

(3) Auxiliary power ratio. It refers to the proportion of auxiliary power consumption of a power plant in total generating capacity of this power plant in the same period and it is one of the economic indicators that reflect the operation management level of a power plant or nuclear power plant. Its calculation formula is as follows:

$$\text{Auxiliary power ratio} = \frac{\text{auxiliary power consumption}}{\text{total generating capacity}} \times 100\%$$

(4) Transmission line loss. It refers to the power loss generated on the line resistance when the load current passes through the line.

Its calculation formula is as follows:

a. For single line, the active power loss is

$$\Delta P = I^2 R$$

where

$\Delta P$ lost power, W;
$I$ load current, A;
$R$ conductor resistance, $R_0$

b. For three-phase power line, the active power loss is

$$\Delta P = \Delta PU + \Delta PV + \Delta PW = 3I^2 R$$

(5) Power generation water consumption indicator. Also known as comprehensive water consumption rate of power generation, it refers to the amount of freshwater consumed by thermal power generation enterprises for generating 1 kWh of electricity (unit: kg/kWh or $m^3$/MWh). Its calculation formula is

$$\text{Water consumption} = \frac{\text{amount of fresh water used for power generation}}{\text{generating capacity}}$$

a. Notes: The amount of freshwater used for power generation refers to total amount of all freshwater used by thermal power generation enterprises for power generation in the statistical period including the amount of water extracted from various conventional water resources, such as surface water, groundwater, urban water supply engineering, and other forms of water (including reclaimed water) and water products (steam, hot water, geothermal water,) purchased from the market, but excluding the amount of reused water, non-production water and water consumed by raw water pretreatment system and reclaimed water advanced treatment system. Water intake using a DC cooling system does not include the amount of water which is taken from river, lake and other water bodies, used for the starting cooling of steam condenser and other heat exchangers and finally discharged into the original water body. The water taken from the DC cooling water (excluding seawater) system for other purposes should be included in the amount of water intake.

Generating capacity refers to total electricity generated by power generation enterprises in the statistical period.

b. Indicator descriptions. Comprehensive water consumption rate of power generation reflects the freshwater consumption per unit generating capacity and is a comprehensive indicator of water utilization efficiency. Through this indicator, we can understand the water conservation situation of power generation enterprises and promote power generation enterprises to full tap their water conservation potentials.

(6) Comprehensive utilization rate of solid waste. It refers to the proportion of the comprehensive utilization of solid waste (fly ash and gypsum) of power generation enterprises in the statistical period in total amount of produced solid waste. Its calculation formula is

$$\text{Comprehensive utilization rate of solid waste} = \frac{\begin{array}{c}(\text{comprehensive utilization of fly ash}\\ +\,\text{comprehensive utilization of gypsum})\end{array}}{(\text{amount of produced fly ash} + \text{amount of produced gypsum})} \times 100\%$$

a. Notes: Comprehensive utilization rate of solid waste refers to the amount of solid waste which enterprises extract useful resources, energy and other raw materials from or convert into useful resources, energy and other raw materials, for example, being used for producing building materials and road construction, through recycling, processing, circulation, exchange, etc., in the statistical period. For thermal power generation enterprises, solid waste mainly includes fly ash and gypsum produced by desulfurization.

Comprehensive utilization rate of fly ash refers to the ratio of comprehensive utilization of fly ash to amount of produced fly ash and its calculation formula is

$$\text{Comprehensive utilization rate of fly ash} = \frac{\text{comprehensive utilization of fly ash}}{\text{amount of produced fly ash}} \times 100\%$$

Amount of produced fly ash refers to the amount of the fine solid particles collected from flue gas generated in the process of coal combustion in the statistical period; comprehensive utilization of fly ash includes the amount of fly ash used for producing building materials, constructing roads and export, etc., as well as the utilization of previously stored fly ash.

Utilization rate of desulfurized gypsum refers to the ratio of comprehensive utilization of desulfurized gypsum to amount of produced desulfurized gypsum and its calculation formula is

$$\text{Utilization rate of desulfurized gypsum} = \frac{\text{utilization of desulfurized gypsum}}{\text{amount of produced desulfurized gypsum}} \times 100\%$$

Amount of produced desulfurized gypsum refers to the amount of by-products produced by the reaction of the absorbent and sulfur dioxide in flue gas in the wet limestone-gypsum flue gas desulphurization process. The utilization of desulfurized gypsum includes the amount of desulfurized gypsum used for producing building materials and export as well as the utilization of previously stored desulfurized gypsum.

b. Indicator descriptions. Comprehensive utilization rate of solid waste reflects the comprehensive utilization of solid waste by power generation enterprises. It is one of the important energy conservation and environmental protection indicators of thermal power generation enterprises. Through this indicator, we can understand enterprises' solid waste utilization level in order to reduce consumption of raw materials, realize the sustainable recycling and reusing of resources, reduce environmental pollution caused by solid waste and eliminate the hidden security risk for destroying the ecological system.

(7) Mine water utilization rate. The mine water refers to the water generated in the coal construction and coal mining process and pumped to the ground, including mine inflow and waste water generated in the downhole production process. According to the industry statistics, about 1.8 t mine water is discharged for mining 1 t of raw coal on average. Its calculation formula is

$$\text{Mine water utilization rate} = \frac{\text{total utilization of mine water in the statistical period}}{\text{total amount of produced mine water in the statistical period}} \times 100\%$$

(8) Coal mine land reclamation rate. In the process of underground coal mining, the mining of coal resources leads to the vertical and horizontal movement of the overlying strata and the surface and surface subsidence. Land excavating, collapsing, occupying which occur in the production and construction activities conducted by human beings are called land destruction.

Land reclamation refers to the activity of taking comprehensive measures to restore the destroyed land to the usable state.

The land reclamation rate is the ratio of the reclaimed land area to the destroyed land area, namely

$$\text{Land reclamation rate} = \frac{\text{reclaimed land area}}{\text{destroyed land area}} \times 100\%$$

(9) Utilization rate of coal gangue. Coal gangue is the collective term for all the rocks generated in the coal production process including the rocks mixed in coal, rocks discharged in tunnel excavation, falling rocks in mined-out areas, rocks appearing at the working surface and the carbonaceous rocks discharged in the coal preparation process.

Coal gangue includes washing gangue and tunneling gangue. According to industry statistics, the proportion of washing gangue and tunneling gangue in 1 t of mined raw coal on average is 18 and 9%, respectively.

The calculation formula of utilization rate of coal gangue is

$$\begin{aligned}&\text{Utilization rate of coal gangue}\\&\quad = \frac{\text{Utilization of produced coal gangue in the statistical period}}{\text{total amount of produced coal gangue in the statistical period}} \times 100\%\end{aligned}$$

(10) Utilization rate of coalbed gas. Coalbed gas is hydrocarbon gas in coal seam and coal measure strata and its main component is methane. Its calorific value is equivalent to that of conventional natural gas. It is a valuable energy resource.

The coalbed gas industry is an emerging energy industry. Development the coalbed gas industry is of important significance to ensuring the safety of coal production, optimizing the energy structure and protecting the ecological environment.

The calculation formula of utilization rate of coalbed gas is

$$\begin{aligned}&\text{Utilization rate of coalbed gas}\\&\quad = \frac{\text{utilization of coalbed gas in the statistical period}}{\text{extraction of coalbed gas in the statistical period}} \times 100\%\end{aligned}$$

(11) Utilization rate of mine gas. The main component of mine gas is methane $CH_4$. Mine gas is the associated gas mainly present in coal mine and the self-stored natural gas that exists in the coal seam and solid rocks in the adsorption or free state through the biochemical pyrolysis in the coal formation process.

Mine gas is a kind of unconventional natural gas and excellent chemical and energy raw material.

The calculation formula of utilization rate of mine gas is

$$\text{Utilization rate of mine gas} = \frac{\text{utilization of mine gas in the statistical period}}{\text{total amount of gas extracted from mines in the statistical period}} \times 100\%$$

(12) Mine water discharge compliance rate. Industrial waste water standard discharge means the indicators of industry characteristic pollutants in the waste water reach the national or local standard for discharged industrial waste water. According to the assessment requirements, industrial waste water discharge compliance rate should be higher than 95%.

The discharge of mine water as industrial waste water should meet national or local discharge standards. However, as a precious freshwater resource, mine water should be purified and comprehensively treated before being discharged and the part which cannot be completely utilized should be discharged in accordance with the standards.

The calculation formula of mine water discharge compliance rate is

$$\text{Mine water discharge compliance rate} = \frac{\text{standard discharge of mine water in the statistical period}}{\text{total discharge of mine water in the statistical period}} \times 100\%$$

(13) Compliance rate of sulfur dioxide emissions from electric power generation. It refers to the proportion of the time when the concentration of sulfur dioxide emissions is in line with that stipulated in Standard for Air Pollutants Emitted from Thermal Power Generation Plants (GB 13223—2011) (if the local standard is stricter than the national standard, the local standard shall prevail) in the boiler running time.

(14) Total annual sulfur dioxide emissions from electric power generation. Total annual sulfur dioxide emissions from electric power generation refers to total amount of sulfur dioxide discharged into the air by power generation enterprises (mainly thermal power generation enterprises) in the statistical period (one year).

(15) Sulfur dioxide emissions from electric power generation per unit thermal power generating capacity. Sulfur dioxide ($SO_2$) emissions from electric power generation per unit thermal power generating capacity refers to the amount of $SO_2$ discharged by power generation enterprises for generating 1 kWh of electricity on average and is used to reflect the $SO_2$ emission intensity per unit generating capacity (unit: g/kWh). Its calculation formula is

$$\text{SO}_2\text{ emission performance} = \frac{\text{SO}_2\text{ emissions}}{\text{total thermal power generating capacity}}$$

a. Notes: Total $SO_2$ emissions refer to total amount of $SO_2$ emitted from thermal power generating units of power generation enterprises in the statistical period. Total thermal power generating capacity refers to total amount of electricity generated by thermal power generating units of power generation enterprises in the statistical period.
b. Indicator descriptions. Sulfur dioxide emissions from electric power generation per unit thermal power generating capacity reflects the emissions of $SO_2$ per unit electricity generated by thermal power generation enterprises and is an important indicator measure the level of $SO_2$ emission control technology of thermal power generation enterprises.

(16) Proportion of flue gas desulfurization units. Proportion of flue gas desulfurization units refers to the proportion of thermal power generating units equipped with and using desulfurization devices in total thermal power generating units in the statistical range. Its calculation formula is

$$\text{Proportion of flue gas desulfurization units} = \frac{\text{thermal power generating units using desulfurization devices}}{\text{total thermal power generating units}}$$

(17) Compliance rate of nitrogen oxide emissions from electric power generation. It refers to the proportion of the time when the concentration of nitrogen oxide emissions is in line with that stipulated in Standard for Air Pollutants Emitted from Thermal Power Generation Plants (GB 13223—2011) (if the local standard is stricter than the national standard, the local standard shall prevail) in the boiler running time.
(18) Total annual nitrogen oxide emissions from electric power generation. Total annual nitrogen oxide emissions from electric power generation refers to total amount of nitrogen oxide discharged into the air by power generation enterprises (mainly thermal power generation enterprises) in the statistical period (one year).
(19) Nitrogen oxide emissions per unit thermal power generating capacity. Nitrogen oxide ($NO_x$) emissions per unit thermal power generating capacity refers to the amount of $NO_x$ discharged by power generation enterprises for generating 1 kWh of electricity on average and is used to reflect the $NO_x$ emission intensity per unit generating capacity (unit: g/kWh). Its calculation formula is

$$\text{NO}_x\,\text{emission performance} = \frac{\text{NO}_x\,\text{emissions}}{\text{total thermal power generating capacity}}$$

a. Notes: $NO_x$ emissions refers to total amount of $NO_x$ discharged by the thermal power generating units of power generation enterprises in the statistical period.
b. Indicator descriptions. Nitrogen oxide emissions per unit thermal power generating capacity reflects the emissions of $NO_x$ per unit electricity generated by thermal power generation enterprises and is an important indicator measure the level of $NO_x$ emission control technology of thermal power generation enterprises.

(20) Proportion of flue gas denitration units. Proportion of flue gas denitration units refers to the proportion of thermal power generating units equipped with and using denitration devices in total thermal power generating units in the statistical range. Its calculation formula is

$$\begin{aligned}&\text{Proportion of flue gas denitrification units}\\&=\frac{\text{thermal power generating units using denitrification devices}}{\text{total thermal power generating units}}\end{aligned}$$

(21) Compliance rate of smoke and dust emissions from electric power generation. It refers to the proportion of the time when the concentration of smoke and dust emissions is in line with that stipulated in Standard for Air Pollutants Emitted from Thermal Power Generation Plants (GB 13223—2011) (if the local standard is stricter than the national standard, the local standard shall prevail) in the boiler running time.

(22) Total annual smoke and dust emissions from electric power generation. Total annual smoke and dust emissions from electric power generation refers to total amount of smoke and dust discharged into the air by power generation enterprises (mainly thermal power generation enterprises) in the statistical period (one year).

(22) Smoke and dust emissions from electric power generation per unit thermal power generating capacity. Smoke and dust emissions from electric power generation per unit thermal power generating capacity refers to the amount of smoke and dust discharged by thermal power generating units of power generation enterprises for generating 1 kWh of electricity and is used to reflect the smoke and dust emission intensity per unit electricity (unit: g/kWh). Its calculation formula is

$$\text{Smoke and dust emission performance} = \frac{\text{smoke and dust emissions}}{\text{total thermal power generating capacity}}$$

a. Notes: Smoke and dust emissions refers to total amount of smoke and dust discharged by thermal power generating units of power generation enterprises in the statistical period.
b. Indicator descriptions. Smoke and dust emissions from electric power generation per unit thermal power generating capacity refers to the emissions of smoke and dust per unit electricity and reflects the application of smoke and dust treatment technology and application and treatment efficiency of smoke and dust treatment equipment of power generation enterprises.

(24) Carbon dioxide emission intensity per unit generating capacity. Carbon dioxide ($CO_2$) emission intensity per unit generating capacity refers to the amount of $CO_2$ discharged by thermal power generating units of power generation enterprises for generating 1 kWh of electricity and is used to reflect $CO_2$ emission intensity per unit electricity (unit: g/kWh). Its calculation formula is

$$CO_2 \text{ emission performance} = \frac{CO_2 \text{ emissions}}{\text{total thermal power generating capacity}}$$

a. Notes: $CO_2$ emissions refers to total amount of $CO_2$ discharged by thermal power generating units of power generation enterprises in the statistical period.

b. Indicator descriptions. Carbon dioxide emission intensity per unit generating capacity refers to the emissions of carbon dioxide per unit electricity and reflects the overall efficiency and carbon dioxide capture, utilization and storage of power generation enterprises. Through this indicator, we can promote power generation enterprises to strengthen carbon dioxide emission management and reduce carbon dioxide emissions.

(25) Proportion of non-fossil energy power generating capacity in total power generating capacity. It refers to the proportion of non-fossil energy power generation in total power generating capacity (unit: %). Its calculation formula is

**Table 1.6** Comparison of some main constraint indicators of energy development

| Category | Indicator | Indicator limit by the end of the "12th Five-Year Plan" period | Compared with 2005 | Attribute |
|---|---|---|---|---|
| Comprehensive | Energy intensity reduction rate in 5 years (%) | A decrease of 15–20 as compared with 2010 | A decrease of 32–36 | Level I indicator |
| Energy and resource conservation | Coal consumption for power supply (g/kWh) | 330 | A decrease of 40 | Level I indicator |
| | Auxiliary power ratio (%) | 5.4 (auxiliary power ratio of thermal power plants decreased by 6.3) | A decrease of 0.47 (auxiliary power ratio of thermal power plants decreased by 0.5) | Level II indicator |
| | Transmission line loss (%) | 6.5 | A decrease of 0.71 | Level I indicator |
| | Power generation water consumption indicator (g/kWh) | 2 | A decrease of 1.1 | Level I indicator |
| | Comprehensive utilization rate of fly ash (%) | 70 | An increase of 1 | Level II indicator |
| Energy and resource conservation | The comprehensive utilization rate of desulfurized byproduct (%) | 60 | | Reference indicator |
| | Mine water utilization rate (%) | 75 in North China, East China, Northeast China, and Central China; 95 in Northwest China; 70 in Southwest China | | Reference indicator |
| | Coal mine land reclamation rate (%) | 40 in North China, East China, and Northwest China; 45 in Northeast China and Central China; 50 in Southwest China | | Reference indicator |
| | Utilization rate of coal gangue (%) | 75 in North China, East China, Northeast China, Central China, Northwest China, and Southwest China | | Reference indicator |

(continued)

**Table 1.6** (continued)

| Category | Indicator | Indicator limit by the end of the "12th Five-Year Plan" period | Compared with 2005 | Attribute |
|---|---|---|---|---|
| | Utilization rate of coalbed gas (%) | 40 in North China, East China, and Northwest China; 45 in Northeast China and Central China; 50 in Southwest China | | Reference indicator |
| | Utilization of mine gas (100 million $m^3$) | 45 | | Reference indicator |
| | Utilization rate of mine gas (%) | 55 | | Reference indicator |
| Environmental protection | Compliance rate of sulfur dioxide emissions from electric power generation (%) | $\geq 98$ | | Level I indicator |
| | Total annual sulfur dioxide emissions from electric power generation (10,000 t) | 800 | A decrease of 38% | Level I indicator |
| | Sulfur dioxide emissions from electric power generation per unit thermal power generating capacity (g/kWh) | 1.8 | A decrease of 4.6 | Level II indicator |
| | Proportion of flue gas desulfurization units (%) | 100 | An increase of 86 | Reference indicator |
| | Compliance rate of nitrogen oxide emissions from electric power generation (%) | $\geq 90$ | | Level I indicator |
| | Total annual nitrogen oxide emissions from electric power generation (10,000 t) | 800 | | Level I indicator |
| | Proportion of denitration units at coal-fired power plants (%) | 50 | An increase of 49 | Reference indicator |

(continued)

**Table 1.6** (continued)

| Category | Indicator | Indicator limit by the end of the "12th Five-Year Plan" period | Compared with 2005 | Attribute |
|---|---|---|---|---|
| | Compliance rate of smoke and dust emissions from electric power generation (%) | 100 | | Level I indicator |
| | Total annual smoke and dust emissions from electric power generation (10,000 t) | 310 | A decrease of 14% | Level II indicator |
| | Smoke and dust emissions from electric power generation per unit thermal power generating capacity (g/kWh) | 0.68 | A decrease of 1.2 | Level II indicator |
| | Mine water discharge compliance rate (%) | 100 | | Reference indicator |
| Climate change adaptations | Carbon dioxide emission intensity per unit generating capacity (g/kWh) | 643 | A decrease of 131 (down 16.9%) | Level II indicator |
| | Proportion of non-fossil energy power generating capacity in total power generating capacity (%) | 24.84 | An increase of 6.75 | Level II indicator |

*Notes* Except as otherwise indicated, all the indicators are the national average values

$$\text{Proportion of non-fossil energy power generating capacity in total power generating capacity} = \frac{\text{non-fossil energy power generating capacity}}{\text{total power generating capacity}} \times 100\%$$

a. Notes: Non-fossil energy refers to energy resources other than those disposable energy forms such as coal, oil and natural gas that have been formed through long time geological changes. Non-fossil energy includes new energy and renewable energy resources including nuclear energy, wind energy, solar energy, hydroenergy, biomass energy, geothermal energy, ocean energy.

Non-fossil energy power generating capacity refers to total amount of electricity generated by generating units in service that use non-fossil energy for power generation owned by power generation enterprises in the statistical period.

Total generating capacity refers to total amount of electricity generated by generating units in service owned by power generation enterprises in the statistical period.

b. Indicator descriptions. Proportion of non-fossil energy power generating capacity in total power generating capacity reflects the using of non-fossil energy for power generation of power generation enterprises.

See Table 1.6 for some main constraint indicators of energy development.

# Chapter 2
# Energy Development Trend

**Li Pan and Jingjie Zhang**

## 2.1 Overview of World Energy Resources

In 1930, more than 95% of the world's energy consumption depended on the supply of coal, oil, natural gas, and traditional biomass, while in 1990, the energy supply included other energy forms such as nuclear energy, hydro energy, wind energy, solar energy, geothermal energy and biofuel. Coal, oil and natural gas are the driving force for world economic development. With the increasing energy dependence in social and economic development, ecological environment is increasingly constraining energy development. In the future, world energy consumption will continue to grow and the contradiction between supply and demand of world energy resources will become increasingly prominent.

Column Interpretation
Main categories of energy
Primary energy and secondary energy

Primary energy is the energy embodied in natural resources prior to undergoing any human-made conversions or transformations. Examples of primary energy resources include coal, oil, natural gas, and running rivers. Secondary energy is the energy contained in energy forms directly or indirectly converted from primary energy. Secondary energy resources include electricity, gas, steam, and all kinds of petroleum products.

Renewable energy and non-renewable energy

Primary energy can be further categorized into renewable and non-renewable energy. Renewable energy sources can be continuously replenished or regenerated in a relatively short period of time such as wind energy, hydro energy, ocean

L. Pan (✉) · J. Zhang
China Electricity Council, Beijing, People's Republic of China
e-mail: panli@cec.org.cn

 
Z. Wang (ed.), *Constraints and Solutions for Energy and Electricity Development*,
https://doi.org/10.1007/978-981-10-8304-4_2

energy, tidal energy, solar energy, and biomass energy; otherwise, they are non-renewable energy sources such as coal, oil, and natural gas. Renewable energy is defined in different ways in different Chinese regulations and policies. In *Renewable Energy Law of the People's Republic of China* (2009 Amendment), renewable energy refers to non-fossil energy such as wind energy, solar energy, hydro energy, biomass energy, geothermal energy, and ocean energy; it is stipulated in *Renewable Energy Law of the People's Republic of China* that the application of hydraulic power generation should be regulated by relevant departments of the State Council that are in charge of energy and approved by the State Council while this law is not applicable to the utilization of straw and firewood, manure, etc., in the low-efficiency modes such as direct combustion using stove. According to the *Medium and Long-Term Development Plan for Renewable Energy in China*, forms of renewable energy include hydro energy, biomass energy, wind energy, solar energy, geothermal energy, and ocean energy; characterized by great resource potential, low environmental pollution, and sustainable use, renewable energy is an important form of energy conducive to the harmonious development of human and nature.

Fossil energy and non-fossil energy

Fossil energy is a kind of hydrocarbon or one of its derivatives contained in natural resources such as coal, oil, and natural gas. Non-fossil energy refers to wind energy, solar energy, hydro energy, biomass energy, geothermal energy, ocean energy, and other forms of renewable energy and nuclear energy.

Conventional energy and new energy

Conventional energy sources include non-renewable primary energy resources such as coal, oil, and natural gas and renewable water resources. New energy resources include solar energy, wind energy, geothermal energy, ocean energy, bioenergy, and nuclear fuel which can be used for nuclear power generation. Most forms of new energy are renewable energy. In addition, the United Nations Development Program categorizes the new energy resources into three categories: large- and medium-sized hydropower, new renewable energy forms (including small hydropower, solar energy, wind energy, modern biomass energy, geothermal energy, ocean energy), and traditional biomass energy.

Clean energy

According to China National Committee for Terms in Sciences and Technologies, clean energy refers to any form of energy that does not emit any harmful substances in the production and utilization processes. It can be renewable energy sources that can be replenished or reproduced after being consumed or non-renewable energy sources (such as clean coal and oil) that have been treated with clean energy technologies.

Studies suggest that as shown in Fig. 2.1, energy can be categorized into the following categories.

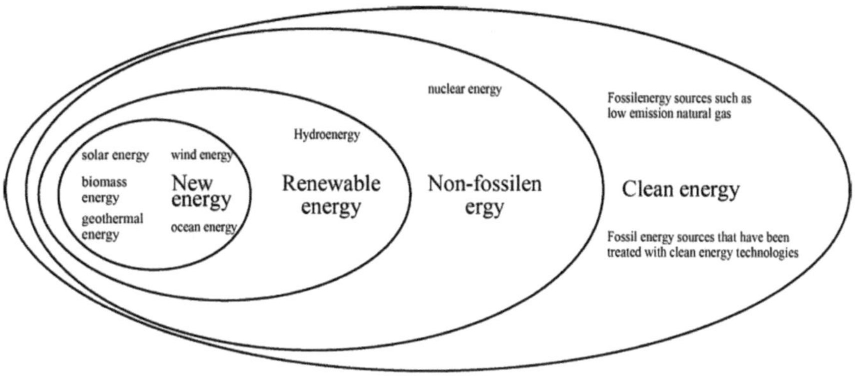


**Fig. 2.1** Energy categories

## *2.1.1 Fossil Energy Resources*

1. Raw coal

Coal is a kind of cheap, common, and accessible resource. Meanwhile the coal resources are finite, non-renewable, and not sustainable in the long term. Over 100 countries in the world have access to coal resources, and those that do not are mainly OPEC countries (Organization of Petroleum Exporting Countries). According to *BP (British Petroleum) Statistical Review of World Energy*, by the end of 2013 the world's proved raw coal reserves were 891.531 billion t and the reserves-to-production ratio (also known as extraction rate) was 113. The world's top ten countries with the largest proved raw coal reserves were America, Russia, China, Australia, India, Germany, Ukraine, Kazakhstan, South Africa, and Turkey successively. The proved raw coal reserves of the top ten countries accounted for 88.9% of the world's total. America's proved raw coal reserves were 237.295 billion t, the largest in the world, accounting for 26.6% of the world's total. Russia's reserves-to-production ratio was 452, the highest in the world. See Table 2.1 for the world's top ten countries with the largest proved raw coal reserves by the end of 2013.

2. Crude oil

Crude oil is a fossil fuel. The exploitation and utilization of crude oil are closely related to its reserves. According to BP statistics, by the end of 2013, the world's proved crude oil reserves were 238.2 billion t (about 1687.9 billion barrels) and the reserves-to-production ratio was 53.3. The world's top ten countries with the largest proved crude oil reserves were Venezuela, Saudi Arabia, Canada, Iran, Iraq, Kuwait, the United Arab Emirates, Russia, Libya, and America successively. The proved crude oil reserves of the top ten countries accounted for 84.8% of the world's total. Venezuela's proved crude oil reserves were up to 46.6 billion t, the largest in the world, accounting for 17.7% of the world's total. The

**Table 2.1** World's top ten countries with the largest proved raw coal reserves by the end of 2013

| Country | Proved raw coal reserves (100 million t) | Proportion in the world's proved raw coal reserves (%) | Reserves-to-production ratio |
|---|---|---|---|
| World | 8915.31 | 100 | 113 |
| America | 2372.95 | 26.6 | 266 |
| Russia | 1570.10 | 17.6 | 452 |
| China | 1145.00 | 12.8 | 31 |
| Australia | 764.00 | 8.6 | 160 |
| India | 606.00 | 6.8 | 100 |
| Germany | 405.48 | 4.5 | 213 |
| Ukraine | 338.73 | 3.8 | 384 |
| Kazakhstan | 336.00 | 3.8 | 293 |
| South Africa | 301.56 | 3.4 | 117 |
| Turkey | 87.02 | 1.0 | 141 |

reserves-to-production ratio of Venezuela, Canada, Iran, and Iraq was higher than 100. See Table 2.2 for the world's top ten countries with the largest proved crude oil reserves by the end of 2013.

3. Natural gas

Natural gas is a fossil fuel with the highest energy efficiency and the least pollution. It has abundant reserves and a wide range of applications. According to BP

**Table 2.2** World's top ten countries with the largest proved crude oil reserves by the end of 2013

| Country | Proved crude oil reserves (100 million t) | Proportion in the world's proved crude oil reserves (%) | Reserves-to-production ratio |
|---|---|---|---|
| World | 2382 | 100 | 53.3 |
| Venezuela | 466 | 17.7 | >100 |
| Saudi Arabia | 365 | 15.8 | 63.2 |
| Canada | 281 | 10.3 | >100 |
| Iran | 216 | 9.3 | >100 |
| Iraq | 202 | 8.9 | >100 |
| Kuwait | 140 | 6.0 | 89.0 |
| The United Arab Emirates | 130 | 5.8 | 73.5 |
| Russia | 127 | 5.5 | 23.6 |
| Libya | 63 | 2.9 | >100 |
| America | 54 | 2.6 | 12.1 |

statistics, by the end of 2013, the world's proved natural gas reserves were 185.7 trillion $m^3$ and the reserves-to-production ratio was 55.1. The world's top ten countries with the largest proved natural gas reserves were Iran, Russia, Qatar, Turkmenistan, America, Saudi Arabia, the United Arab Emirates, Venezuela, Nigeria, and Algeria. The proved natural gas reserves of the top ten countries accounted for 78.5% of the world's total. Iran's proved natural gas reserves were 33.8 trillion $m^3$, the largest in the world, accounting for 18.2% of the world's total. The reserves-to-production ratio of Iran, Qatar, Turkmenistan, United Arab Emirates, Venezuela, and Nigeria was higher than 100. See Table 2.3 for the world's top ten countries with the largest proved natural gas reserves by the end of 2013.

4. Combustible ice

It is estimated that the world's combustible ice reserves are two times the existing natural gas and oil reserves. Only the distribution area of the combustible ice in seabed area is up to 40 million $km^2$, accounting for 1/4 of total area of the earth's oceans. At present, combustible ice is found to be distributed in 116 places and has broad development prospects. Combustible ice is distributed in deep-sea sediments. The combustible ice resources in South China Sea are about 100 billion t (of standard coal), about 1/2 of total amount of China's onshore oil and natural gas resources. The prospective combustible ice resources in Qinghai–Tibet plateau are more than 50 billion t (of standard coal). Combustible ice is a kind of important strategic reserve resources, but at present still in the stage of resource investigation and technology research and development, and there is still a long way to go before its commercial utilization. In March 2013, the Japanese government announced

**Table 2.3** World's top ten countries with the largest proved natural gas reserves by the end of 2013

| Country | Proved natural gas reserves (100 million t) | Proportion in the world's proved natural gas reserves (%) | Reserves-to-production ratio |
|---|---|---|---|
| World | 2382 | 100 | 53.3 |
| Iran | 466 | 17.7 | >100 |
| Russia | 365 | 15.8 | 63.2 |
| Qatar | 281 | 10.3 | >100 |
| Turkmenistan | 216 | 9.3 | >100 |
| America | 202 | 8.9 | >100 |
| Saudi Arabia | 140 | 6.0 | 89.0 |
| The United Arab Emirates | 130 | 5.8 | 73.5 |
| Venezuela | 127 | 5.5 | 23.6 |
| Nigeria | 63 | 2.9 | >100 |
| Algeria | 54 | 2.6 | 12.1 |

Japan succeeded in separating methane gas from combustible ice deposits in Japan's offshore formation and became the first country in the world to master the seabed combustible ice mining technology.

Column Interpretation

Principles of Converting Various Energy Carriers into Standard coal
It is specified in *General Rules for the Calculation of Comprehensive Energy Consumption* (GB/T 2589—2008), "In calculating comprehensive energy consumption, various energy carriers should be converted to energy calorific value of standard coal, the unit of primary energy. The energy consumption unit should convert the inferior calorific value of its actually consumed fuel energy into standard coal. The fuel whose inferior calorific value is 29,307 kJ is called 1 kg standard coal (1 kgce). For the energy carriers and energy-consumed working media purchased by the energy consumption unit, the data released by National Bureau of Statistics of China can be used as a reference for the energy transformation coefficient. For the energy produced by the energy consumption unit and energy consumed by the working medium, the energy consumption unit can calculate the energy transformation coefficient based on the actual input and output. When the measured inferior calorific value of various fuel energy and the energy consumption per unit of working medium cannot be obtained, see Appendix A and Appendix B for reference." The reference coefficients for converting various energy carriers into standard coal are as follows:

Reference Coefficients for Converting Various Energy Carriers into Standard Coal

| Energy | | Mean inferior calorific value | Coefficient for conversion into standard coal |
|---|---|---|---|
| Raw coal | | 20,908 J/kg (5000 kcal/kg) | 0.7143 kgce/kg |
| Cleaned coal | | 26,344 J/kg (6300 kcal/kg) | 0.9000 kgce/kg |
| Other types of washed coal | Middlings | 8363 J/kg (5000 kcal/kg) | 0.2857 kgce/kg |
| | Slime | 8363–12,545 J/kg (2000–3000 kcal/kg) | 0.2857–0.4286 kgce/kg |
| Coke | | 28,435 J/kg (6800 kcal/kg) | 0.9714 kgce/kg |
| Crude oil | | 41,816 J/kg (10,000 kcal/kg) | 1.4286 kgce/kg |
| Fuel oil | | 41,816 J/kg (10,000 kcal/kg) | 1.4286 kgce/kg |
| Oil field natural gas | | 3893 kJ/m$^3$ (9310 kcal/m$^3$) | 1.3300 kgce/m$^3$ |
| Gas field natural gas | | 35,544 kJ/m$^3$ (8500 kcal/m$^3$) | 1.2143 kgce/m$^3$ |
| Mine gas | | 14,636–16,726 kJ/m$^3$ (3500–4000 kcal/m$^3$) | 0.5000–0.5714 kgce/m$^3$ |
| Thermal power (energy calorific value) | | —— | 0.03412 kgce/MJ |
| Electric power (energy calorific value) | | 3600 kJ/Wh (860 kcal/kWh) | 0.1229 kgce/kWh |
| Electric power (energy equivalent value) | | Calculated based on standard coal consumption of thermal power generation in the same year | |

## 2.1.2 *Non-fossil Energy Resources*

1. Nuclear resources

Nuclear energy resources are the resources that generate energy in nuclear reaction and can be divided into nuclear fission energy resources and nuclear fusion energy resources. The main raw material for nuclear fission is uranium, and the raw material of nuclear fusion is hydrogen isotope. Uranium is the most important natural nuclear fuel, and its reserves in the earth's crust reserves total several billion tons, but its distribution is very scattered. Most of the uranium ore is deposited in the land in the form of oxide state, and the average uranium content of ore is 0.1–0.5%. Although the seawater contains a lot of uranium, the concentration is very low and cannot be used at present. According to the International Atomic Energy Agency (IAEA) statistics, by the end of 2011, total amount of uranium resources with the proved mining cost below $269/kg uranium, $269/kg uranium, $269/kg uranium, and $269/kg uranium was 7.0966, 5.3272, 3.0785, and 0.6809 million t, respectively. The world's top five countries with the largest proved uranium reserves with the mining cost below $130/kg uranium were Australia, Kazakhstan, Russia, Canada, and Niger, respectively. Total amount of proved uranium reserves of the top five countries accounted for 68.8% of the world's total. Australia's proved uranium reserves with the mining cost below $130/kg uranium were 1.6616 million t, the largest in the world, accounting for 31.2% of the world's total. China's proved uranium reserves with the mining cost below $130/kg uranium were 0.1661 million t, accounting for 3.1% of the world's total. See Table 2.4 for the world's top five countries with the largest proved uranium reserves (with the mining cost below $130/kg uranium) in 2011.

2. Hydropower resources

Hydropower resources generally refer to the electric energy generated by the hydrogenerator which is driven by the potential energy and kinetic energy of

**Table 2.4** World's top five countries with the largest proved uranium reserves (with the mining cost below $130/kg uranium) in 2011

| Country | Proved uranium reserves (10,000 t) | Proportion (%) |
|---|---|---|
| World | 532.72 | |
| Australia | 166.16 | 31.2 |
| Kazakhstan | 62.91 | 11.8 |
| Australia | 48.72 | 9.1 |
| Canada | 46.87 | 8.8 |
| Niger | 42.10 | 7.9 |

running rivers. According to the World Energy Council statistics, the world's theoretical reserves of hydropower resources are about 40 trillion kWh/a of which 16 trillion kWh/a were technically exploitable hydropower resources. The development and utilization of hydropower resources in developed countries have reached a high level in general. The utilization rate of technically exploitable hydropower resources of Italy, France, Norway, Japan, and Sweden is higher than 40%; that of Canada and Austria, 40–50%; that of Mexico and Spain, 20–40%. China's utilization rate of technically exploitable hydropower resources is 23.4%, and there is still a lot of room for development. The world's top five countries with the largest technically exploitable hydropower resources are China, Russia, America, Brazil, and Canada, respectively. The amount of technically exploitable hydropower resources of the top five countries accounts for 47.4% of the world's total. See Table 2.5 for the world's top five countries with the largest hydropower resources.

3. Wind energy resources

Wind energy is one of the most important energy sources, and wind energy resources on the earth are very rich. According to the estimates of World Energy Council, the annual average wind speed in 27% of the earth's land area is higher than 5 m/s (10 m above the ground). In 1981, World Meteorological Organization published World-wide Wind Energy Resource Assessment and Distribution Map in which worldwide wind energy resources are categorized into ten grades by average wind energy density and corresponding annual average wind speed. Continental wind energy resources are generally below Grade 7, and the grade of wind energy resources in windy regions such as the Western United States, coastal areas in Northwestern Europe, peak of Ural Mountains, and Black Sea region is high. See Table 2.6 for world wind energy resource distribution.

4. Solar energy resources

**Table 2.5** World's top five countries with the largest hydropower resources

| Country | Technically exploitable hydropower resources (TWh/a) | Technically exploitable hydropower resources per capita (MWh) | Economically exploitable hydropower resources (TWh/a) | Economically exploitable hydropower resources per capita (MWh) |
|---|---|---|---|---|
| World | 15,955 | | | |
| China | 2474 | 1.8 | 1753 | 1.3 |
| Russia | 1670 | 11.8 | 852 | 6.0 |
| America | 1339 | 4.3 | 376 | 1.2 |
| Brazil | 1250 | 6.4 | 818 | 4.2 |
| Canada | 827 | 24.0 | 536 | 15.5 |

*Notes* Per capita data are calculated based on the IEA 2011 population data
*Data source* World Energy Council; 2010 Survey of Energy Resources

**Table 2.6** World wind energy resource distribution

| Region | Land area ($km^2$) | Area of grades 3–7 wind energy resources ($km^2$) | Proportion of area of grades 3–7 wind energy resources (%) |
|---|---|---|---|
| North America | 19,339 | 7876 | 41 |
| Latin America and Caribbean Sea | 18,482 | 3310 | 18 |
| Western Europe | 4742 | 1968 | 42 |
| Eastern Europe and CIS | 23,049 | 6783 | 29 |
| Middle East and North Africa | 8142 | 2566 | 32 |
| Sub-Saharan Africa | 7255 | 2209 | 30 |
| Pacific Region | 21,354 | 4188 | 20 |

*Data source* www.kepu.neti.cn

Regions with optimal duration of sunshine in the world include North Africa, the Middle East, the Southwestern United States and Mexico, southern Europe, Australia, South Africa, South America east coast, east coast of South America, west coast of South America, and western regions of China. North Africa is one of the regions in the world with the most annual solar radiation. Technically exploitable solar energy resources in Algeria are about 169.44 trillion kWh; the annual solar radiation in southern European exceeds 7200 MJ/$m^2$; technically exploitable solar energy resources in Spain are about 1646 billion kWh.

## 2.2 Overview of China's Energy Resources

### *2.2.1 Total Amount of Energy Resources*

#### 2.2.1.1 Energy Distribution Characteristics

China's proved reserves of conventional primary energy are 925.8 billion tons of standard coal (China Energy Development Report 2007). 56.03% of the proved reserves are distributed in 12 provinces in Central and Western China; 34.15%, in 6 provinces in Central China; less than 10%, in eastern China (4.13% in the three northeastern provinces); only 5.69%, in 10 provinces in eastern coastal areas. See Fig. 2.2 for distribution of China's conventional primary energy resources.

China's coal resources are widely distributed. Rich in West China and North China and poor in East China and South China, coal resources are distributed in a reverse of the regional economic development. Oil resources are relatively

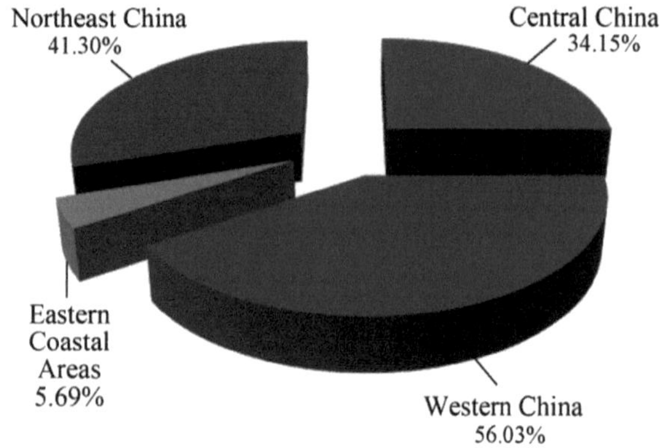


**Fig. 2.2** Distribution of China's conventional primary energy resources

concentrated. Conventional oil resources are mainly distributed in the eastern and western regions; heavy oil resources, in the eastern and coastal areas; low permeability oil resources, in the eastern, central, and western regions. The distribution of natural gas resources is relatively concentrated, and its depth distribution is characterized by deep burial, shallow in the east, deep in the west, shallow in offshore areas, and deep in high seas. See Fig. 2.3 for China's conventional primary energy structure.

Hydropower resources are rich in the western regions and poor in the eastern regions. Particularly, southwestern regions have the richest hydropower resources. As a result, hydropower resources are very unevenly distributed all over China. Due to the impact of the monsoon climate, the annual runoff and intra-annual runoff of most rivers are unevenly distributed. The difference in seasonal water flow is quite large, and water flow is unevenly distributed in time.

#### 2.2.1.2 Fossil Energy Resources

China has abundant energy resources. However, the energy resource endowment is not high and the per capita share is far below the world average. In recent years, the external energy and resource dependence has risen. China's energy resources are

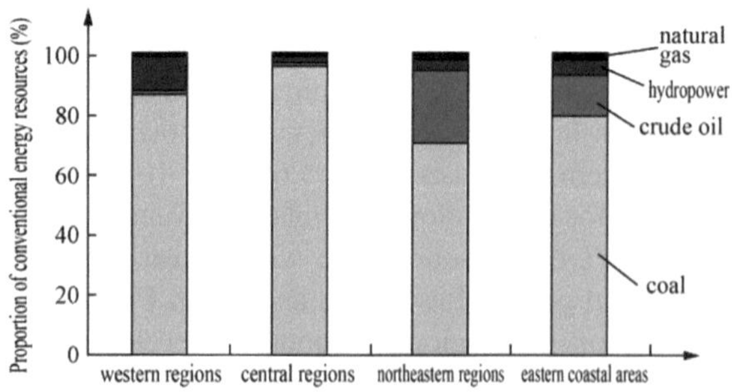


**Fig. 2.3** China's conventional primary energy structure

characterized by abundant coal, lacking oil, and shortage of gas, and coal is China's main energy source. China's coal reserves are ranked third in the world, while China's proved oil and natural gas reserves are relatively inadequate, ranked the 14th and 15th in the world, respectively. The combustible ice (methane gas) discovered on the Qinghai–Tibet Plateau is recognized as one of the largest new energy sources on earth that have not yet been developed.

Due to China's large population and rapid economic and social development, energy demand continues to grow rapidly and supplies of energy resources are relatively inadequate. China's population accounts for nearly 20% of the world's population, but the per capita ownership of major fossil fuels is much lower than the world's average. Oil reserves per capita are 1.78t, only accounting for 5.15% of the world's average; natural gas reserves per capita are 2301 $m^3$, only accounting for 8.38% of the world's average; coal reserves per capita are 85.13t, accounting for 67.49% of the world's average. The reserves-to-production ratio is far below the world's average. The reserves-to-production ratio of oil is 21.6% of the world's average; natural gas, 51.9%; and coal, 28.4%. See Table 2.7 for comparison of remaining recoverable reserves of fossil energy resources in China and the world in 2013.

Comparison of primary energy consumption structure of major countries in the world in 2013 is shown in Fig. 2.4. It can be seen from it that China's coal consumption proportion is the highest in the world, while oil and natural gas consumption proportions are the lowest in the world.

China's proved coal reserves are 114.5 billion t, ranked the third in the world, only preceded by America and Russia. China has a great variety of widely

**Table 2.7** Comparison of remaining recoverable reserves of fossil energy resources in China and the world in 2013

| Category of resource | | Unit | China | World | China/world (%) |
|---|---|---|---|---|---|
| Remaining proved recoverable reserves | Oil | 100 million t | 25 | 2382 | 1.05 |
| | Natural gas | Trillion $m^3$ | 3.30 | 185.70 | 1.78 |
| | Coal | 100 million t | 1145 | 8915 | 12.84 |
| Population (2011) | | 100 million people | 13.61 | 70.57 | 19.29 |
| Remaining recoverable reserves per capita | Oil | t | 1.84 | 33.75 | 5.44 |
| | Natural gas | $m^3$ | 2424.69 | 26,314.30 | 9.21 |
| | Coal | t | 84.13 | 126.33 | 66.60 |
| Reserves-to-production ratio | Oil | Year | 11.90 | 53.30 | |
| | Natural gas | Year | 28.00 | 55.10 | |
| | Coal | Year | 31 | 113 | |

*Data source* BP Energy Outlook 2014; http://www.stats.gov.cn/

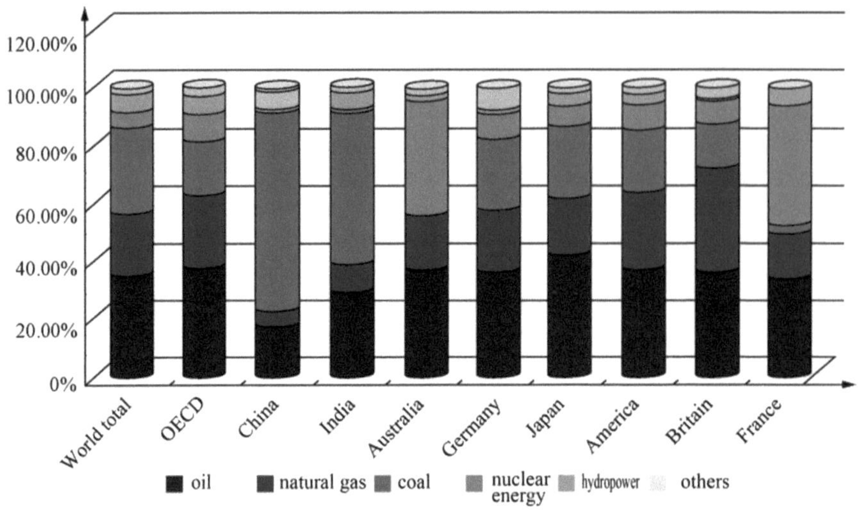


**Fig. 2.4** Comparison of primary energy consumption structure of major countries in the world in 2013

distributed rich coal resources. Geographically, China's coal resources are unevenly distributed, rich in the north and west and poor in the south and east. Coal resources are mainly distributed in North China and Northwest China, especially in Shanxi, Shaanxi and Inner Mongolia. The coal resources in Shanxi, Shaanxi, and Inner Mongolia account for 60% of China's total, while those in Sichuan, Yunnan, Guizhou, and Chongqing, 9%.

China's proved crude oil reserves are 2.5 billion t, only accounting for 1.1% of the world's total. China's oil reserves have continued to rise. Since 2005, the growth rate of the remaining proven recoverable reserves of oil in western land and seas has reached 9.6 and 4.6%, respectively. The growth of oil reserves is mainly attributed to the growth in Ordos, Junggar, Tarim, and other basins. Offshore oil reserves are mainly concentrated in Bohai Gulf and Pearl River Mouth Basin.

China's proved natural gas reserves are 3.3 trillion $m^3$, accounting for 1.8% of the world's total. Onshore proven natural gas reserves in West China are mainly concentrated in Tarim, Ordos, Sichuan, and other basins and account for 85% of China's total, and the annual average growth rate has reached 7.8% since 2005. In 2012, China's onshore natural gas exploration continued to maintain a good momentum of development and major progresses were made in the Sichuan Basin Mill Creek Sinian, Xujiahe Formations in Central Sichuan, middle shallow layer in Western Sichuan, northwestern part of Ordos Basin, southern part of Songliao Basin, Changyuan, and other areas. The natural gas reserves in the sea areas have remained stable since 2005, and the deep-seated gas potential in Pearl River Mouth Basin is huge.

#### 2.2.1.3 Non-fossil Energy Resources

China is rich in hydropower resources, and China's technically exploitable hydropower resources are ranked the first in the world. According to the results of review of China's hydropower resources in 2005, China's economically exploitable hydropower installed capacity was 400k MWk MW and technically exploitable hydropower installed capacity was 540k MW. In addition, China's small hydropower resources are widely distributed and technically exploitable small hydropower installed capacity reached 128k MWk MW. China's hydropower resources are extremely unevenly distributed, rich in the west and poor in the east. Particularly the hydropower resources in Southwest China are the most abundant. Due to the impact of the monsoon climate, the annual runoff and intra-annual runoff of most rivers are unevenly distributed. The difference in seasonal water flow is quite large, and water flow is unevenly distributed in time.

China's wind energy resources are mainly concentrated in the onshore areas of North China, Northwest China, and Northeast China and eastern coastal areas. According to the results of the National Wind Energy Resource Assessment 2009 issued by China Meteorological Bureau, the potentially exploitable wind energy resources above Grade 3 (with the wind energy density no less than 300 $W/m^2$) 50 m above the ground in onshore areas were about 2.38 billion kW; the potentially exploitable wind energy resources above Grade 3 50 m above the ground in offshore areas of 5–25 m in depth were about 200k MWk MW; China's technically exploitable wind energy resources totaled 0.7 billion to 1.2 billion kW.

China has abundant solar energy resources, and its annual solar radiation is 1050–2450 kWh/($m^2$ a). The characteristic line with an annual average daily radiation intensity of 180 $W/m^2$ runs southwestward from the central Inner Mongolia to central Yunnan along the eastern edge of the Qinghai–Tibet Plateau, dividing China into two parts, Northwest China with solar energy resources above 180 $W/m^2$ and Southeast China with solar energy resources below 180 $W/m^2$. See Table 2.8 for distribution of China's solar energy resources.

### 2.2.2 *Energy Production and Consumption*

Energy consumption is the end of energy demand which drives the whole energy system. According to BP statistics, in 2013, the world's primary energy production totaled 18.04 billion tons of standard coal, a year-on-year increase of 2.3%; total consumption increased by 2.1%, lower than the average annual growth rate of 2.7% in the past 10 years (BP). In 2013, China's primary energy production totaled 3.33 billion tons of standard coal, a year-on-year increase of 4.8%. In a breakdown, the production of coal, oil, and natural gas was 2.539 billion tons of standard coal, 295 million tons of standard coal, and 143 million tons of standard coal, respectively, a year-on-year increase of 2.6, 2.1, 4.4%; China's primary energy consumption

**Table 2.8** Distribution of China's solar energy resources

| Name | Sign | Index [kWh/($m^2$ a)] | Proportion in China's land area (%) | Regions |
|---|---|---|---|---|
| Extremely rich zone | I | ≥ 1750 | 17.4 | Most parts of Tibet, southern Xinjiang, Qinghai, Gansu, and Western Inner Mongolia |
| Very rich zone | II | 1400–1750 | 42.7 | Most parts of Xinjiang, eastern Qinghai and Gansu, Ningxia, Shaanxi, Shanxi, Hebei, northeastern Shandong, eastern Inner Mongolia, southwestern Northeast China, Yunnan, western Sichuan |
| Rich zone | III | 1050–1400 | 36.3 | Heilongjiang, Jilin, Liaoning, Anhui, Jiangxi, southern Shaanxi, northeastern Inner Mongolia, Henan, Shandong, Jiangsu, Zhejiang, Hubei, Hunan, Fujian, Guangdong, Guangxi, eastern Hainan, Sichuan, Guizhou, southeast corner of Tibet and Taiwan area |
| Average zone | IV | <1050 | 3.6 | Central Sichuan, northern Guizhou, northwestern Hunan |

*Data source* Solar Energy Volume of China's Renewable Energy Development Strategy Research Series published by China Electric Power Press in 2010

totaled 3.91 billion tons of standard coal, a year-on-year increase of 4.7%. See Table 2.9 for China's primary energy production and consumption.

#### 2.2.2.1 Characteristics of China's Energy Production and Consumption

(1) Relative shortage of resources
(2) Rapid growth of energy production. In China's industrialization process since the reform and opening up, between 1981 and 2006, with a GDP growth rate of 9.7%, China was one of the countries with the fastest growing GDP in the world and in the first 20 years (between 1981 and 2000) China made significant stage achievement of "doubling energy consumption and redoubling total volume of economy." Since the "11th Five-Year Plan" period, both China's total energy production and consumption have been increasing; China's primary energy production has been ranked the first in the world for years in a row, and it reached 3.4 billion tons of standard coal in 2013. See Table 2.10 and Fig. 2.5 for China's total energy production and structure between 1980 and 2013. Figure 2.6 shows the comparison of total energy production between China and other major countries.

**Table 2.9** China's primary energy production and consumption

| Indicator | | Unit | 1990 | 2000 | 2010 | 2011 | 2012 | Data source |
|---|---|---|---|---|---|---|---|---|
| Population | | 10,000 people | 114,333 | 126,743 | 134,091 | 134,735 | 135,404 | China Statistical Yearbook |
| Land area | | 10,000 km$^2$ | 960 | | | | | China Statistical Yearbook |
| Energy production | | One million tons of standard coal | 1039.22 | 1289.78 | 2969.16 | 3179.87 | 3318.48 | China Statistical Yearbook |
| Breakdown | Coal | | 771.1 | 988.55 | 2274.38 | 2473.93 | 2538.64 | China Statistical Yearbook |
| | Oil | | 197.45 | 232.28 | 290.98 | 289.37 | 295.34 | China Statistical Yearbook |
| | Natural gas | | 20.78 | 36.46 | 124.70 | 136.73 | 142.69 | China Statistical Yearbook |
| Primary energy supply | | One million tons of standard coal | 879.32 | 1174.75 | 2530.57 | 2742.62 | – | IEA Energy Balances of non-OECD Countries |
| Breakdown | Coal | | 533.1 | 694.31 | 1696.42 | 1866.38 | – | IEA Energy Balances of non-OECD Countries |
| | Oil | | 122.03 | 227.46 | 431.81 | 446.15 | – | IEA Energy Balances of non-OECD Countries |
| | Natural gas | | 12.8 | 23.2 | 91.7 | 110.24 | – | IEA Energy Balances of non-OECD Countries |
| Primary energy consumption | | | 664.6 | 980.3 | 2338 | 2540.8 | 2735.2 | BP Statistical Review of World Energy 2013 |
| Final energy consumption | | | 669.43 | 824.97 | 1536.04 | 1643.34 | – | IEA Energy Balances of non-OECD Countries |
| Energy self-sufficiency rate | | | 1.01 | 0.96 | 0.91 | 0.89 | – | IEA statistics over the years |

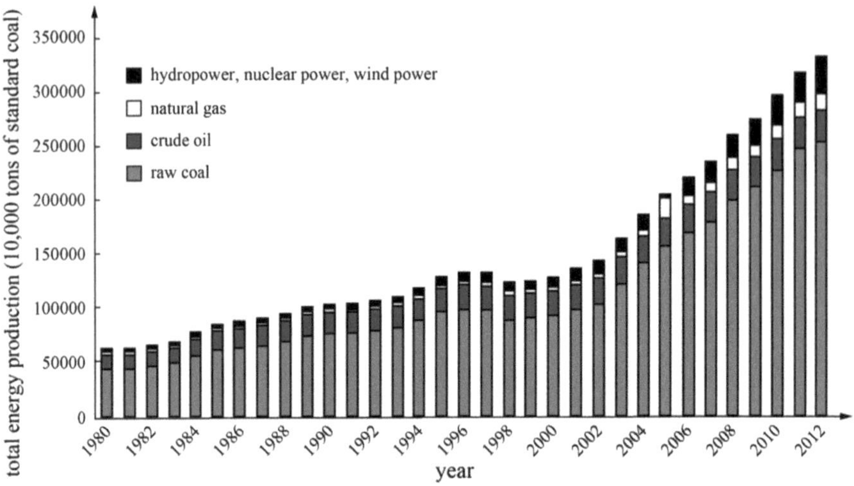


**Fig. 2.5** China's total energy production and structure between 1980 and 2013

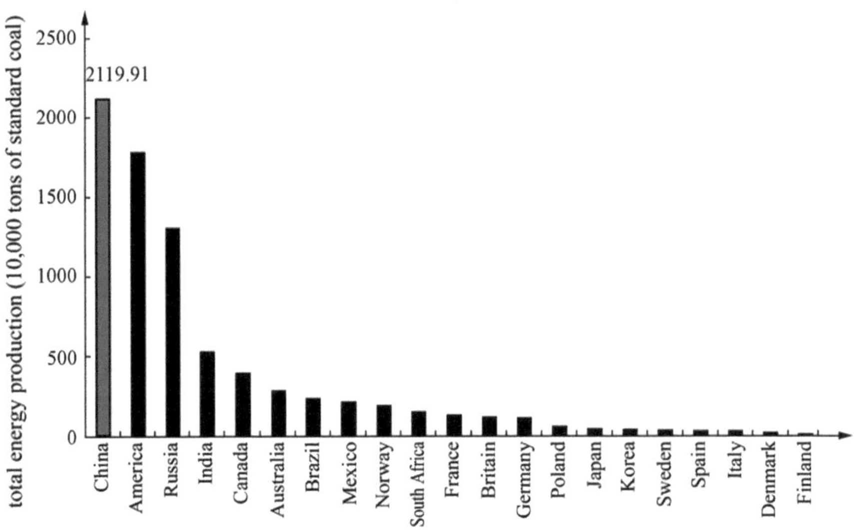


**Fig. 2.6** Comparison of total energy production between China and other major countries in 2011. *Data source* China's data are converted from *China Statistical Yearbook 2012* issued by National Bureau of Statistics of the People's Republic of China; the data of other countries are based on the statistics of IEA in 2012

In terms of energy production growth, China's energy production can be roughly divided into three stages: the slow growth stage before 1997; relatively flat growth stage between 1998 and 2001 which is characterized by violent fluctuations of total energy production, the entering of energy products into the buyers' market, and the relative surplus of some energy sources; the rapid growth stage since 2001 which is characterized by significantly enhanced energy supply capacity. China's energy production in 2013 is 2.4 times as much as that in 2001, and with an annual average growth rate of 7.8, China became the world's largest energy producer.

In terms of structure, China's primary energy production has long been dominated by coal production. The proportion of raw coal production in China's total energy production has long been above 70%, which is determined by the natural endowment of China's energy resources. The proportion of non-fossil energy production has been increasing and reached 10.9% in 2013.

(3) Rapid expansion of energy demand. China has become the fastest growing energy consumer in the world. According to BP Statistical Review of World Energy 2014, in 2013 the world's primary energy consumption increased by 2.3%, almost equal to the historical average, but much lower than the growth rate of 5.1% in 2010. China contributed 22.4% of the world's energy consumption increment. In 2013, China's primary energy consumption reached 3.91 billion tons of standard coal and China became the world's largest primary energy consumer. See Fig. 2.7 for China's total energy production and consumption in the past 30 years. See Figs. 2.8 and 2.9 for comparison of primary energy consumption and final energy consumption between China and other major countries.
(4) Increasing external energy dependence. In the first 10 years of China's rapid economic growth (in the 1980s), energy self-sufficiency could be basically ensured. Since 1992, China's total energy consumption has exceeded its total energy production and the difference between energy production and energy consumption has become increasingly large. In 2009, China's net coal imports appeared for the first time and its external crude oil dependence exceeded the 50% guard line for the first time. According to data released by the General Administration of Customs of the People's Republic of China, in 2012 China's

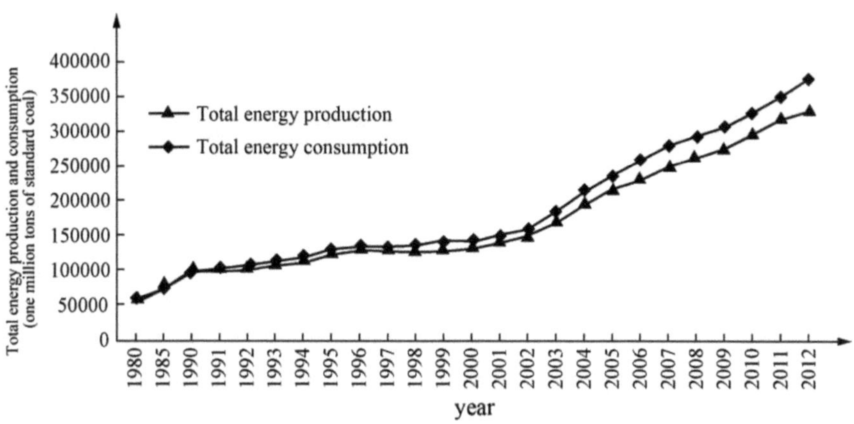


**Fig. 2.7** Comparison of China's total energy production and consumption in the past 30 years

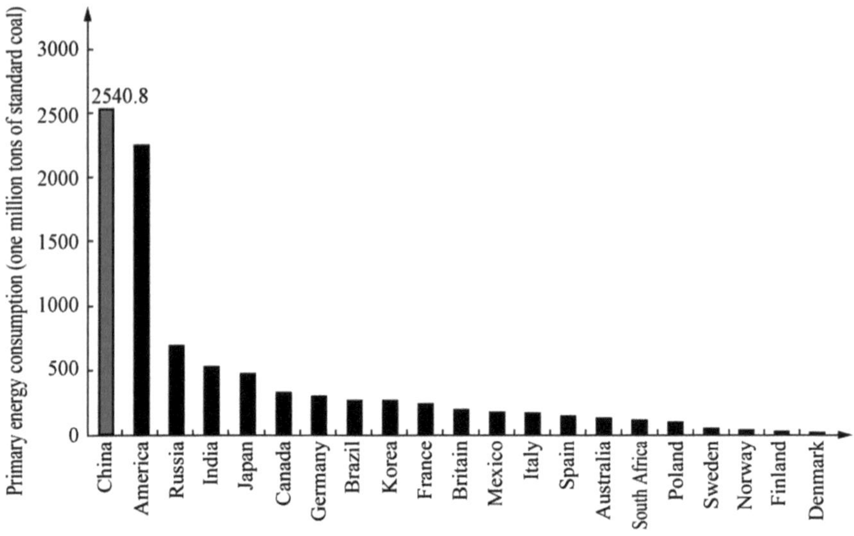


**Fig. 2.8** Comparison of primary energy consumption between China and other major countries in 2013. *Data source* BP Statistical Review of World Energy 2014

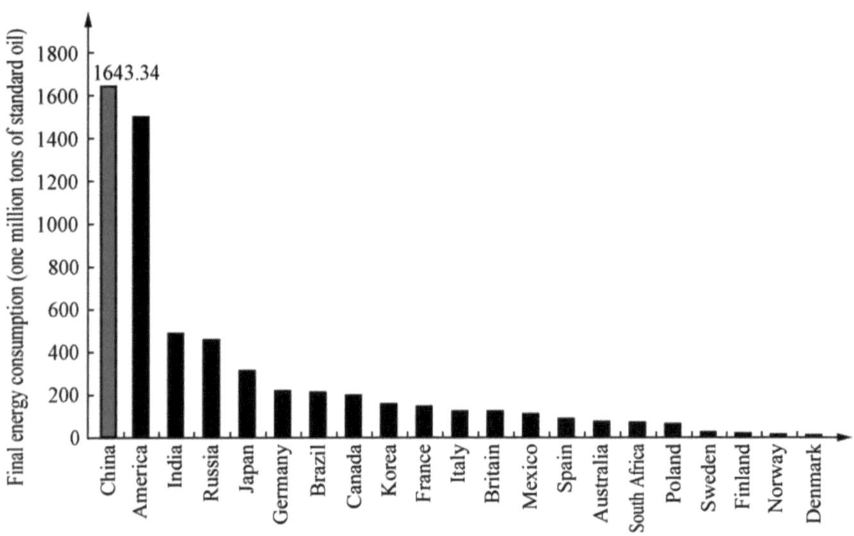


**Fig. 2.9** Comparison of final energy consumption between China and other major countries in 2013. *Data source* The data of China, India, and Japan are based on IEA Energy Balances of non-OECD Countries; the data of other countries are based on the statistics of IEA in 2012

crude oil imports reached 271 million tons, up 6.8% year-on-year and China's coal imports reached 290 million tons, up 29.8% year-on-year. According to BP Statistical Review of World Energy 2013, in 2012 China imported

**Table 2.10** China's total energy production and structure between 1980 and 2013

| Year | Total energy production (10,000 tons of standard coal) | Growth rate (%) | Proportion in total energy production (%) | | | |
|---|---|---|---|---|---|---|
| | | | Raw coal | Crude oil | Natural gas | Hydropower, nuclear power and wind power |
| 1980 | 63,735 | −1.28 | 69.4 | 23.8 | 3 | 3.8 |
| 1981 | 63,227 | −0.80 | 70.2 | 22.9 | 2.7 | 4.2 |
| 1982 | 66,778 | 5.62 | 71.3 | 21.8 | 2.4 | 4.5 |
| 1983 | 71,270 | 6.73 | 71.6 | 21.3 | 2.3 | 4.8 |
| 1984 | 77,855 | 9.24 | 72.4 | 21 | 2.1 | 4.5 |
| 1985 | 85,546 | 9.88 | 72.8 | 20.9 | 2 | 4.5 |
| 1986 | 88,124 | 3.01 | 72.4 | 21.2 | 2.1 | 4.3 |
| 1987 | 91,266 | 3.57 | 72.6 | 21 | 2 | 4.4 |
| 1988 | 95,801 | 4.97 | 73.1 | 20.4 | 2 | 4.5 |
| 1989 | 101,639 | 6.09 | 74.1 | 19.3 | 2 | 4.6 |
| 1990 | 103,922 | 2.25 | 74.2 | 19 | 2 | 4.8 |
| 1991 | 104,844 | 0.89 | 74.1 | 19.2 | 2 | 4.7 |
| 1992 | 107,256 | 2.3 | 74.3 | 18.9 | 2 | 4.8 |
| 1993 | 111,059 | 3.55 | 74.0 | 18.7 | 2 | 5.3 |
| 1994 | 118,729 | 6.91 | 74.6 | 17.6 | 1.9 | 5.9 |
| 1995 | 129,034 | 8.68 | 75.3 | 16.6 | 1.9 | 6.2 |
| 1996 | 132,616 | 2.78 | 75.2 | 17 | 2 | 5.8 |
| 1997 | 132,410 | −0.16 | 74.1 | 17.3 | 2.1 | 6.5 |
| 1998 | 124,250 | −6.16 | 71.9 | 18.5 | 2.5 | 7.1 |
| 1999 | 125,935 | 1.36 | 72.6 | 18.2 | 2.7 | 6.6 |
| 2000 | 128,978 | 2.42 | 72.0 | 18.1 | 2.8 | 7.2 |
| 2001 | 137,445 | 6.56 | 71.8 | 17 | 2.9 | 8.2 |
| 2002 | 143,810 | 4.63 | 72.3 | 16.6 | 3 | 8.1 |
| 2003 | 163,842 | 13.93 | 75.1 | 14.8 | 2.8 | 7.3 |
| 2004 | 187,341 | 14.34 | 76.0 | 13.4 | 2.9 | 7.7 |
| 2005 | 205,876 | 9.89 | 76.5 | 12.6 | 3.2 | 7.7 |
| 2006 | 221,056 | 7.37 | 76.7 | 11.9 | 3.5 | 7.9 |
| 2007 | 235,415 | 6.5 | 76.6 | 11.3 | 3.9 | 8.2 |
| 2008 | 260,552 | 10.68 | 76.8 | 10.5 | 4.1 | 8.6 |
| 2009 | 274,618 | 5.4 | 77.3 | 9.9 | 4.1 | 8.7 |
| 2010 | 296,916 | 8.2 | 76.6 | 9.8 | 4.2 | 9.4 |
| 2011 | 317,987 | 7.1 | 77.8 | 9.1 | 4.3 | 8.8 |
| 2012 | 331,848 | 4.4 | 76.5 | 8.9 | 4.3 | 10.3 |
| 2013 | 340,000 | 2.46 | 75.6 | 8.9 | 4.6 | 10.9 |

*Notes* Coefficient for converting power into standard coal is calculated based on the current annual average coal consumption for power generation; based on *China Statistical Yearbook 2007* issued by National Bureau of Statistics of the People's Republic of China. The data of 2009–2013 are calculated based on the *International Energy and Power Statistical Manual*

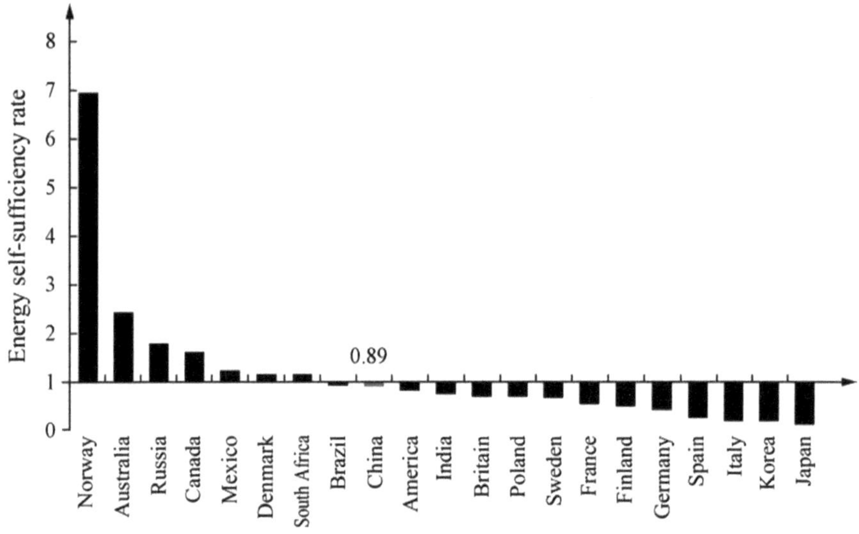


**Fig. 2.10** Comparison of energy self-sufficiency rate between China and other major countries in 2011. *Data source* The statistics of IEA in 2012

41.426 billion m$^3$ of natural gas. See Fig. 2.10 for comparison of energy self-sufficiency rate between China and Other Major Countries.

(5) Over-reliance on fossil fuels. China's energy production and consumption have long been dominated by coal. Between the late 1950s and the early 1960s, the proportion of coal in China's total energy production was higher than 90% and afterward it declined. In 2013, coal accounted for 75.6% of total energy production and 66.0% of total energy consumption. With the strengthened energy restructuring, the proportion of coal in energy consumption has gradually decreased, while the proportion of natural gas and hydropower in energy consumption has constantly increased.

See Table 2.11 for China's total energy consumption and structure between 1980 and 2013. See Fig. 2.11 for China's main energy consumption structure changes.

#### 2.2.2.2 Oil Production and Consumption

According to *BP Statistical Review of World Energy 2014*, in 2013 China's oil production was 4.18 million barrels/day, ranked the fourth in the world, and only preceded by Saudi Arabia whose oil production was 11.525 million barrels/day, Russia whose oil production was 10.788 barrels/day, and America whose oil production was 10.003 barrels/day. By strengthening key production capacity construction, promoting the application of new technologies and fine development of

**Table 2.11** China's total energy consumption structure and structure between 1980 and 2013

| Year | Total energy consumption (10,000 tons of standard coal) | Proportion in total energy consumption (%) | | | |
|---|---|---|---|---|---|
| | | Raw coal | Crude oil | Natural gas | Hydropower and other energy sources |
| 1980 | 60,275 | 72.2 | 20.7 | 3.1 | 4.0 |
| 1985 | 76,682 | 75.8 | 17.1 | 2.2 | 4.9 |
| 1990 | 98,703 | 76.2 | 16.6 | 2.1 | 5.1 |
| 1992 | 109,170 | 75.7 | 17.5 | 1.9 | 4.9 |
| 1993 | 115,993 | 74.7 | 18.2 | 1.9 | 5.2 |
| 1994 | 122,737 | 75.0 | 17.4 | 1.8 | 5.7 |
| 1995 | 131,176 | 74.6 | 17.5 | 1.8 | 6.1 |
| 1996 | 138,948 | 74.7 | 18.0 | 1.7 | 5.5 |
| 1997 | 137,798 | 71.7 | 20.4 | 2.2 | 6.2 |
| 1998 | 132,214 | 69.6 | 21.5 | 2.2 | 6.7 |
| 1999 | 133,831 | 68.0 | 23.2 | 2.2 | 6.6 |
| 2000 | 138,553 | 66.1 | 24.6 | 2.50 | 6.8 |
| 2001 | 143,199 | 65.28 | 24.27 | 2.70 | 7.9 |
| 2002 | 151,797 | 66.32 | 23.41 | 2.56 | 7.7 |
| 2003 | 174,990 | 68.38 | 22.21 | 2.58 | 6.8 |
| 2004 | 203,227 | 67.99 | 22.23 | 2.60 | 7.1 |
| 2005 | 223,319 | 69.11 | 21.00 | 2.58 | 7.1 |
| 2006 | 246,000 | 69.40 | 20.44 | 3.00 | 7.2 |
| 2007 | 266,000 | 69.50 | 19.70 | 3.50 | 7.3 |
| 2008 | 291,448 | 70.3 | 18.3 | 3.7 | 7.7 |
| 2009 | 306,647 | 70.4 | 17.9 | 3.9 | 7.8 |
| 2010 | 324,939 | 68.0 | 19.0 | 4.4 | 8.6 |
| 2011 | 348,002 | 68.4 | 18.6 | 5.0 | 8.0 |
| 2012 | 390,740 | 68.5 | 17.7 | 4.7 | 9.1 |
| 2013 | 375,000 | 66.0 | 18.4 | 5.8 | 9.8 |

*Notes* data are from *China Energy Statistical Yearbook* over the years published by China Statistics Press (which are different from those in BP Statistical Review of World Energy)

old oil fields and improving unconventional development technologies, China has achieved the steady growth of crude oil production. In 2013, China's crude oil consumption was 10.756 barrels/day, ranked the second in the world, and only preceded by the USA after whose crude oil consumption was 18.887 barrels/day. See Fig. 2.12 for China's oil production and consumption between 2000 and 2011; see Fig. 2.13 for comparison of oil production between China and other major countries in 2011; see Fig. 2.14 for oil supply comparison.

In 2012, affected by macroeconomy and the operation of major oil consumption industries, China's crude oil consumption growth presented the falling and rising pattern; product oil consumption growth generally slowed down; diesel consumption growth dropped significantly; gasoline consumption continued to increase

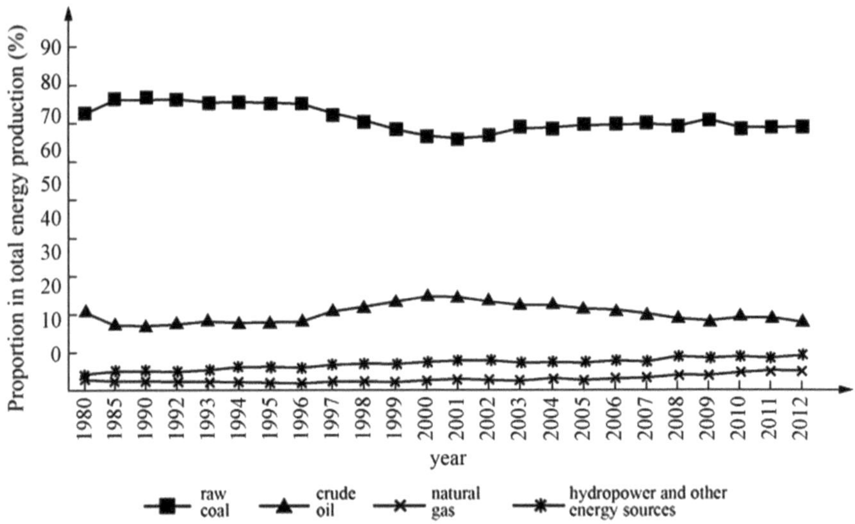


**Fig. 2.11** China's main energy consumption structure changes

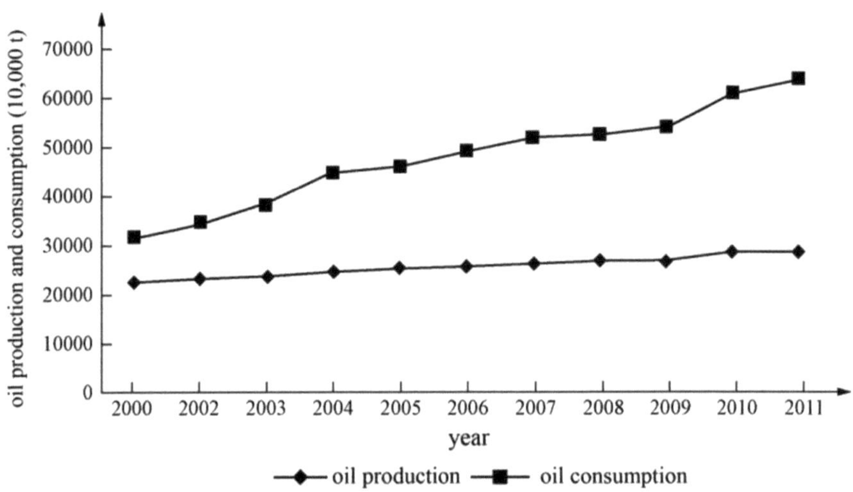


**Fig. 2.12** China's oil production and consumption between 2000 and 2011

substantially; kerosene consumption grew steadily. In 1993, China became a net oil importer. In 1995, China's oil import dependence increased to 5.3%. In 2000 driven by the increasing economic demand, China's dependence on imported oil approached 34%. Between 2000 and 2010, China's net oil imports increased by 13.67%. In 2012, China's oil imports reached 285 million t and its oil import

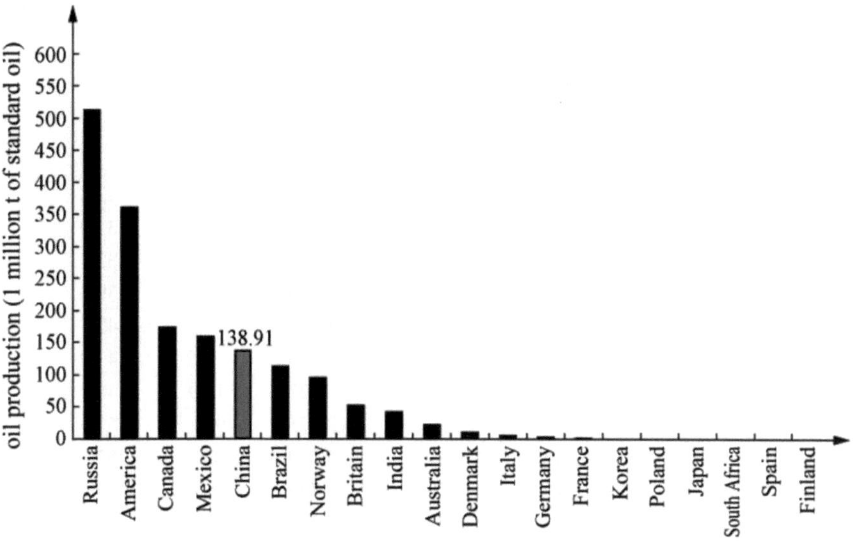


**Fig. 2.13** Comparison of oil production between China and other major countries. *Data source* China's data are converted from *China Statistical Yearbook 2012* issued by National Bureau of Statistics of the People's Republic of China; the data of other countries are based on the statistics of IEA in 2012

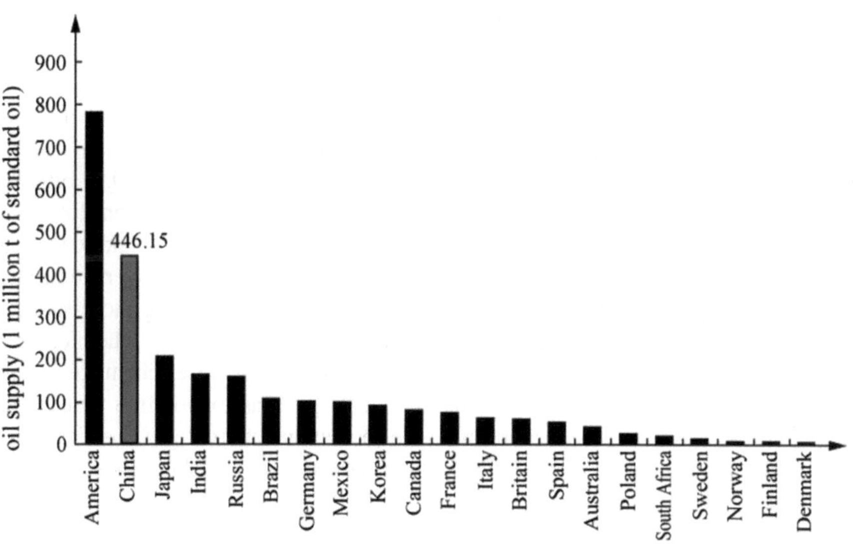


**Fig. 2.14** Comparison of oil supply between China and other major countries. *Data source* China's data are based on IEA *Energy Balances of non-OECD Countries*; the data of other countries are based on the statistics of IEA in 2012

**Table 2.12** China's oil consumption and import and export volume between 2000 and 2012

| Year | Consumption (10,000 t) | Import volume (10,000 t) | Export volume (10,000 t) | Net import volume (10,000 t) | Net import growth (%) | Import dependence (%) |
|---|---|---|---|---|---|---|
| 2000 | 22,439 | 9749 | 2172 | 7577 | – | 33.8 |
| 2001 | 22,838 | 9118 | 2047 | 7071 | −6.7 | 31.0 |
| 2002 | 24,780 | 10,269 | 2139 | 8130 | 15.0 | 32.8 |
| 2003 | 27,126 | 13,190 | 2541 | 10,649 | 31.0 | 39.3 |
| 2004 | 31,700 | 17,291 | 2240 | 15,051 | 41.3 | 47.5 |
| 2005 | 32,535 | 17,163 | 2888 | 14,275 | −5.2 | 43.9 |
| 2006 | 34,876 | 19,453 | 2626 | 16,827 | 17.9 | 48.2 |
| 2007 | 36,570 | 21,139 | 2664 | 18,475 | 9.8 | 50.5 |
| 2008 | 37,303 | 23,016 | 2946 | 20,070 | 8.6 | 53.8 |
| 2009 | 38,385 | 25,642 | 3917 | 21,725 | 8.2 | 56.6 |
| 2010 | 43,770 | 29,437 | 4079 | 25,358 | 16.7 | 57.9 |
| 2011 | 46,180 | 25,378 | 378 | 25,000 | 5.8 | 56.6 |
| 2012 | 47,341 | 27,102 | 498 | 26,604 | 10.4 | 58.3 |

*Data source* China Statistical Yearbook 2012

dependence reached 58.3%. See Table 2.12 for China's oil consumption and import and export volume between 2000 and 2012.

China's oil industry has formed a complete industrial system with certain international competitiveness. China has become the world's second largest oil and chemical product producer, and China's crude oil production ranks the fifth in the world. At the same time, China is also the world's second largest consumer of oil and chemical products and has become the oil and chemical product market center in the Asia-Pacific region. In 2011, the import and export value of China's petroleum and chemical industry totaled \$607.15 (about RMB 3789.5 billion), accounting for 16.7% of China's total value of foreign trade. Between 2005 and 2011, total value of imports and exports of China's petroleum and chemical industry increased from \$199.5 billion to \$607.15 billion, with an average annual growth rate of 34.1%. The whole petrochemical industry, covering more than 20 sub-sectors including the production of oil and gas resources, basic chemical raw materials, synthetic materials, agricultural chemicals, defense chemicals, new materials, specialty chemicals and fine chemicals such as exploration and development of oil and natural gas, oil refining, petrochemical, natural gas chemical, coal chemical, salt chemical, biochemical, can produce over 40,000 products in all categories and varieties.

#### 2.2.2.3 Coal Production and Consumption

China is the world's largest coal producer. In 2013, China's coal production reached 1840 million t of standard oil, accounting for 47.4% of the world's total. China's coal production has continued to maintain the growth momentum, but due to the impact of the macroeconomic slowdown, the growth rate fell sharply. China's primary energy consumption is dominated by coal, and China is the world's largest coal consumer. In 2013, China's coal consumption was 1925.3 million t of standard oil, accounting for 50.3% of the world's total, but the coal consumption growth slowed down significantly, the lowest since 2009. China's economic growth is heavily dependent on resource consumption, and the coal production capacity is mainly concentrated in Western China. The sustainable development of the coal industry is constrained by environmental protection. See Fig. 2.15 for China's coal production and consumption between 2000 and 2011; see Fig. 2.16 for comparison of coal production between China and other major countries; see Fig. 2.17 for comparison of coal supply between China and other major countries.

During the "11th Five-Year Plan" period, guided by the thought of "big base, big group and big integration," the industrial distribution of the coal industry became more concentrated; the proportion of coal production in Shanxi, Shaanxi, Inner Mongolia, Ningxia, and Xinjiang area increased, while most provinces (cities and regions) south of Yangtze River withdrew from coal development. The coal development mode of giving priority to the development of medium- and small-sized coal mines was changed into focusing on developing large- and medium-sized coal mines and the coal production of large- and medium-sized coal mines accounted for about 60% of total coal production. In terms of coal development means, manual operation was changed into modern mechanized production and the mining mechanization rate of key state-owned coal mines exceeded 80%. In

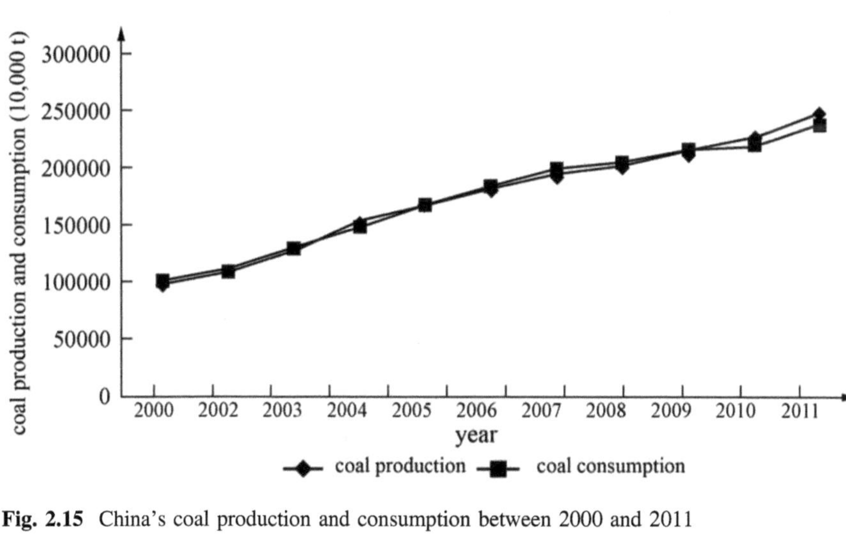


**Fig. 2.15** China's coal production and consumption between 2000 and 2011

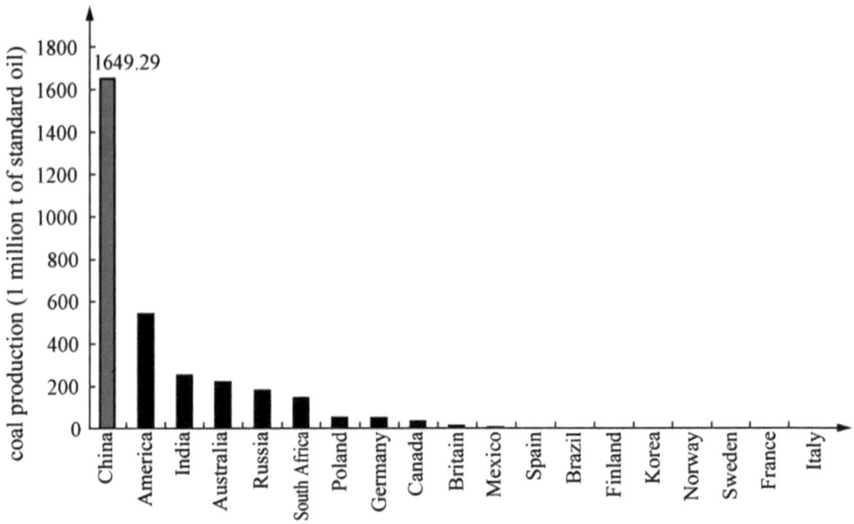


**Fig. 2.16** Comparison of coal production between China and other major countries. *Data source* China's data are converted from *China Statistical Yearbook 2012* issued by National Bureau of Statistics of the People's Republic of China; the data of other countries are based on the statistics of IEA in 2012

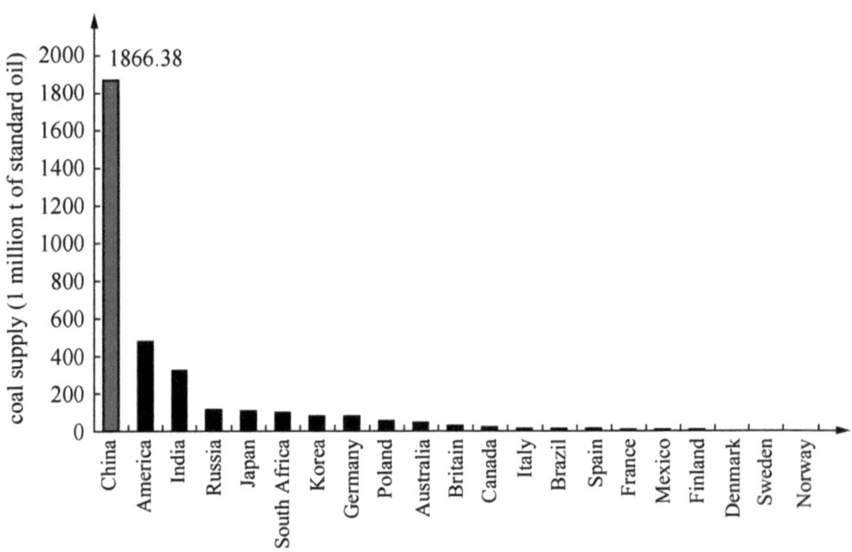


**Fig. 2.17** Comparison of coal supply between China and other major countries. *Data source* China's data are based on IEA *Energy Balances of non-OECD Countries*; the data of other countries are based on the statistics of IEA in 2012

terms of industrial structure, single coal production was changed into diversified development; in terms of product utilization, coal sales was changed into combination of coal sales and consumption; in terms of product structure, simple screening was changed into deep processing. In terms of nature of coal mines, the combination of state-owned, collective, and private coal mines changed into focusing on state holding coal mines and private coal mines. Thanks to the changes in the development layout, development mode, and economic structure, the coal industry has entered a new development age.

Since the "11th Five-Year Plan" period, the coal industry has adhered to relying on scientific and technological progress to enhance the capability of independent innovation. In terms of coal strategic planning and research, "controlling East China, Stabilizing Central China and Developing West China" has become the guiding ideology of "12th Five-Year Plan" for the Coal Industry. In terms of coal sector industrial restructuring, the structural optimization and transformation of the coal industry have been achieved by promoting the integration of coal resources and accelerating the elimination of backward production capacity and developing the coal power generation, coal carbonization, and coal gasification industry chain. In terms of ecological coal mine construction, the ecological environment in the mining areas has been improved through the development of the coal resource comprehensive utilization and green mining technologies and control of emissions of pollutants.

Guided by the national policy of constructing large coal bases and fostering large coal enterprise groups, 14 large coal bases made remarkable construction achievements; large coal enterprise groups developed rapidly; national coal supply capacity improved significantly. In 2013, China's coal production reached 3.68 t, up 0.8% year-on-year. The production of 14 large coal bases was about 3.4 billion t, accounting for about 92% of China's total, up 2% year-on-year. The raw coal production of each of the eight enterprises including Shenhua Group, China Coal, Datong Coal Mine Group, Shandong Energy Group, Jizhong Energy Group, Shaanxi Coal and Chemical Industry Group, Henan Energy and Chemical Industry Group, and Shanxi Coking Coal Group exceeded 100 million tons, and the raw coal production of these eight enterprises totaled 1.43 tons, accounting for about 38% of China's total. The coal industry concentration increased, and China's long-term stable coal supply capacity improved greatly. In terms of consumption, China's annual coal consumption reached 3.61 tons, while the coal consumption growth reduced to 2.6%. In terms of import, in 2013 China's coal imports reached 327 million tons and exports reached 7.51 million tons. The net coal imports reached 320 million tons, increasing by about 40 million tons as compared to 2012. Coal imports which were mainly concentrated in China's economically developed areas with large coal consumption had a significant impact on China's coal price.

By increasing the integration of resources and eliminating backward production capacity, the number of coal mines in China reduced from 25,000 in 2005 to 12,000 at the end of 2013, eliminating about 650 million tons of backward production capacity. With the successful experiment on the domestic production of fully mechanized coal mining equipment with an annual production capacity of 6 million tons and the industrialization of fully mechanized mining equipment with an annual production capacity of the tens of millions of tons as well as progress made in the R&D of key parts, China's coal mining machinery equipment manufacturing level has improved significantly, effectively promoting the construction of large-scale modern coal mines and optimizing the structure of coal production.

Accounting for the largest proportion in China's energy consumption, coal will remain China's dominating energy source for a long time and China's coal industry will shoulder the responsibility for ensuring safe and stable energy sully. In recent years, China's coal inventory has increased significantly and has remained over 300 million tons for 30 months in a row, resulting in the downturn in the coal market. There will be a long way to go for the Chinese coal market to digest the excess production capacity, reduce the coal inventory to a reasonable level, and achieve balanced coal supply and demand.

In 2012, China's total coal consumption was 3.52 billion tons and coal consumption was concentrated in East China, North China, Central China, and Northwest China, mainly in Shanxi, Shaanxi, Inner Mongolia, Hebei, Anhui, Zhejiang, Jiangsu, and Guizhou. Due to the impact of resource distribution difference and economic development demand, a large amount of coal was transferred inter-provincially and mainly transferred to East China and North China. China's coal consumption was mainly concentrated in the industrial sector including the power, metallurgy, construction material, and chemical industries.

#### 2.2.2.4 Natural Gas Production and Consumption

According to *BP Statistical Review of World Energy 2014*, in 2013 China's natural gas production was 117.1 billion $m^3$, up 9.5% year-on-year, accounting for 3.5% of the world's total; China's natural gas consumption was 161.6 billion $m^3$, up 10.8% year-on-year, accounting for 4.8% of the world's total; China's natural gas production ranked the sixth in the world, and consumption ranked the fourth in the world. See Fig. 2.18 for China's natural gas production and consumption between 2000 and 2011; see Fig. 2.19 for comparison of natural gas production between China and other major countries; see Fig. 2.20 for comparison of natural gas supply between China and other major countries.

#### 2.2.2.5 Electric Power Production and Consumption

1. Electric power production

In 2013, China's power generation equipment capacity reached 1257.68k MWk MW, up 9.7% over 2012. In a breakdown, hydropower capacity was 280.44k MW, up 12.4% over 2012; thermal power capacity was 870.09k MW, up 6.1% over 2012; nuclear power capacity was 14.66k MW, up 16.6% over 2012; grid-connected wind power capacity was 76.52k MW, up 24.6% over 2012;

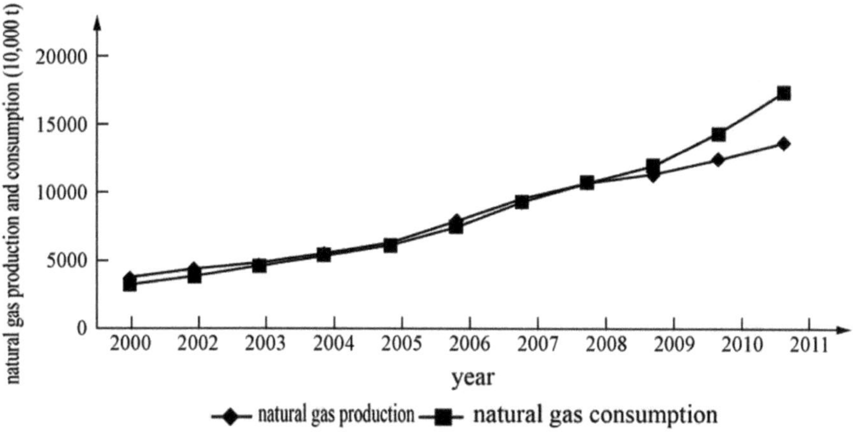


**Fig. 2.18** China's natural gas production and consumption between 2000 and 2011

grid-connected solar power capacity was 15.89k MW, up 365.8% over 2012. The loop length of transmission line of 220 kV and above was 54,390,000 km, up 7.57% over 2012; capacity of substation equipment of 220 kV and above was 2.782 billion kVA, up 9.03% over 2012. See Fig. 2.21 for China's installed capacity structure in 2013; see Fig. 2.22 for China's installed capacity changes between 2001 and 2013; see Fig. 2.23 for China's installed capacity structure changes between 2001 and 2013.

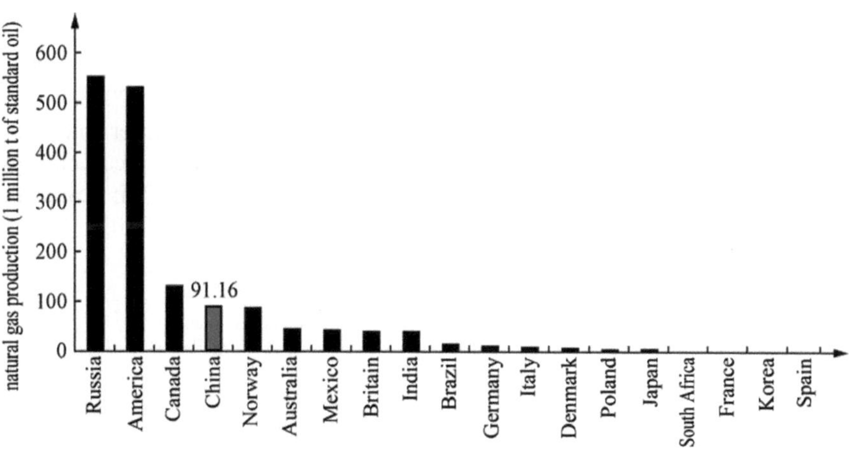


**Fig. 2.19** Comparison of natural gas production between China and other major countries. *Data source* China's data are converted from *China Statistical Yearbook 2012* issued by National Bureau of Statistics of the People's Republic of China; the data of other countries are based on the statistics of IEA in 2012

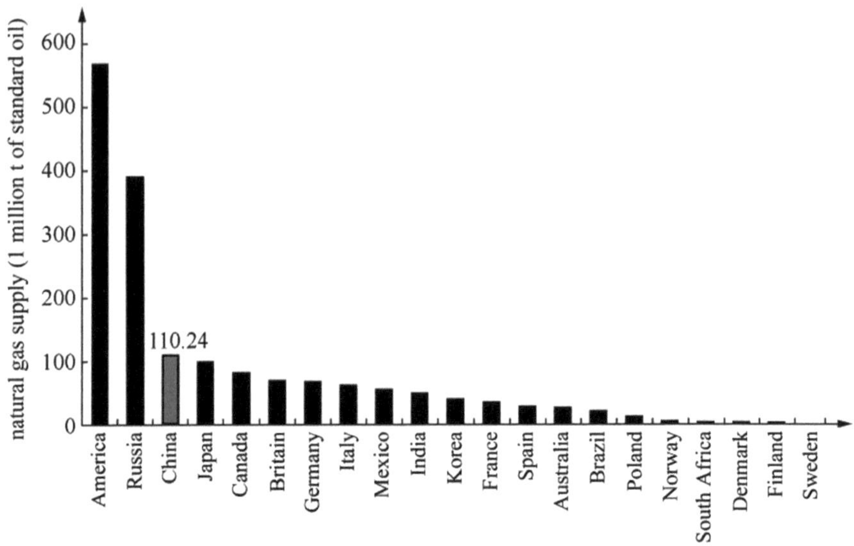


**Fig. 2.20** Comparison of natural gas supply between China and other major countries. *Data source* China's data are based on *IEA Energy Balances of non-OECD Countries*; the data of other countries are based on the statistics of IEA in 2012

In 2013, China's generating capacity was 5372.1 billion kWh, up 7.73% over 2012. In a breakdown, hydropower generating capacity was 892.1 billion kWh, up 4.27% over 2012; thermal power generating capacity was 4221.6 billion kWh, up 7.54% over 2012; nuclear power generating capacity was 111.5 billion kWh, up 13.41% over 2012; wind power generating capacity was 138.3 billion kWh, up 34.17% over 2012. See Fig. 2.24 for China's generating capacity structure in 2013; see Fig. 2.25 for generating capacity and composition of major countries in 2013; see Fig. 2.26 for generating capacity structure of major countries in 2013.

China's electric power industry has become increasingly larger and more powerful. In the early stage of new China, China's installed power capacity was 1.73k MW; annual generating capacity was only 4.3 billion kWh; residential power consumption per capita was less than 1 kWh. By the end of 2013, China's

**Fig. 2.21** China's installed capacity structure in 2013

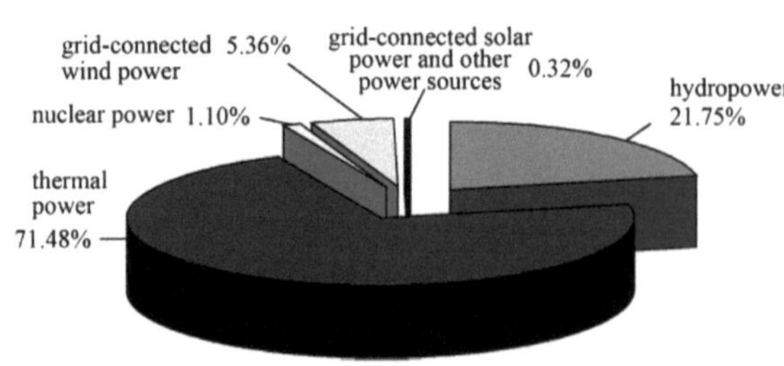

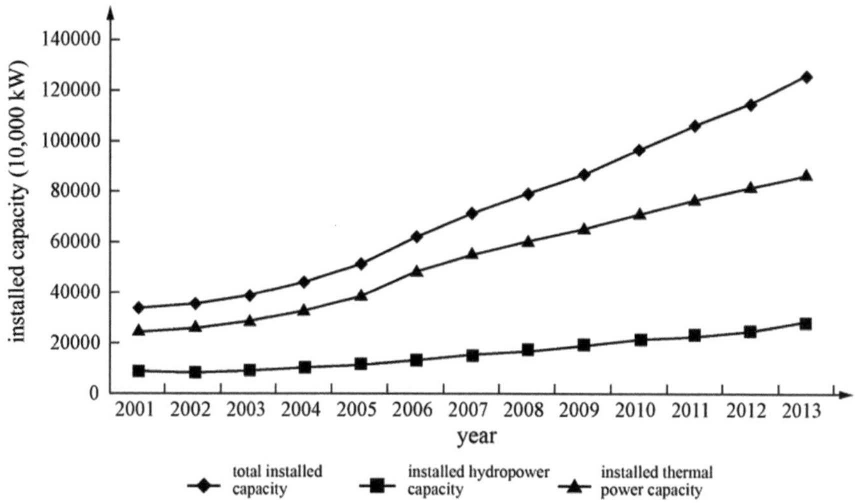


**Fig. 2.22** China's installed capacity changes between 2001 and 2013. *Data source* Annual Development Report of China's Electric Power Industry 2014 issued by China Electricity Council

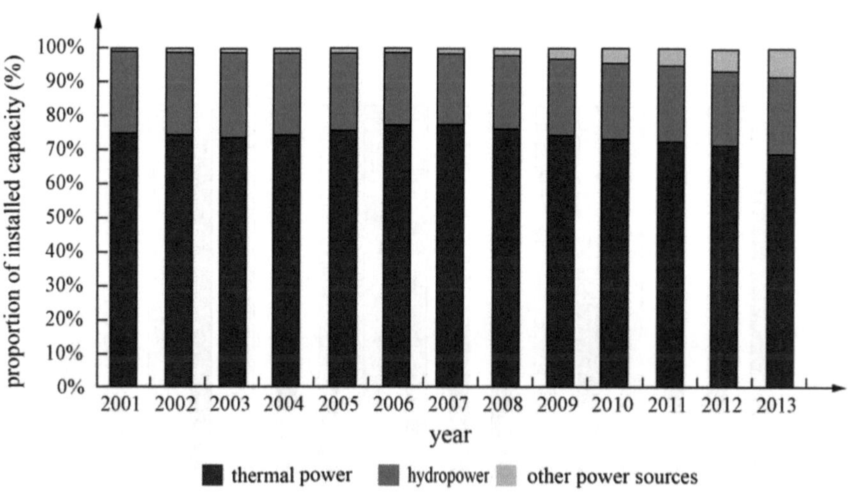


**Fig. 2.23** China's installed capacity structure changes between 2001 and 2013. *Data source* Annual Development Report of China's Electric Power Industry 2014 issued by China Electricity Council

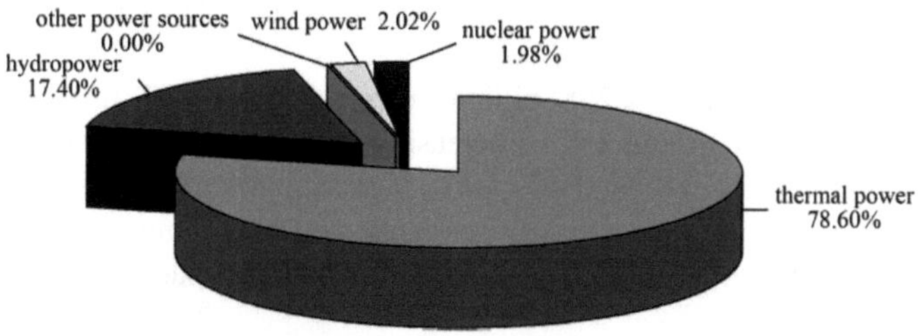


**Fig. 2.24** China's generating capacity structure

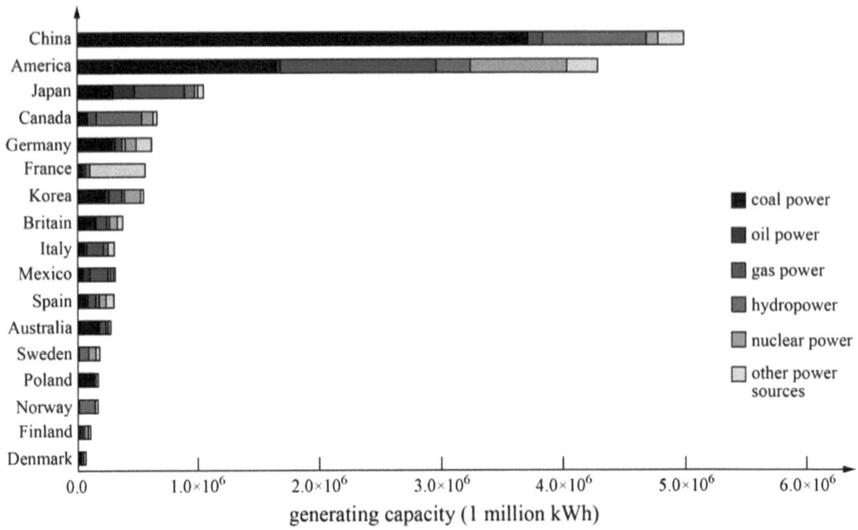


**Fig. 2.25** Generating capacity and composition of major countries in 2013. *Data source* Statistics of IEA

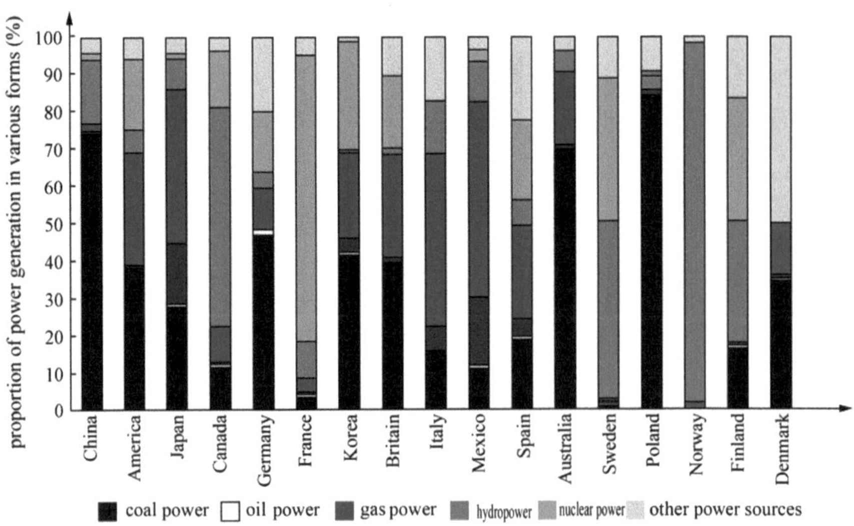


**Fig. 2.26** Generating capacity structure of major countries in 2013. *Data source* Statistics of IEA

installed capacity reached 1.258 billion, increasing by 727 times; generating capacity reached 5.37 trillion kWh, increasing by 1249 times; residential power consumption per capita exceeded 400 kWh, increasing by more than 400 times. At present both China's installed capacity per capita and generating capacity per capita were quite low. See Figs. 2.27 and 2.28 for installed capacity growth rate and

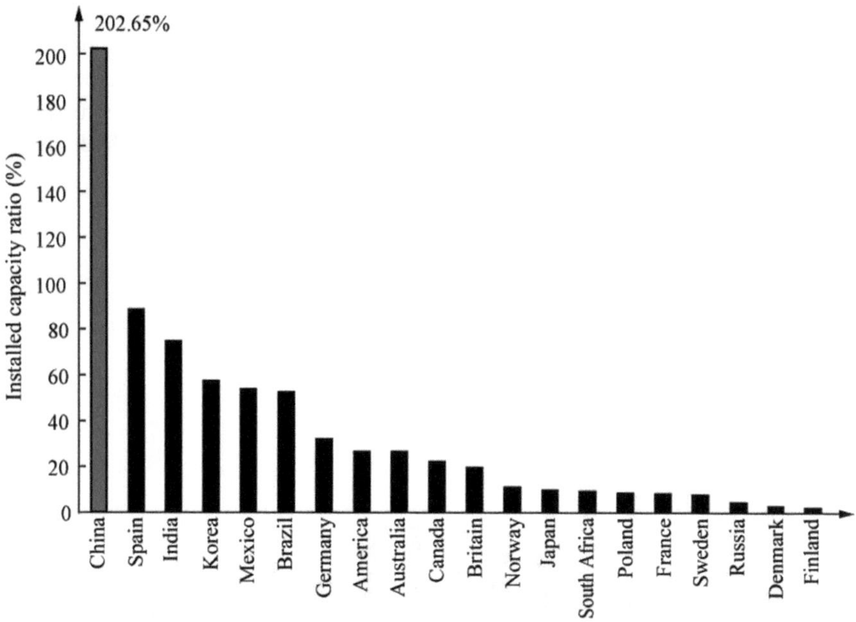


**Fig. 2.27** Installed capacity growth rate of major countries between 2000 and 2010. *Data source* Statistics of IEA

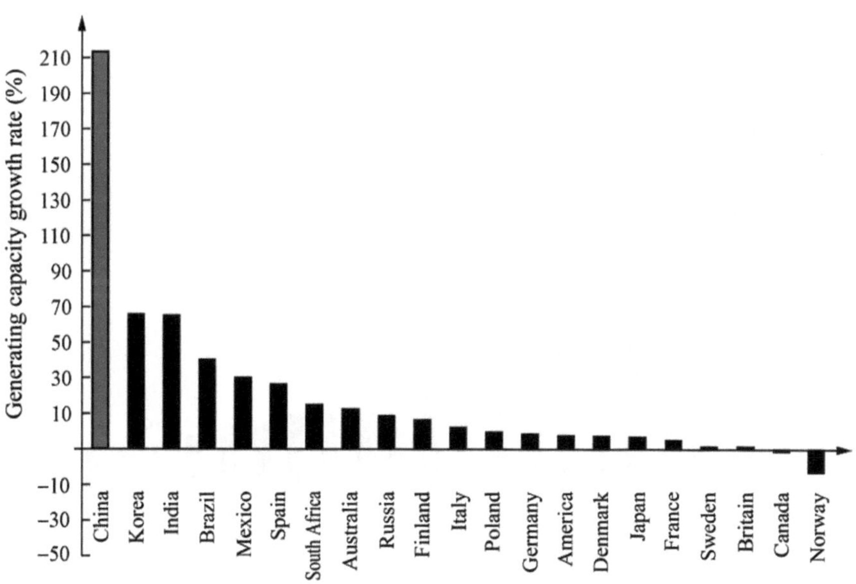


**Fig. 2.28** Generating capacity growth rate of major countries between 2000 and 2010. *Data source* Statics of IEA

generating capacity growth rate of major countries between 2000 and 2010, respectively; see Figs. 2.29 and 2.30 for comparison of installed capacity per capita and generating capacity per capita between major countries in 2011, respectively.

In recent years, China's power industry has strengthened industrial restructuring, resulting in optimization of thermal power, accelerated development of hydropower

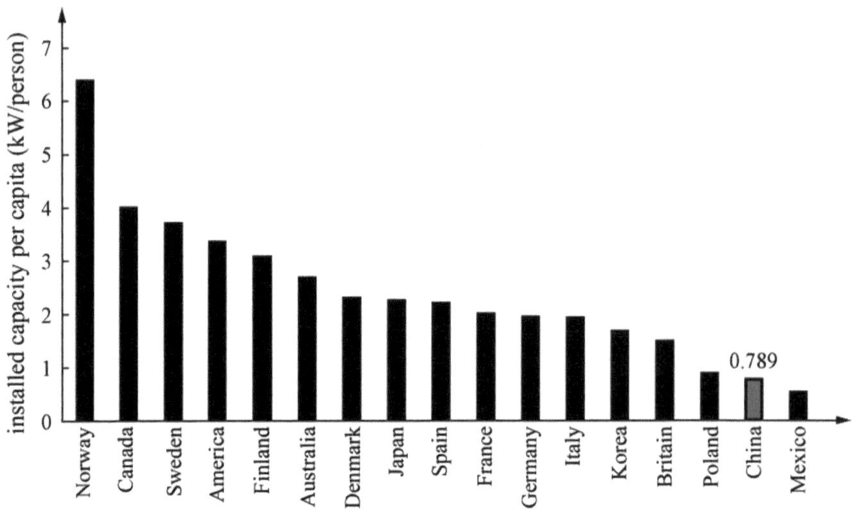


**Fig. 2.29** Comparison of installed capacity per capita between major countries in 2011. *Data source* Statistics of IEA

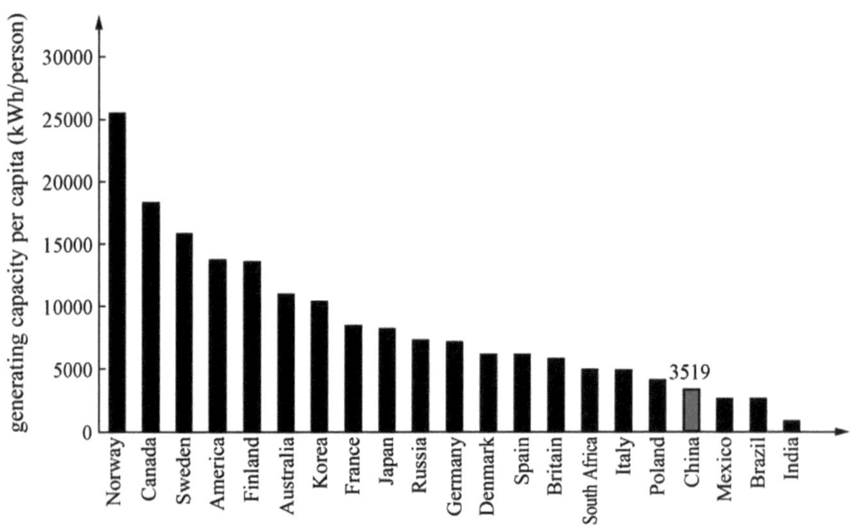


**Fig. 2.30** Comparison of generating capacity per capita between major countries. *Data source* Statistics of IEA

and nuclear power, leaping development of wind power, and coordinated development of power grid construction and power supply construction. By the end of 2013, the capacity of hydroelectric generating units included in the statistic survey of generating units of 6000 kW and above was 236.17k MW, accounting for 94.19% of the capacity of hydroelectric generating units of 6000 kW and above in China. Among all the hydroelectric generating units included in the survey, the proportion of hydroelectric generating units of 600 MW and above increased by 4.41% over 2012; the proportion of hydroelectric generating units of 300–600 MW (excluding 600 MW), 200–300 MW (excluding 300 MW), 100–200 MW (excluding 200 MW), 50–100 MW (excluding 100 MW), and below 50 MW decreased by 0.22, 1.37, 1.19, 0.84, and 0.8% over 2011, respectively. See Fig. 2.31 for the proportion of hydroelectric generating units of various capacities in the national industry statistic survey.

By the end of 2013, the capacity of thermal power generating units included in the statistic survey of generating units of 6000 kW and above was 848.91k MW, accounting for 98.17% of the capacity of thermal power generating units of 6000 kW and above in China. In the statistic survey, the average unit capacity of thermal power generating units was 118 MW, almost equal to that in 2012. Among all the thermal power generating units included in the survey, the proportion of thermal power generating units of 600 MW and above reached 41.13%, up 1% over 2012 and up 29.45% over 2005, reflecting the rapid development of thermal power generating units of large capacity and high parameters since the "11th Five-Year Plan" period; the proportion of thermal power generating units of 300–600 MW (excluding 600 MW), 200–300 MW (excluding 300 MW), and 100–200 MW (excluding 200 MW) decreased by 0.3, 0.39, and 0.44% over 2012, respectively. By the end of 2012, China had 54 ultra-supercritical thermal power generating units of one k MW in operation, ranked the first in the world. See Table 2.13 for the proportion of thermal power generating units of various capacities in the national industry statistic survey. See Fig. 2.32 for China's thermal power generating unit structure changes between 1995 and 2013.

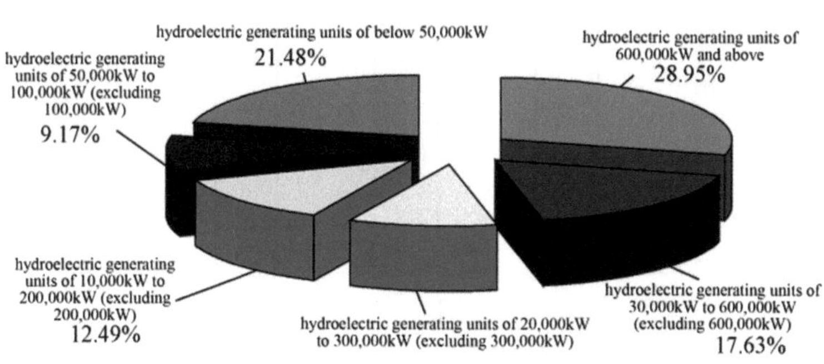


**Fig. 2.31** Proportion of hydroelectric generating units of various capacities in the national industry statistic survey

**Table 2.13** Proportion of thermal power generating units of various capacities in the national industry statistic survey

| Thermal power generating units of various capacities | | Unit | Total number of thermal power generating units | Proportion in total thermal power capacity in the statistic survey (%) |
|---|---|---|---|---|
| 6000 kW and above | | Unit | 7223 | 100 |
| | | 10 MW | 84,891 | |
| Breakdown | Generating units of 600 MW and above | Unit | 529 | 41.13 |
| | | 10 MW | 34,916 | |
| | Generating units of 300,000–600 MW (excluding 600 MW) | Unit | 919 | 35.12 |
| | | 10 MW | 29,810 | |
| | Generating units of 200,000–300 MW (excluding 300 MW) | Unit | 246 | 6.10 |
| | | 10 MW | 5180 | |
| | Generating units of 100,000–200 MW (excluding 200 MW) | Unit | 468 | 7.36 |
| | | 10 MW | 6251 | |
| | Generating units of below 100 MW | Unit | 5068 | 10.29 |
| | | 10 MW | 8735 | |

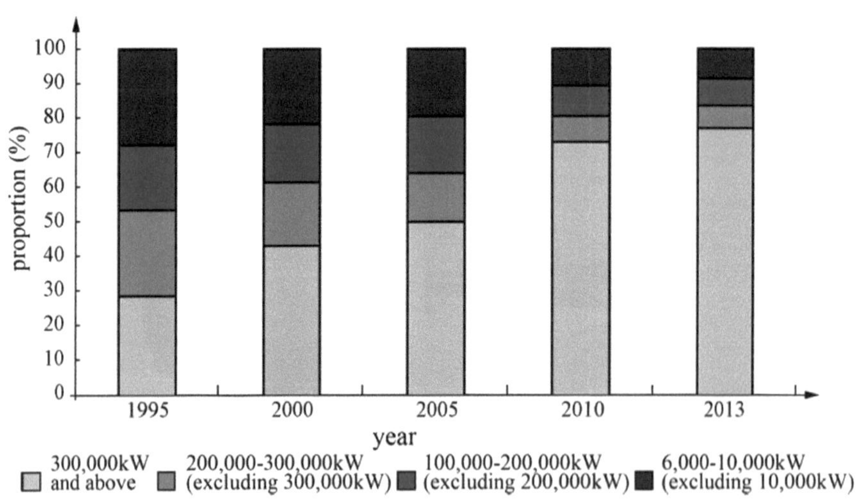


**Fig. 2.32** China's thermal power generating unit structure changes between 1995 and 2013

**Table 2.14** Loop length of transmission lines of 35 kV and above and capacity of substation equipment at the end of 2013 in China

| Transmission lines of various voltage classes | | Loop length of transmission lines | | Capacity of substation equipment | |
|---|---|---|---|---|---|
| | | Length (km) | Growth rate (%) | Capacity (10,000 kVA) | Growth rate (%) |
| 35 kV and above | | 1,554,236 | 5.03 | 483,427 | 8.42 |
| 220 kV and above | | 543,896 | 7.57 | 278,166 | 9.03 |
| Breakdown | 1000 kV | 1936 | 202.78 | 3900 | 116.67 |
| | ±800 kV | 6904 | 29.93 | 4654 | 6.74 |
| | 750 kV | 12,666 | 25.56 | 6500 | 22.18 |
| | ±600 kV | 1400 | 0.00 | 948 | 0.00 |
| | 500 kV | 146,166 | 6.61 | 90,112 | 8.05 |
| | ±500 kV | 10,653 | 16.48 | 7637 | 5.64 |
| | ±400 kV | 1031 | 0.00 | 141 | 0.00 |
| | 330 kV | 24,065 | 6.01 | 8575 | 11.16 |
| | 220 kV | 339,075 | 6.55 | 155,699 | 7.95 |

China has constantly expanded the scale and improved the technology of power grid construction. By the end of 2013, the loop length of power transmission line of 35 kV and above was 1.5542 million km, up 5.03% over 2012. In a breakdown, the loop length of power transmission line of 220 kV and above was 543,900 km, up 7.57% over 2012. The capacity of substation equipment of 35 kV and above was 4.834 billion kVA, up 8.42% over 2012. In a breakdown, the capacity of substation equipment of 220 kV and above was 2.782 billion kVA, up 9.03% over 2012. The inter-provincial and inter-regional backbone transmission network dominated by 500 kV transmission lines was gradually formed; the 750 kV transmission demonstration project was successfully completed; 1000 kV UHV and ±800 kV DC pilot demonstration project was put into operation. See Table 2.14 for the loop length of transmission lines of 35 kV and above and capacity of substation equipment at the end of 2013 in China.

2. Electric power consumption

In 2012 driven by the power consumption growth in developing countries such as China and India, the world's electric power consumption continued to achieve a year-on-year growth rate of 2.4%. According to the statistics of China Electricity Council, in 2012 consuming a total of 4959.1 billion kWh of electricity, China exceeded America and became the world's largest electric power consumer. See Fig. 2.33 for electricity consumption growth rate of major countries between 2000 and 2010; see Fig. 2.34 for comparison of electricity consumption per capita between major countries in 2011.

In 2013, China's total electricity consumption was 5342.3 billion kWh, up 7.58% over 2012, 1.98% lower than 2012. See Fig. 2.35 for China's electricity consumption and growth rate between 2006 and 2013.

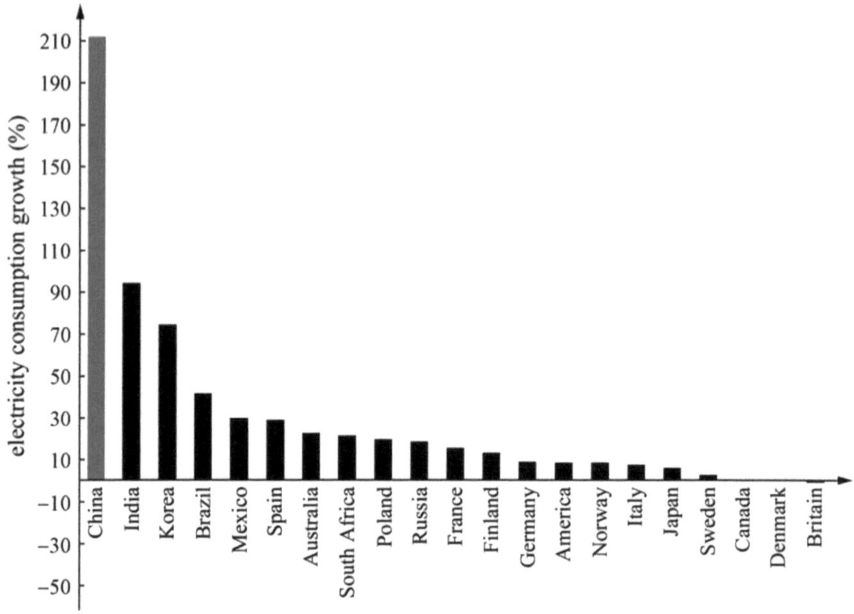


**Fig. 2.33** Electricity consumption growth rate of major countries between 2000 and 2010. *Data source* Statistics of IEA

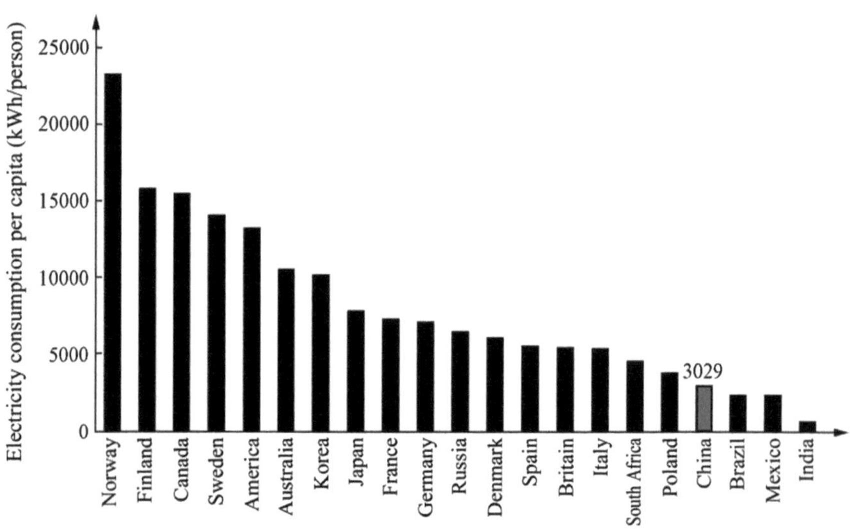


**Fig. 2.34** Comparison of electricity consumption per capita between major countries in 2011. *Data source* Statistics of IEA

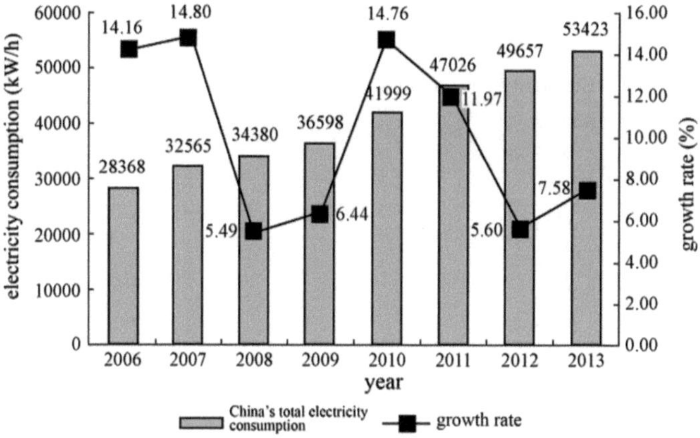


**Fig. 2.35** China's electricity consumption and growth rate between 2006 and 2013. *Data source* Statistics of IEA

In 2013, the proportion of electricity consumed by the tertiary industry and urban and rural residents in China's total electricity consumption increased by 0.28 and 0.15% over 2012, respectively; the proportion of electricity consumed by the secondary industry was 73.62%, down 0.35% over 2012; the proportion of electricity consumed by the primary industry was only 1.92%, down 0.10% over 2012. See Fig. 2.36 for China's electric power consumption structure.

In 2013, the electricity consumption growth rate of 17 provinces, municipalities, and autonomous regions in China was higher than the national average level, 7.58% and 7 provinces, municipalities, and autonomous regions, namely Xinjiang

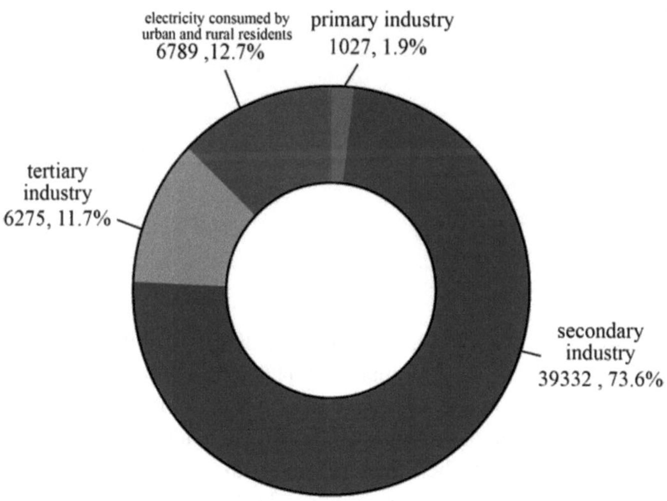


**Fig. 2.36** China's electric power consumption structure

**Table 2.15** Electricity consumption of China's provinces, municipalities, and autonomous regions in 2013

| Region | Electricity consumption (100 k MWh) | Growth rate over 2012 (%) |
|---|---|---|
| China | 53,423 | 7.58 |
| Beijing | 913 | 4.44 |
| Tianjin | 774 | 7.20 |
| Hebei | 3251 | 5.63 |
| Shanxi | 1832 | 3.77 |
| Inner Mongolia | 2182 | 8.19 |
| Liaoning | 2008 | 5.72 |
| Jilin | 654 | 2.64 |
| Heilongjiang | 845 | 2.09 |
| Shanghai | 1411 | 4.22 |
| Jiangsu | 4957 | 8.20 |
| Zhejiang | 3453 | 7.55 |
| Anhui | 1528 | 12.27 |
| Fujian | 1701 | 7.68 |
| Jiangxi | 947 | 9.16 |
| Shandong | 4083 | 7.60 |
| Henan | 2899 | 5.51 |
| Hubei | 1630 | 8.08 |
| Hunan | 1423 | 5.69 |
| Guangdong | 4830 | 4.56 |
| Guangxi | 1238 | 7.27 |
| Hainan | 232 | 10.04 |
| Chongqing | 813 | 12.37 |
| Sichuan | 1949 | 6.46 |
| Guizhou | 1126 | 7.60 |
| Yunnan | 1460 | 10.94 |
| Tibet | 31 | 10.43 |
| Shaanxi | 1152 | 8.01 |
| Gansu | 1073 | 7.91 |
| Qinghai | 676 | 12.30 |
| Ningxia | 811 | 9.35 |
| Xinjiang | 1540 | 33.72 |

(33.72%), Chongqing (12.37%), Qinghai (12.30%), Anhui (12.27%), Yunnan (10.94%), Tibet (10.43%) and Hainan (10.04%), had an electricity consumption growth rate of higher than 10%. Except Guangxi and Sichuan, all the other provinces, municipalities, and autonomous regions in Western China had an electricity consumption growth rate higher than the national average level. The electricity consumption growth rate of Heilongjiang (2.09%), Jilin (2.64%), and Shanxi (3.77%) was lower than 4%.

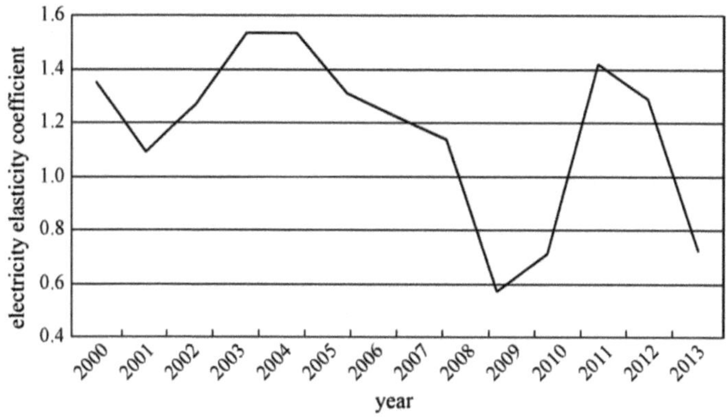


**Fig. 2.37** China's electricity elasticity coefficient

In 2013, the electricity consumption of eight provinces and autonomous regions including Jiangsu, Guangdong, Shandong, Zhejiang, Hebei, Henan, Inner Mongolia, and Liaoning exceeded 200 billion kWh (the electricity consumption of Liaoning exceeded 200 billion kWh for the first time). The electricity consumption of eight provinces and autonomous regions totaled 2766.4 billion kWh, up 6.61% over 2012, 0.97% lower than the national average growth rate. Except Jiangsu and Inner Mongolia, all the other five provinces had an electricity consumption growth rate lower than the national average level. Total electricity consumption of eight provinces and autonomous regions accounted for 51.78% of China's total electricity consumption, down 0.47% over 2012, and contributed 45.56% to China's electricity consumption growth rate, up 2.10% over 2012. See Table 2.15 for details.

According to the preliminary estimates of National Bureau of Statistics of the People's Republic of China, in 2013 China's GDP was RMB 56,884.5 billion, up 7.7% over 2012 in terms of comparable price, equal to that in 2012. In 2013, China's electricity elasticity coefficient was 0.98, up 0.25% over 2012. See Fig. 2.37 for China's electricity elasticity coefficient.

# Chapter 3
# Analysis of Environmental Protection Constraint of Energy Development

**Fan Yang, Jie Tan and Xin Sui**

## 3.1 Coal Development

### *3.1.1 Environmental Impacts and Present Situation of Control*

#### 3.1.1.1 Major Environmental Impacts

In the process of coal development, it is inevitable that the natural environment will be changed and destroyed, producing a large amount of pollutants. Main environmental impacts include land subsidence, water loss and soil erosion, land occupation, river system damage, discharge of solid waste such as coal gangue, discharge of mine water, gas emissions, destruction of vegetation, smoke and dust emissions, impact of biodiversity, autoignition caused by mining.

(1) Land subsidence. Due to coal mining, especially underground mining, the coal layer of tens of meters thick that was originally buried hundreds of and even tens of hundreds of meters under the ground is lifted to the ground. Without effective filling measures, the rocks around the coal mining face will collapse due to unbalanced forces. The collapse will extend to the earth's surface and result in earth surface collapse, namely land subsidence. According to

F. Yang (✉)
China Electricity Council, Beijing, People's Republic of China
e-mail: yangfan@cec.org.cn

J. Tan
China Coal Processing & Utilization Association, Beijing
People's Republic of China

X. Sui
China Institute of Water Resources and Hydropower Research, Beijing
People's Republic of China



Z. Wang (ed.), *Constraints and Solutions for Energy and Electricity Development*,
https://doi.org/10.1007/978-981-10-8304-4_3

incomplete statistics, on average the land subsides 0.2 $hm^2$ for mining 10,000 tons of coal. What is more, land subsidence is the main cause of environmental impacts such as geological disasters, soil erosion, and ecological destruction. In addition, owing to the limitation of natural endowment of coal resources, China's coal is mainly mined in the way of underground mining. As a result, land subsidence is an important environmental impact in China's current coal development.

(2) Water loss and soil erosion. In the coal development process, activities such as underground mining, open-air strip, and pumping of mine water lead to land subsidence, topsoil stripping, and decline of groundwater level in mining areas, which increases the local water loss and soil erosion, result in soil depletion and vegetation degradation, and further exacerbate water loss and soil erosion, forming a vicious cycle. Therefore, water loss and soil erosion are another environmental impact in China's coal development.
(3) Land occupation. It mainly refers to the land occupied by solid waste such as coal gangue and open-air stripping soil in the coal development process.
(4) River system damage. It includes water pollution and destruction of underground hydrogeological conditions. The discharge of industrial wastewater produced in the coal development process will result in water pollution in the mining areas as well as surrounding water areas. What is worse, the coal development might result in more serious water system damage without good solutions. The pumping of groundwater or discharge of drainage leads to the change in underground hydrogeological conditions in mining areas and the decline of groundwater level. As a result, the ground vegetation further degrades due to inadequate water supply and soil becomes increasingly barren, increasing water loss and soil erosion.
(5) Discharge of solid waste such as coal gangue, mine water, gas, smoke, and dust. In the coal development process, solid waste such as coal gangue, mine water or production wastewater, gas, smoke, and dust will be produced. At present, in addition to comprehensive utilization, the "three wastes" (wastewater, waste gas, and solid waste) generated in the coal development process will be discharged into the environment through reasonable treatment and make certain impacts on the environment.
(6) Impacts of vegetation destruction and biodiversity. The above-mentioned environmental impacts will definitely lead to vegetation destruction and changes in ecological function and biodiversity in mining areas.
(7) Autoignition caused by mining. When coal is exposed to the air, due to oxidative heat release, the temperature gradually rises to 70–80 °C and then rises sharply. When it reaches the coal ignition temperature, coal is ignited, which is called spontaneous combustion of coal seam. If in the coal mining process the coal resources stored deep down underground are exposed to the air, the possibility of the spontaneous combustion of coal will increase. The autoignition caused by mining is also related to coal quality. For example, coal which is easy to be oxidized and has intense exothermic oxidation reaction and low ignition temperature is prone to autoignition in the mining process. The

spontaneous combustion of coal seam is the incomplete oxidation of coal and produces a large amount of harmful gases including CO, $SO_2$, and $NO_x$, polluting the environment.

#### 3.1.1.2 Present Situation of Control Technology

In terms of the “three wastes” problem, improving the comprehensive utilization rate of resources can fundamentally reduce the pollution of “three wastes”. For the earth’s surface vegetation destruction, reclamation plans can be made in accordance with local conditions. For example, land suitable for tree planting or fishing should be reclaimed for tree planting or fishing. Sources of domestic water for local residents should be developed in a protective way, and it is strictly forbidden to destroy them. When the decline of the groundwater level does affect water for production and living of local residents, local residents should be compensated according to the actual situation or be provided with well water or tap water. For areas where geological disasters might occur, local residents should be relocated. The relocation site and schedule should be arranged by the local government and funded by enterprises. See Table 3.1 for pollution control of coal enterprises.

#### 3.1.1.3 Control Effectiveness and Existing Problems

1. Control effectiveness

China’s coal industry has done a lot of work in environmental protection and has made positive progress. In terms of management scope, it is changed from focusing on key state-owned coal mines to the whole coal industry; in terms of management means, the administrative management in the past is changed into the comprehensive services integrating supervision according to laws, industry guidance, technology, and information guidance. In terms of work focus, it is changed from focusing on solving environmental pollution problems in coal development process to focusing on solving both pollution problems and ecological problems. In terms of technology roadmap, it is changed from end-of-pipe treatment and focusing on discharging pollutants in accordance with standards to process control, ecological renovation, and comprehensive utilization of resources. In terms of economic policy, it is changed from relying on sole financing of enterprises to joint financing of the central government, local governments, and enterprises for comprehensive control of environmental pollution. The investment has increased year by year, and channels have gradually expanded. These changes have effectively promoted the environmental protection of the coal industry. In recent years, China’s coal industry has mainly made the following achievements:

(1) Positive progress has been made in green mining and ecological mine construction. Since the “12th Five-Year Plan” period, more and more coal

**Table 3.1** Pollution control of coal enterprises

| Pollutant | Pollution producing process | Pollution control measure | Process and technology used | Control effect |
|---|---|---|---|---|
| Coal gangue | Coal gangue produced in the coal mining and excavation process; washery rejects produced in the coal washing process | Comprehensive utilization | Coal gangue power plant, coal gangue brick field, filling subsidence area, road paving, backfilling mined-out areas, filling the mined-out areas when gangue is directly crushed without being lifted out of the well | Coal gangue can be comprehensively utilized to eliminate pollution, develop circular economy, and achieve certain economic benefits |
| Mine water | Mine drainage discharged in the coal mining process | Treat mine water in different ways based on the usage, for example, purification, in-depth treatment and simple treatment | Precipitation → filtration → disinfection → utilization | Mine water can be comprehensively utilized to improve the environment, save resources, and develop circular economy; if excess water yield of mine cannot be used otherwise, it can be discharged; highly mineralized mine water treatment has a high cost, but direct discharge might have a negative impact on environment |
| | | | Precipitation → filtration → reverse osmosis → disinfection → utilization | |
| | | | Precipitation → agricultural irrigation and dust reduction | |
| Mine gas | Coal seam gas extracted in the coal mining process | Comprehensive utilization | The coal mine gas with concentration higher than 6% can be used as fuel for low concentration gas power station; the coal mine gas with concentration higher than 30% can be used as fuel for gas power station or household fuel | The coal mine gas with concentration higher than 6% can be used as clean fuel to eliminate pollution. Coal mine gas with low concentration is mainly discharged into the atmosphere currently, but with the improvement of |

(continued)

**Table 3.1** (continued)

| Pollutant | Pollution producing process | Pollution control measure | Process and technology used | Control effect |
|---|---|---|---|---|
| | | | | low concentration coal mine gas comprehensive utilization technology, low concentration coal mine gas can be gradually realized |
| Sulfur dioxide | Coal-fired boiler flue gas | Desulfurization with calcium content in furnace | Co-firing of coal and limestone $SO_2 + CaO \rightarrow CaSO_3 + CaSO_4$ | Low desulfurization efficiency |
| | | Calcium injection desulfurization in furnace | Injection of limestone powder to a circulating fluidized bed furnace; desulfurization reaction is the same as above | Desulfurization efficiency can reach 70% |
| | | Semi-dry flue gas desulfurization | Flue gas is put into reactor and mixed with water +desulfurization agent; $SO_2$ is absorbed | Desulfurization efficiency is 80–90% |
| | | Wet flue gas desulfurization | Use desulfurization agent to spray, and wash flue gas containing $SO_2$ in the absorption tower; $SO_2$ is absorbed | Desulfurization efficiency is higher than 95% |
| Chemical oxygen demand | Coal mine drainage is mainly mine water.<br>The COD concentration of the mine water is generally not high. If the displacement is large, total amount of COD will increase accordingly | Mine water is used after treatment and is not discharged | Precipitation → filtration → disinfection → utilization<br>Precipitation → filtration → reverse osmosis → disinfection → utilization<br>Precipitation → agricultural irrigation and dust reduction | Mine water can be comprehensively used to completely eliminate COD. If the mine water in large quantity cannot be completely utilized, some of the mine water will be discharged, forming discharge COD |

enterprises have carried out the green mine and ecological mine construction activities. Coal enterprises have actively explored and implemented green mining technologies such as filling mining and water-preserved mining which not only improve the recovery rate of coal resources, save valuable resources of coal and water, but also reduce the impact on and damage to environment of coal mining.

(2) New progress has been made in energy conservation and consumption reduction. In recent years through technical progress, management improvement, industry and product structure adjustment, and other measures, the coal industry has improved the energy efficiency of enterprises and the unit product energy consumption has been constantly reduced. According to statistics, in 2012 comprehensive energy consumption of raw coal production and power consumption of raw coal production of large state-owned coal mines reached 13.1 kg standard coal/t and 22.5 kWh/t, respectively, down 2.6 and 3.0% over 2011, respectively.

(3) Continuous improvement of comprehensive utilization level. With the continuous development of the coal gangue, mine water, and coal gas comprehensive utilization technology, remarkable achievements have been made in the development and utilization of "three wastes" of mines and associated resources of coal measures. In 2012, the mine inflow was 6.75 billion tons and the mine inflow utilization rate was 62%; coal gangue production was 720 million tons, and the utilization rate was 62.7%; the mine gas production was 18.25 billion $m^3$; underground drainage was 9.5 billion $m^3$, and the utilized quantity and utilization rate are 4 billion $m^3$ and 42%, respectively. The scale of using coal gangue and fly ash to produce new types of buildings materials has steadily expanded. The technology of exchanging coal gangue for coal and filling mining has constantly improved. The utilized quantity of coal gangue and fly ash has gradually increased.

(4) Construction of circular economy park has been steadily promoted. At present, in the coal industry the circular economy development concept has been deeply rooted in everybody's mind. A lot of coal enterprises have particularly prepared circular economy development plans and explored and created distinctive circular economy development patterns.

(5) Ecological environment governance has promoted harmonious mining area construction. Ecological environment governance of mining areas is not only an important part of the construction of ecological civilization, but also closely related to the construction of harmonious mining area. The governance of coal mining subsidence area and the renovation of the shanty towns are steadily advanced. In 2012, the average annual income of large coal enterprises was about RMB 68,000, up 5.7% year-on-year. Combining ecological environment governance and harmonious mining area construction, the state-owned enterprises better fulfilled their due social responsibility. Eighteen large coal enterprises issued the social responsibility report, and 156 coal enterprises carried out the credit rating evaluation.

2. Major existing problems

Since the "12th Five-Year Plan" period, with the growing demand for social development, coal mining areas have made gratifying achievements in environmental protection. However, compared with the advanced world level, there are the following problems:

(1) High strength development and rapid growth of production have made a great impact on environment, and ecological damage has been constantly exacerbated. China's total coal production in the past 30 years totals 45 billion tons, and coal washing and processing capacity is about 11 billion tons. However, emissions of various wastes have not been synchronously used and treated. China's coal mining subsidence area increases by 20,000–25,000 $hm^2$ annually, and the area of open pit mining and land occupation is about 10,000 $hm^2$ annually. In the development mode of mining and discharging followed by governance or without governance, the discharge and reserves of pollutants also increase with the production, processing capacity, and utilization. Besides, the Chinese government has issued the pollution charge policy. Coal enterprises have difficulties bearing the double pressures.
(2) China has a large number of coal mines covering a broad area. As a result, coal mines have a great impact on environment. The strategy of the westward movement of coal development has made the ecological environment problem more prominent. In the twentieth century, China's coal mines were distributed in 1458 counties (cities) and at most there were over 80,000 coal mines. The ecological damage caused by coal mining affected more than 50% of these counties (cities), increasing the difficulty of environmental governance. Moreover, due to the serious shortage of the Chinese government's investment in environmental governance of mining areas, the governance effect is affected.
(3) Due to the low proportion of coal processing and raw coal-oriented product structure, environmental pollution has become increasingly exacerbated. Washing and processing raw coal are the effective ways to reduce emissions of pollutants, improve product quality, and obtain the best economic benefit. At present, most coal mines in China have not constructed the supporting coal washeries and the proportion of raw coal washing is quite low. In 2012, the raw coal washing proportion was 52.6%, but compared with the proportion of 80–90% of the world's major coal producers, it is quite low.

As the ecological environment protection has played an increasingly important role in coal development, the ecological environment protection system for coal development has been basically formed in China's newly built mining areas, improving the ecological environment around the mining areas. China is gradually advancing toward green mining.

## *3.1.2 Analysis of Major Environmental Protection Constraints*

### 3.1.2.1 Analysis of Laws, Regulations, and Policies

1. Requirements of existing laws, regulations, and policies

Since the beginning of the twenty-first century, China's coal industry has entered the unprecedented rapid development stage of increasing production and improving efficiency. The Chinese government has identified the "coal-dominated and electric power-centered" development strategy and put forward the development idea of "constructing large coal production bases, adjusting and transforming medium and small coal mines, developing and utilizing coal seam gas, and encouraging coal and electricity production joint venture." In addition, it has made more than 20 economic and technical policies.

With the continuous development of China's economy, the progress of science and technology, and the gradual improvement of the coal industry's environmental protection mechanism, environmental governance effect has gradually improved. Especially in recent years, relevant state ministries and commissions have developed policies such as Township Coal Mine Environmental Protection Management Regulations, Notice on Strengthening Soil and Water Conservation in Coal Mine Production, and Construction Projects and Land Reclamation Management Ordinance and issued standards, laws, and regulations such as Clean Production Standard Coal Mining Industry, Emission Standard for Pollutants from Coal Industry and Rules for the Construction of Coal Pillars in Buildings, Water Bodies, Railways and Main Mine Lanes and Pressed Coal Mining to lay a solid foundation for further standardizing environmental protection in coal industry.

2. Judgement of laws, regulations, and policies during the "12th Five-Year Plan" period

Since the establishment of environmental protection mechanism, the coal industry has achieved remarkable results and effectively curbed the further deterioration of the mining area environment. However, with the progress of society and the development of science and technology, the existing environmental protection policies and measures cannot meet the increasingly higher environmental requirements. During the "12th Five-Year Plan" period, China should issue supporting laws, regulations, and policies on changing the development mode, environmental compensation mechanism, ecological environmental standards, environmental supervision system, and broadening channels of funds and meanwhile put forward development goals to promote the coordinated development of economy and environment.

#### 3.1.2.2 Analysis of Major Environmental Protection Constraint Factors

From the point of view of environmental impact factors in a single mining area, coal development will inevitably lead to surface subsidence, damage of ground vegetation and groundwater, geological disasters such as landslide and ground cracks, secondary disasters such as surface vegetation destruction, soil degradation, and water pollution. Through comprehensive analysis, we find environmental protection constraints in coal development are mainly reflected in the following aspects:

1. Water resource destruction constrains coal mining

Coal mining causes serious damage to the groundwater and surface water resources, exacerbating the loss of water resources. Especially in water-deficient areas, due to the loss of groundwater and pollution of surface water, local residents have difficulty having access to production and living water. According to the survey, about 1.87 $m^3$ water resources are destroyed for producing 1t of raw coal on average. China's raw coal production in recent 7 years totals 13.1 billion tons and about 25,152 billion $m^3$ water resources are lost. According to incomplete statistics, the area of water resources destroyed by coal mining in Shanxi province reaches 20,352 $km^2$. Therefore, relevant Chinese ministries and commissions have made strict requirements for discharge of pollutants and required that the comprehensive utilization rate of mine water and domestic sewage should be higher than 75% and achieve zero discharge preferably.

2. Solid waste discharge constrains coal mining and selection

It is inevitable that some solid wastes including coal gangue, coal slime, and fly ash are discharged in the coal mining and washing process. These solid wastes have three distinct characteristics: large quantity, containing a large amount of low-grade energy, and low man-made (chemical) pollution degree. With the increase of coal production, the production of coal gangue, coal slime, and fly ash has also increased sharply, exacerbating the destruction of the ecological environment in mining areas. Moreover, the accumulation of a large amount of coal gangue and fly ash directly results in the unorganized discharge of $SO_2$ and dust.

3. Land and vegetation damage constrains coal mining

Large-scale energy base construction and coal mining leads to large area land subsidence in mining areas. According to statistics, the earth's surface subsides 0.2 $hm^2$ (about 3 acres) for China's coal mines mining every 10,000 tons of coal. It has been more than 60 years since the founding of the new China, and about 50 billion tons coal have been produced to meet the demand of rapid development of national economy. However, a cumulative area of 960,000 $hm^2$ of the earth's surface has subsided, causing damage to the land and surface vegetation.

Coal mining leads to the destruction of forest vegetation and exacerbates water loss and soil erosion. China's 13 large coal bases are distributed in 100 mining areas in 14 provinces and have a mining area of 355,130 $km^2$. The coal production of the eight large coal bases in Shanxi, Shaanxi, and Inner Mongolia accounts for 60% of China's total coal production. These eight large coal bases cover an area of 33,000 $km^2$. With the water and soil loss area reaching 1.14 million $km^2$, these areas have the most serious problem of water loss and soil erosion in China. Shanxi, Shaanxi, and Inner Mongolia lose about 1.6 billion tons of water and soil annually.

#### 3.1.2.3 Analysis of Main Environmental Protection Indicator Limit

1. Limit of environmental protection indicators during the "12th Five-Year Plan" period

In terms of efficient utilization of waste resources, China's coal industry has focused on the efficient utilization of coal gangue, mine water, and gases, vigorously reformed the backstopping process and utilization means, strengthened the control of the discharge and storage of pollutants, and implemented land reclamation in mining areas. The Chinese central government, local governments, and enterprises have joined hands to establish unified standards, improve laws, regulations and policies, increase investment, and construct a batch of resource-saving and environment-friendly mining areas. See Table 3.2 for development goals and major measures for efficient utilization of waste resources in all regions in 2015.

With the adjustment of coal production structure and production capacity, China's coal industry aims to, with large coal bases as the leader, improve management level and technical equipment, adjust and optimize the industrial structure, use clean production technologies, reduce resource and energy consumption, and construct a batch of ecologically harmonious mining areas during the "12th Five-Year Plan" period. By 2015, the coal industry will basically achieve the goal that the discharge of all pollutants meets the standards; the ecological environment of mining areas improves significantly; a batch of large and medium coal mines are established as clean coal production demonstration mine areas.

2. Long-term environmental protection indicator limit

Based on the coal environmental control during the "12th Five-Year Plan" period, by 2020, environment quality of China's coal mining areas and key coal producing cities (counties and districts) will be comprehensively improved; the system of sustainable development of coal economy and ecological benign cycle will be established; 80% of coal mining areas will reach the level of clean production demonstration coal mining areas. See Table 3.3 for development goals for efficient utilization of waste resources in all regions in 2020.

**Table 3.2** Development goals and major measures for efficient utilization of waste resources in all regions in 2015

| Region | Main indicators (%) | | | | Major measures |
|---|---|---|---|---|---|
| | Mine water utilization rate | Coal gangue utilization rate | Coalbed gas utilization rate | Land reclamation rate | |
| North China and East China | 75 | 75 | 60 | 40 | All large, medium, and small coal enterprises have perfect environmental protection facilities to realize the coordinated development of regional economy and ecological mining areas |
| Northeast China | 75 | 75 | 55 | 45 | In restoring and developing production implement the clean production strategy, strengthen environmental protection, and carry out the governance and restoration of ecological environment after the shutting down of coal mines |
| Central China | 75 | 75 | 55 | 45 | Actively solve the ecological and environmental problems in coal production, and carry out the promotion of clean production and circular economy |
| Northwest China | 95 | 75 | 60 | 40 | The coal mine development and construction layout, development scale and development strength must be identified; prior to the completion of new mining areas, coal mining water conservation test must be conducted and detailed planning and measures for soil and water conservation must be made; coal environmental control must be strengthened to achieve the goal that the discharge of all pollutants meets standards and the comprehensive utilization of associated resources and wastes of coal must be improved |
| Southwest China | 70 | 75 | 55 | 50 | Comprehensively promote clean production, realize the combination of clean production and clean coal technologies and close integration of comprehensive utilization of associated resources of coal measures and environmental and geological disaster prevention and control and change the present situation of backward environmental protection |

**Table 3.3** Development goals for efficient utilization of waste resources in all regions in 2020

| Region | Main indicator (%) | | | |
|---|---|---|---|---|
| | Mine water utilization rate | Coal gangue utilization rate | Coalbed gas utilization rate | Land reclamation rate |
| North China and South China | 80 | 80 | 65 | 50 |
| Northeast China | 80 | 80 | 60 | 45 |
| Central China | 85 | 85 | 55 | 45 |
| Northwest China | 95 | 80 | 55 | 45 |
| Southwest China | 70 | 85 | 55 | 70 |

### *3.1.3 Solutions and Policy Suggestions*

Coal development, especially large-scale coal base construction, should be based on the protection of the ecological environment. In the long run, there are many factors affecting and constraining the large-scale development and utilization of coal bases, especially the shortage of water supply. As 13 large coal bases are located in areas characterized by fragile environment and water shortage or in densely populated and economically developed areas, the highly centralized and large-scale coal mine-dominated coal mining and utilization mode will definitely result in the large area underground water-level decline, farmland subsidence, house cracking and depression and relocation of villages and lack of comprehensive planning might lead to social instability.

During the "12th Five-Year Plan" period, China's coal industry should take into consideration its own characteristics and balance industrial development and environmental protection based on the principle of protective development and increasing income and reducing expenditure. Specific measures include: ① base production on water supply and make reasonable planning to control the coal production of areas with fragile ecological environment. ② Vigorously adjust economic structure; strictly limit the development of heavy chemical industry enterprises that use a large amount of water; actively develop circular economy and ecological economy; protect environment and promote the benign development of coal development. ③ Explore underground water and surface water resources in a protective way. ④ Vigorously improve the comprehensive utilization of waste resources.

The guiding ideology is as follows: To continue to thoroughly implement the scientific outlook on development, take a new road to industrialization, speed up the industrial and product structure adjustment of coal enterprises, and utilize and dispose of the "three-waste" resources of coal mines in a concentrated, scientific, reasonable, and effective way based on the scientific mode of developing mining areas into circular economy parks; carry out "exchanging coal gangue for coal" to realize terminal solid waste disposal and vertical ecological renovation of the well underground and above the ground in the subsidence area, develop the vegetation greening of mining areas in the carbon sequestration model and set the producing of thermal and clean products by the circular economy parks as the goal to achieve the basic balance between coal development and environmental capacity, promote the healthy and sustainable development of the coal industry, and construct ecologically civilized mining areas.

### *3.1.4 Summary of Environmental Protection Constraints in Coal Development*

Environmental problems in coal mining areas are in essence development problems and will finally be solved through economic laws and market mechanism. In recent 5 years, China's coal industry has developed very rapidly and the development layout, development mode, development scale, economic structure, and management methods of the coal industry have undergone fundamental changes. However, at present the coal industry still has a lot of problems; for example, the preferential tax policy for comprehensive utilization of waste resource is not implemented; environmental protection funds are not in place; environmental protection technology research is not in place; relevant environmental protection standards are not in place. The delayed formulation of environmental protection policies and measures affects the successful promotion of environmental protection work conducted by enterprises. In order to solve environmental protection constraint factors in coal development, we should carry out work in the following aspects:

(1) Speed up the comprehensive control of subsidence areas to construct ecological economy. A lot of resource-based cities in China, especially many mining areas where resources will soon be exhausted, are built up based on coal development. From the perspective of vigorously developed ecological economy, we should carry out all-round rectification and governance including gangue dump management, land reclamation, and renovation of the shanty towns, actively promote the experience of constructing ecological economy of Fuxin, Huainan and Kailuan mining areas, and compile unified ecological construction plan. The Chinese central government and all provincial governments should make unified land reclamation standards and the standards for the collection of the ecological environment governance deposit and strengthen the research on the subsidence area control technology.

(2) Vigorously implement comprehensive utilization of resources to save energy and reduce emission.

    (1) Promote advanced energy-saving and emission reduction technologies and experiences, develop carbon reduction measures and technologies in the coal industry, constantly increase coal seam gas drainage and utilization,

expand utilization of cooking gas and tar oil, and gradually reduce the venting amount.

(2) Rectify and ban ecological damage and environmental pollution, waste of resources, and the mining of high-sulfur coal mines in accordance with the law, appropriately control total raw coal production, reduce total discharge of the waste, and promote the advanced experience of "zero" discharge of pollutants.

(3) Reform mine development layout and mining process, reduce coal gangue discharge by forbidding the discharge of coal gangue in new mines and encouraging old mines to create conditions to prevent discharge of coal gangue, strengthen the underground filling technology, and greatly reduce surface subsidence.

(4) Adjust the economic structure of the coal industry, and change the economic growth mode. Based on environmental capacity of the coal development regions, increase the strength of macroeconomic regulation and control; in the coal development process, focus on clean production, change the economic growth mode, and formulate economic and technical policies for the development of ecological, circular, and low carbon economy.

(5) Establish the industry environmental protection supervision system, and improve environmental protection level. In recent years, coal production has increased sharply; environmental protection funds have been lacking; relevant regulations have lagged behind; effective unified supervision mechanism of environmental protection in coal mining area has been lacking. During the "12th Five-Year Plan" period, the Chinese government should change the development mode, establish effective ecological compensation mechanism, prepare environmental protection standards for underground mining, surface mining, construction period and operation period and unified and organized environmental protection supervision system, broaden the financing channels, issue supporting laws, regulations, and policies, and enhance environmental protection level.

(6) Achieve sustainable development of circular economy. Establish the scientific outlook on development and the idea of coordinated development of economy, environment, and resources, realize the reduction, reclaiming, and reusing of pollutants in mining areas to maximize resource value and comprehensive benefit, and form the development mode of virtuous cyclic utilization. Specific measures are as follows: ① Vigorously promote energy conservation and emission reduction. ② Comprehensively promote clean production. ③ Vigorously carry out comprehensive utilization of resources, and maximally realize reclamation of wastes and recycling and reusing of renewable resources. ④ Vigorously develop environmental protection industry, focus on the development of environmental protection technology and equipment, and realize the sustainable development of mining areas. For intensive coal production areas, establish laws and regulations as soon as possible to promote the development of circular

economy, incorporate the comprehensive governance of associated resources and wastes of coal produced in coal production (coal gangue, mine water and coal slime) into legal provisions, and mobilize the whole society to participate in it.

## 3.2 Coal-Fired Power Generation

### *3.2.1 Present Situation of Environmental Impacts and Control*

Due to the characteristics of China's energy resource endowment, China's power source structure is dominated by coal-fired power. See Fig. 3.1 for main pollution generating links of coal-fired power plants.

#### 3.2.1.1 Smoke and Dust

1. Main environmental impacts

The smoke and dust pollution of the coal-fired power plant mainly refers to the ash content in coal that is discharged into the air in the form of flying ash, leading to air pollution in the coal combustion process. The smoke and dust emission mainly depends on the amount of ash content in coal, boiler combustion mode and dust removal effect of the smoke and dust treatment device. Smoke and dust have many impacts on environment. Through the dust removal of efficient dust removal device (with dust removal efficiency higher than 99%), generally the smoke and dust particles discharged into the air are particulate matters (PM2.5) and smoke particles usually do harm to the respiratory system. Due to the combined effect of primary particulate matters (smoke and dust) and secondary particulate matters generated from gaseous pollutants, there emerges smog and haze which affect the visibility of the atmosphere, leading to environmental pollution.

2. Main control technology

(1) Technology progress. China's coal-fired power plants have witnessed the development of flue gas dust removal technology. In terms of the application of precipitator, at the very beginning cyclone precipitator, multi-tube precipitator, and water film precipitator were used; in the 1980s, electrostatic precipitator was widely used; in recent years as the pollution discharge standards have becoming increasingly stringent, more and more bag-type precipitator and electric-bag-type precipitator have been used and new dust removal technologies and processes have been further developed and utilized. The smoke and dust control technology of coal-fired power plants has developed into the stage

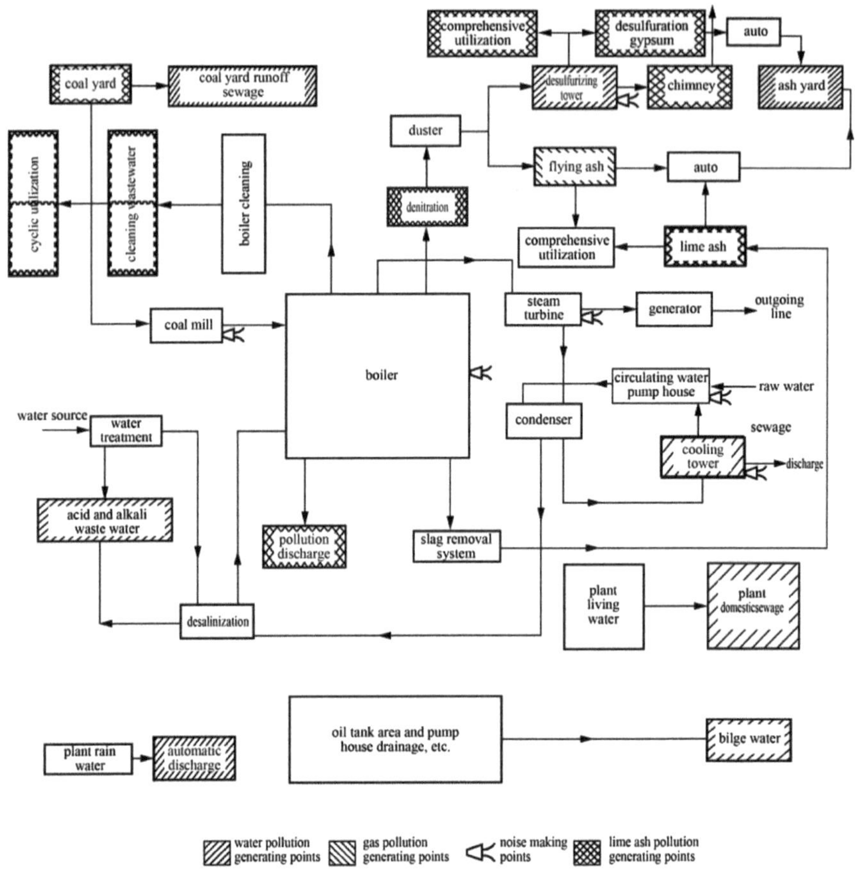


**Fig. 3.1** Main pollution generating points of coal-fired power plants

of using best available techniques (BAT). By the end of 2013, the proportion of the electric precipitator, bag-type precipitator, and electrostatic-fabric integrated precipitator in all precipitators used by coal-fired power plants are 79.5, 9, and 11.5%, respectively. See Fig. 3.2 for the change of dust removal technology of China's coal-fired power plants in various stages.

BAT is defined as the latest development stage of all production activities, processes, and relevant operations. It shows the applicability of a particular technique when it meets the discharge limit or when it cannot meet the discharge limit and no other specified technique is available; this technique can be used to minimize the amount of pollutants discharged into environment. "Techniques" should include the application of techniques and the design, construction, repair, operation, and removal of facilities. "Available techniques" refer to the techniques that have developed based on a certain scale or level and can be applied in relevant industrial fields when economic and process conditions permit and the costs and benefits are

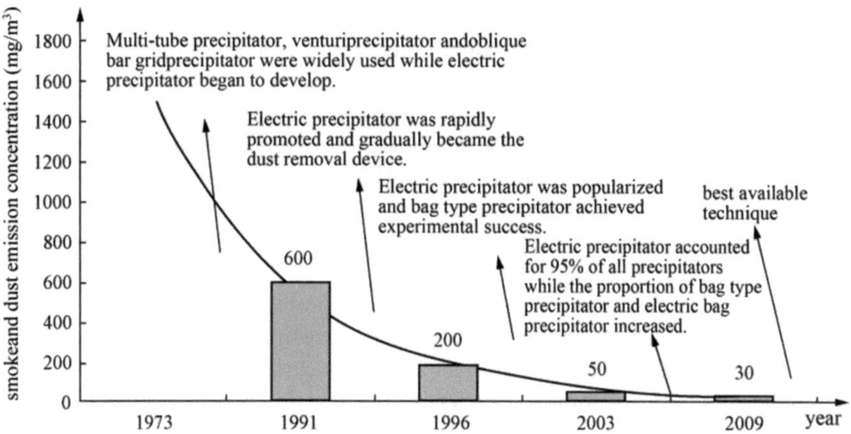


**Fig. 3.2** Change of dust removal technology of China's coal-fired power plants in various stages column interpretation best available techniques

taken into consideration. Whether a technique is used and put into production depends on whether it can be reasonably accepted by the operator. "Best" refers to the maximization of effectiveness based on the comprehensive consideration of environmental protection.

(2) Main existing problems

(1) Electric dust removal technology. As the mainstream dust removal technology in the electric power industry, with the decrease of the limit of the smoke and dust emission concentration, electric dust removal technology is facing some problems to be solved urgently. For example, most of China's coal-fired power plants do not have fixed coal sources, burn coals mixed with inferior, and have poor flue gas working conditions. Electric precipitator is sensitive to smoke and dust, and coal quality changes will reduce the efficiency of dust removal. As the corona wire of conventional electric precipitator largely uses barbed electrode and the matched power supply generally uses the low-power frequency equipment, conventional electric precipitator has high energy consumption; conventional electric precipitator has installation and debugging defects. According to the requirements for the special discharge limit in "key areas" (20 mg/m$^3$), most electric dust removal devices of coal-fired power plants should be upgraded.

(2) Bag-type dust removal technology. Main problems found in the operation of the bag-type precipitator include: ① Generally speaking, due to the short application time, bag-type precipitator used by China's coal-fired power plants needs to be improved in design, manufacturing, and operation. ② Filtering bag has a short service life. The service life of most filtering bags is less than 30,000 h, while the filtering bags used by a few power plants need to be changed every year and even every few months. ③ Dust removal efficiency is

unstable. At the initial stage of the operation, bag-type precipitator has high dust removal efficiency, but some time later, the holes in the bag cannot be easily identified and it is hard to change the bag on the production line. As a result, the actual smoke and dust emission concentration will increase. ④ There are a lot of factors that affect the dust removal effect, and bag-type precipitator has high requirements for design, manufacturing, installation, debugging, maintenance, and operation management. ⑤ Bag-type precipitator has large workload, high operation cost and poor examination and repair work environment.

(3) Electric-bag dust removal technology. Main problems found in the operation of the electrostatic-fabric integrated precipitator are as follows. Electrostatic-fabric integrated precipitator has certain advantages in technical indicators and economical efficiency, but due to its short application time some key technical problems remain to be solved, for example, structure arrangement, impact of cathode discharge on the filtering bag, filtering bag replacement, extremely short service life, low reclamation rate of old bags, and large equipment pressure loss.

3. Present situation of emissions and emission control

With the smoke and dust control technology progress, the dust removal efficiency of China's coal-fired power plants has increasingly improved. Although the installed power capacity and generating capacity have continued to grow rapidly, the annual smoke and dust emissions have been effectively controlled. In 2013, China's annual emissions of electric smoke and dust were about 1.42 million t and smoke and dust emissions per unit thermal power generating capacity were 0.34 g/kWh. See Fig. 3.3 for smoke and dust emissions of China's thermal power plants between 2001 and 2013.

Column Interpretation

Analysis of Control of Particulate Matters Emissions from Thermal Power Plants

I. Emission Characteristics

Particulate matters emitted by coal-fired power plants can be divided into two categories, primary particulate matters directly emitted from power plants and secondary particulate matters that are generated from particulate matters emitted into the air in the form of gas (such as $SO_2$, $NO_x$, $SO_3$ and VOC) through complex atmospheric physical and chemical processes. Primary particulate matters can be categorized into particulate matters directly emitted from power plants in the form of solid (or liquid) and primary condensed particulate matters that are emitted from power plants in the form of gas or vapor at the temperature of flue gas and condensed into solid (or liquid) in the process of dilution and cooling of smoke plume.

II. Impact Estimation

1. Analysis of emissions of primary particulate matters

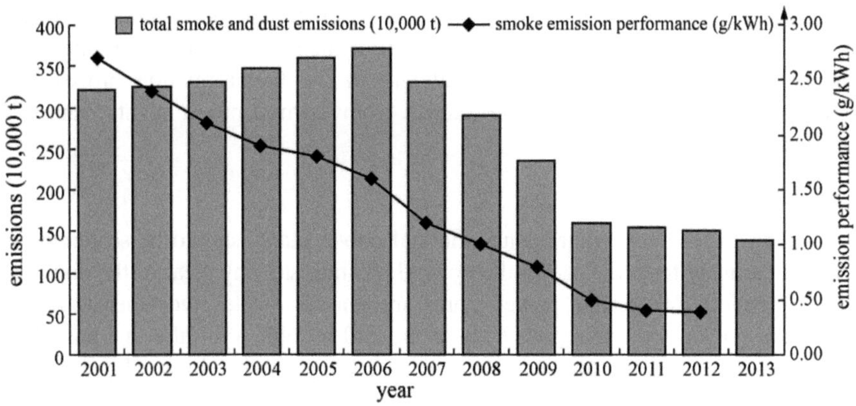


**Fig. 3.3** Smoke and dust emissions of China's thermal power plants between 2001 and 2013

In recent years, coal-fired power plants generally use highly efficient dust removal and desulfurization facilities and the smoke and dust removal efficiency of precipitators has reached 99.6% and even higher. More than 90% of the power plants have used the wet limestone–gypsum flue gas desulfurization technology, and the smoke and dust removal efficiency has also reached 50%. The measured results show that more than 90% of the particulate matters in the flue gas emitted from coal-fired power plants are PM10 in which PM2.5 accounts for a large proportion. For example, State Power Environmental Protection Research Institute measured the smoke and dust, PM10 and PM2.5 emitted from six representative coal-fired power plants in the Yangtze River Delta region, and the results show that after the six coal-fired power plants used electrostatic precipitators and wet limestone–gypsum flue gas desulfurization technology, the proportion of PM10 and PM2.5 in the smoke and dust emitted from these power plants were 87.54–95.90% and 41.22–50.31%, respectively, averaging 91.57 and 46.14%, respectively. In a breakdown, the proportion of PM2.5 in PM10 was 42.99–55.14%, averaging 50.45%. Accordingly, conservative estimation was made on the emissions of primary particulate matters emitted from coal-fired power plants. Suppose all the particulate matters in the smoke and dust emitted from these power plants were PM2.5, the annual emissions of primary particulate matters of China's electric power industry in 2012 would not exceed 1.5 million t.

2. Analysis of emissions of secondary particulate matters

Water soluble ion aerosol formed by sulfate, nitrate, and ammonia ion generated from the homogeneous or heterogeneous oxidation reaction of $SO_2$ and $NO_x$ emitted into the air from thermal power plants is an important source of secondary particulate matters emitted from thermal power plants. In recent years, thermal power plants have installed desulfurization and denitration device on a large scale, significantly improving desulfurization and denitrification efficiency. Total emissions and emissions per unit generating capacity of $SO_2$ and $NO_x$ of the electric power industry have constantly reduced. In the future with technology progress and management innovation, their emissions will further decline; emissions of secondary particulate matters and the proportion of their impact on environment will reduce accordingly. Due to the complexity of the transformation process, the emissions of secondary particulate matters from thermal power plants cannot be estimated accurately.

Ⅲ. Impact level analysis

According to relevant requirements, thermal power plants should be scattered and located far away from the city in the upwind position steering clear of the dominant wind direction. As thermal power plants are sources of elevated emissions, generally the stacks in thermal power plants are 240 m high. With the flue gas lifting height taken into consideration, the effective emission height is above 500 m, having a small impact on the near-surface environment.

In terms of the impact of the emissions of pollutants from power plants on environment, with environmental protection requirements becoming increasingly strict, the coal-fired power plants in cities mainly use large-scale combined heat and power boilers and take efficient dust removal measures to meet the emission requirement of the new standard, 30 $mg/m^3$. According to *Technical Guidelines for Environmental Impact Assessment*, we used the atmospheric steady plume dispersion model recommended by the Ministry of Environmental Protection of the People's Republic of China to predict and analyze the impact of PM2.5 emitted from coal-fired boilers. The maximum daily average concentration of particulate matters emitted from one 1025 t/h boiler in a seaside city was 0.0018 $mg/m^3$; the maximum daily average concentration of particulate matters emitted from one 2060 t/h boiler in a seaside city was 0.0014 $mg/m^3$; the maximum daily average concentration of particulate matters emitted from one 2060 t/h boiler in a hilly city was 0.0011 $mg/m^3$. Through analysis we can make the conclusion that when strict measures are taken to meet the new standard, the concentration of particulate matters emitted from large coal-fired power plants in cities is less than 0.002 $mg/m^3$ and the concentration of PM2.5 is lower, less than 2.6% of the 0.075 $mg/m^3$, the daily average value of the Class 2 standard specified in *Ambient Air Quality Standard* (GB 3095—2012). Therefore, after the implementation of the *Standard for Emissions of Air Pollutants from Thermal Power Plants* (GB 13223—2011), particulate matters emitted from the high altitude point sources of urban coal-fired power plants have a small impact on the PM2.5 in the air.

#### 3.2.1.2 Sulfur Dioxide

1. Major environmental impacts

The atmospheric emissions of $SO_2$ in a small range of spatial scales are an important factor affecting China's air and environment quality. However, as it is also an important factor in the formation of acid rain, its sulfur deposition in a wide range of spatial scales basically shows a linear relationship. In addition, minute particles formed through physical and chemical reactions lead to secondary environmental pollution.

Coal combustion produces a large amount of $SO_2$. The $SO_2$ emissions from low and short surface sources have the greatest impact on environment. The $SO_2$ emissions from low and short surface sources account for 20–30% of total $SO_2$ emissions, but the rate of its contribution to the ambient $SO_2$ concentration is 60–70%. As power plants use high stacks to emit flue gases, the rate of the contribution of $SO_2$ emissions from power plants to the impact on environment quality is 1/4 to 1/3 of the proportion of its emissions.

2. Main control technology

At present, the desulfurization methods widely used in China and foreign countries can be divided into three categories: pre-combustion desulfurization, desulfurization in combustion, and flue gas desulfurization.

(1) Pre-combustion desulfurization includes coal washing, coal gasification, liquefaction and the use of mechanical, electromagnetic, and other physical technologies for desulfurization of coal. This method can only remove some of the sulfur (mainly inorganic sulfur) in coal, which cannot solve the problem of sulfur dioxide pollution.
(2) Desulfurization in combustion mainly includes injection of calcium into furnace and the addition of limestone to fluidized bed (i.e., circulating fluidized bed boiler combustion technology) and so on.
(3) Flue gas desulfurization technologies can be divided into three categories, wet, semi-dry, and dry flue gas desulfurization. The desulfurization technology in which both the desulfurization agent and desulfurization products are in the form of aqueous solution or serous fluids is wet flue gas desulfurization; the desulfurization technology in which desulfurization agent is in the form of aqueous solution or serous fluid and desulfurization products are in the dry state or desulfurization agent is in the dry state and desulfurization products are in the form of aqueous solutions or serous fluid is the semi-dry flue gas desulfurization; the desulfurization technology in which both desulfurization agent and desulfurization products are in the dry state is the dry flue gas desulfurization.

Due to the high desulfurization efficiency and mature technology, wet limestone–gypsum flue gas desulfurization technology has become the mainstream technology both in China and foreign countries. By the end of 2013, the proportion of wet limestone–gypsum flue gas desulfurization technology (including carbide slag method, etc.), sea water desulfurization technology, flue gas circulating fluidized bed desulfurization technology, ammonia desulfurization technology, and other desulfurization technologies in all desulfurization technologies is 92.3, 2.8, 2.0, 1.9, and 1.0% respectively. See Table 3.4 for main characteristics of typical flue gas desulfurization technologies.

**Table 3.4** Main characteristics of typical flue gas desulfurization technologies

| Item | Wet limestone–gypsum flue gas desulfurization | Seawater desulfurization | Flue gas circulating fluidized bed desulfurization | Ammonia desulfurization |
|---|---|---|---|---|
| Sulfur content of applicable coal (%) | A wide range of applications | <1.5 | <1.5 | A wide range of applications |
| Desulfurization efficiency (%) | >90 | >85 | About 90 | >90 |
| Proportion of electricity consumption in total generating capacity (%) | 1.5–2 | About 1 | Relatively small | About 1 |
| Equipment floor area | Large | Large | Relatively small | Relatively large |
| Scaling and blocking | Yes | No | No | Yes |
| Slag state | Wet | – | Dry | Dry |
| Operating cost | High | Relatively low | Relatively low | Relatively high |
| Desulfurization by product | Gypsum | None | Mixtures of dust, $CaSO_4$, $CaSO_3$, and CaO | Ammonium sulfate |
| Technology maturity | Large-scale application in China and foreign countries | Large-scale application in China and foreign countries | Large-scale application in China and foreign countries | Application in China |

Since 2005, China's coal-fired power plants have begun to implement large-scale flue gas desulfurization. By the end of 2013, total capacity of flue gas desulfurization units in thermal power plants that have been put into operation was 720k MW, accounting for 91.6% of the capacity of the in-service coal-fired generating units in China. If the circulating fluidized bed boilers with the function of desulfurization are taken into consideration and the units planned to be shut down are deducted, the proportion of desulfurization units to coal-fired generating units was close to 100%. See Fig. 3.4 for application of flue gas desulfurization units in China's coal-fired power plants between 2005 and 2013.

3. Present situation of emissions and emission control

In 2013, China's sulfur dioxide emissions were 20.439 million tons, down 3.48% over 2012. According to the statistical analysis of China Electricity Council,

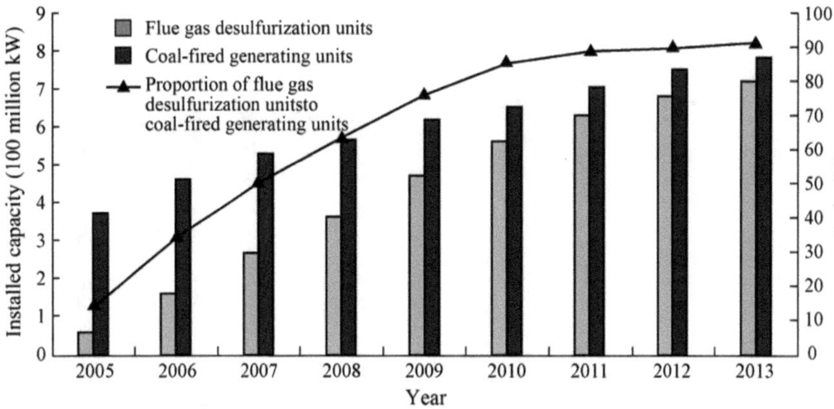


**Fig. 3.4** Application of flue gas desulfurization units in China's coal-fired power plants between 2005 and 2013

in 2013 China's sulfur dioxide emissions from electric power generation were 7.8 million tons, down 11.7% over 2012, equivalent to sulfur dioxide emissions from electric power generation in 1999 and sulfur dioxide emissions from electric power generation accounted for 38.2% of China's total sulfur dioxide emissions. In 2013, China's sulfur dioxide emissions per kilowatt hour of thermal power were 1.86 g, reducing by 0.39 g as compared with 2012, lower than America's (in 2012, America's sulfur dioxide emission performance per unit coal-fired generating capacity was 2.44 g/kWh). See Fig. 3.5 for China's sulfur dioxide emissions and sulfur dioxide emissions from electric power generation between 2001 and 2013.

#### 3.2.1.3 Nitrogen Oxide

1. Major environmental impacts

$NO_x$ is an important factor in the formation of acid rain, precursor of photochemical smog that contains pollutants, mainly ozone and one of the important sources of nutrient nitrogen element in lakes and other water bodies. In addition, the ultrafine particles in secondary pollutants $NO_x$ forms are an important source of smog and haze in city clusters and regions. More than 95% of $NO_x$ emitted from coal-fired power generation are NO, but NO can be easily oxidized by oxidizing substances in the atmosphere into $NO_2$.

2. Main control technology

Methods for controlling nitrogen oxide emissions from thermal power plants can be categorized into two categories, low $NO_x$ combustion technology which is used to control the generation of $NO_x$ in the combustion process and flue gas denitration technology which is used remove the generated $NO_x$ from the flue gas.

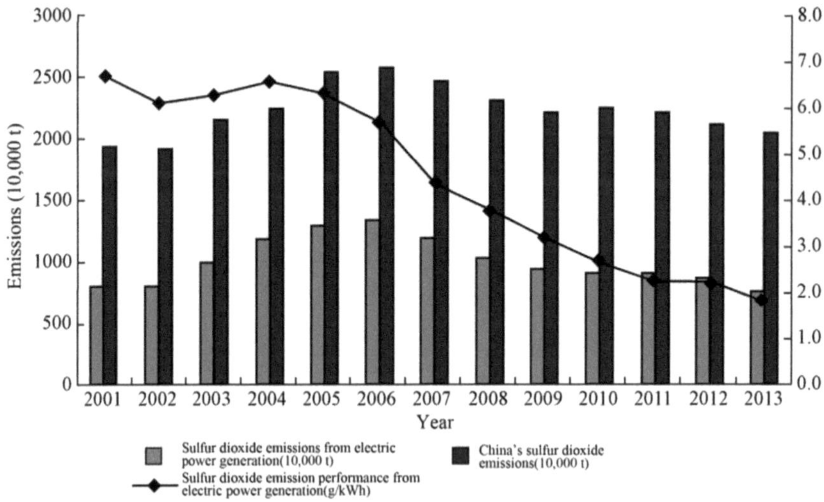


**Fig. 3.5** China's sulfur dioxide emissions and sulfur dioxide emissions from electric power generation between 2001 and 2013. *Notes* China's sulfur dioxide emissions are based on Report on the State of environment of China, while China's sulfur dioxide emission performance from electric power generation is based on the statistical analysis of the electric power industry

Low $NO_x$ combustion technology is the main method widely used to control the formation and emissions of nitrogen oxides from thermal power plant and mainly includes the low $NO_x$ burner (LNB), air classification technology, and fuel-staged combustion (re-combustion).

The flue gas denitration technology widely used by thermal power plants mainly includes selective catalytic reduction (SCR), selective non-catalytic reduction (SNCR), and SNCR/SCR combined denitration technology. SCR technology is currently the most widely used commercial application of flue gas denitration technology. See Table 3.5 for main control technologies for nitrogen oxide emissions from thermal power plants.

Since the "10th Five-Year Plan" period, all the newly built coal-fired generating units use the low $NO_x$ combustion mode as required. A number of existing units are also equipped with low $NO_x$ burners through technical reconstruction. According to statistics of China Electricity Council, by the end of 2013 the capacity of flue gas denitration units that had been put into operation in thermal power plants was 430k MW, accounting for 50% of the capacity of in-service thermal power generating units in China; the proportion of the capacity of denitration units to the capacity of coal-fired generating units reached 55, 5% higher than that in America; most thermal power plants used the SCR technology. See Fig. 3.6 for application of flue gas denitration units in China's thermal power plants between 2005 and 2013.

3. Present situation of emissions and emission control

Since the "12th Five-Year Plan" period, China's electric power industry has constantly strengthened the control of nitrogen oxide emissions. In 2013, China's

**Table 3.5** Main control technologies for nitrogen oxide emissions from thermal power plants

| Technology | Control effect of nitrogen oxides | Advantages | Disadvantages |
|---|---|---|---|
| Air-staged combustion (OFA) | Reducing nitrogen oxide emissions by 30% | Low investment and rich operating experience | It is not applicable to all furnaces, and it is possible to cause furnace corrosion and slagging and reduce the combustion efficiency |
| Reducing the number of burners put into operation | Reducing nitrogen oxide emissions by 15–30% | Low investment, conducive to boiler reconstruction, rich operating experience | It is possible to cause furnace corrosion and slagging and increase carbon content in the fly ash |
| Fuel-staged combustion | Reducing nitrogen oxide emissions by 50–60% | It is applicable to newly built boilers and existing boiler retrofitting; it can reduce the formed nitrogen oxides; it requires medium investment | It might require secondary fuels, may be required, and increases carbon content in fly ash |
| Low oxygen combustion | Reducing nitrogen oxide emissions by 20% | Minimum investment and rich operating experience | It increases carbon content in fly ash |
| Flue gas recirculation (FGR) | Reducing nitrogen oxide emissions by 20% | It can improve mixed combustion and requires medium investment | Unstable flame; increasing the number of recycling fans; not widely used |
| Low $NO_x$ burner (LNB) | When it is combined with air-staged combustion, it can reduce nitrogen oxide emissions by 60% | Suitable for newly built boilers and existing boiler reconstruction; medium investment; rich operating experience | Its structure is more complex than the conventional burner; it is possible to cause the furnace slagging and corrosion and reduce the combustion efficiency |
| Selective non-catalytic reduction (SNCR) | Large units can reduce nitrogen oxide emissions by 25–40%; small units can reduce nitrogen oxide emissions by 80% when it is combined with LNB and OFA | Occupying a small area; not affected by the fuel; medium investment; having a small impact on the air preheater | $NH_3$ escape rate is high, and denitration efficiency is affected by the flue gas flow rate, temperature distribution, and the distribution of nitrogen oxides in the furnace, which is not easy to control |
| Selective catalytic reduction (SCR) | Reducing nitrogen oxide emissions by 95% | High denitrification efficiency | Occupying a large area; high investment and operation cost; the system pressure loss is large; easy to lead to $SO_2$ oxidation |
| SCR/SNCR | Reducing nitrogen oxide emissions by 95% | Moderate investment and high denitration efficiency | Large units are yet to pass industrial tests, and the flow field is not easy to control |

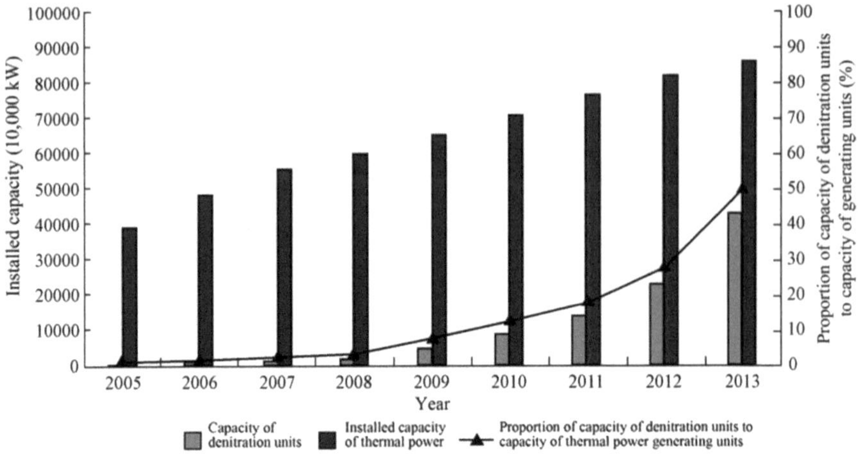


**Fig. 3.6** Application of flue gas denitration units in China's thermal power plants between 2005 and 2013

nitrogen oxide emissions were 22.273 million tons, down 4.72% over 2012. According to the statistical analysis of China Electricity Council, in 2013 China's nitrogen oxide emissions from electric power generation was 8.34 million tons, down 11.7% over 2012 and accounting for about 37.4% of China's total emissions of nitrogen oxides. In 2013, nitrogen oxides emissions per kilowatt hour of thermal power was 1.99 g, reducing by 0.43 g or 17.8% as compared with 2012. See

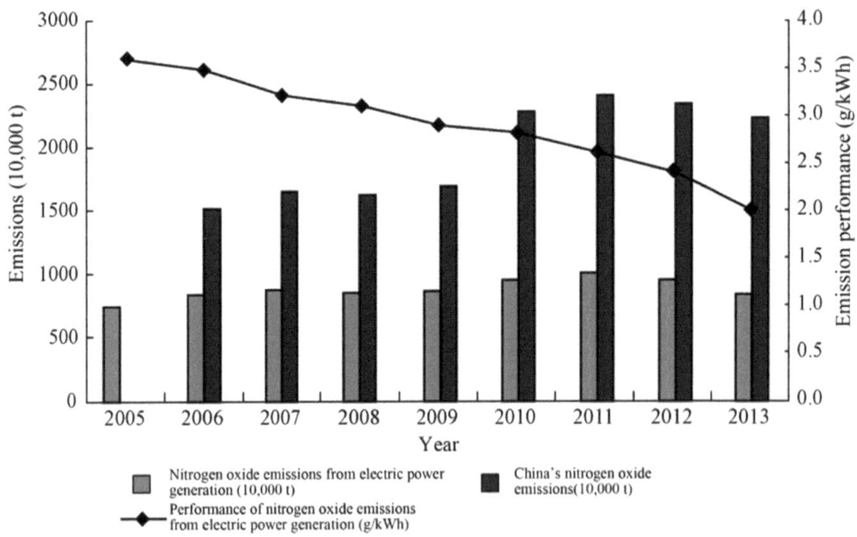


**Fig. 3.7** China's nitrogen oxide emissions and nitrogen oxide emissions from electric power generation between 2005 and 2013

Fig. 3.7 for China's nitrogen oxide emissions and nitrogen oxide emissions from electric power generation between 2005 and 2013.

#### 3.2.1.4 Heavy Metals in Flue Gas

1. Major environmental impacts

Coal contains traces of heavy metal elements. When coal is burned at a high temperature, some of these heavy metal elements will volatilize and change into metal vapor which, in the lower part of the flue pipe, gathers on the surface of submicron particles through nucleation, coagulation, and condensation as the temperature falls and is discharged into the air or polluted water resources, doing great harm to environment and human health. According to the current statistics, China has detected 47 kinds of trace elements and those that have attracted the most attention are mercury, cadmium, and chromium. Especially mercury, even at a very low concentration, is extremely toxic to human beings and wildlife. Emissions of mercury from human activities, mainly from coal-fired power plants, account for 10–30% of total mercury emissions.

2. Main control technology

The methods for controlling mercury emissions from coal combustion can be divided into three categories: pre-combustion control, control in combustion, and post-combustion control. Pre-combustion control mainly includes coal washing technology and chemical desulfurization process; combustion control is mainly achieved through the change of combustion conditions and the injection of the solid adsorbent into the furnace; post-combustion control is mainly achieved through the use of activated carbon, fly ash, calcium-based absorbents, and some new absorbents to reduce mercury emissions. See Table 3.6 for the average demercuration efficiency of existing flue gas pollution control methods.

3. Present situation of emissions and emission control

At present, special technology for removing mercury has not been used in China, but the widely used electrostatic precipitators and wet flue gas desulfurization devices and the rapidly developing flue gas denitration devices can effectively reduce the mercury emissions from coal-fired power plants and are the main means used by China's coal-fired power plants to control mercury pollution. Some

**Table 3.6** Average demercuration efficiency of existing flue gas pollution control methods

| Control measures | Mercury removal rate (%) |
|---|---|
| Electrostatic precipitator (ESP) | 10–30 |
| Fabric filter (FF) | 20–40 |
| ESP/FF + wet flue gas desulfurization (WFGD) | 65–85 |
| SCR + ESP/FF + WFGD | 70–95 |

scholars and research institutes have estimated the mercury emissions from coal combustion in China:

(1) According to estimates made by Tsinghua University, in 2000 China's atmospheric mercury emissions from coal combustion were 219.5 t of which 76.8 t, or 35%, were from electricity production and China's mercury emissions from coal combustion exceeded the USA.
(2) According to the estimates made by Chinese Research Academy of Environmental Sciences in 2006, in 2003 China's atmospheric mercury emissions from coal-fired power plants were 89.4 t of which 18.5 t were Hg0, 1.9 t, particulate mercury (Hgp) and 69.0 t, Hg2.

#### 3.2.1.5 Wastewater

1. Major environmental impacts

Wastewater from power plants mainly includes industrial wastewater, domestic sewage, and ash flushing wastewater. In addition, it also includes non-normal drainage from regular maintenance of power plants, for example, cleaning water for dust collectors and air preheaters and wastewater from boiler pickling. If power plants use the DC cooling system, there will be warm water discharge. The industrial wastewater from power plants and domestic sewage usually can meet the discharge standards through treatment or are recycled and reused, making a small impact on environment.

2. Present situation of discharge and discharge control

In recent years, electric power enterprises have actively adopted dry slag removal technology and direct air cooling technology while most newly built power plants have used the wastewater "zero discharge" technology. Through the use of water-saving technology, water-saving technical reconstruction and water management, the improvement of water reuse rate and external drainage discharge, wastewater discharge per unit generating capacity has reduced year by year. In 2013, water consumption per kilowatt hour of electricity of China's thermal power plants was 2.0 g, reducing by 0.15 g as compared with 2012 and wastewater discharge per kilowatt hour of electricity was 0.10 kg, equivalent to that in 2012. See Fig. 3.8 for water consumption and wastewater discharge per kilowatt hour of electricity of China's thermal power plants between 2001 and 2013.

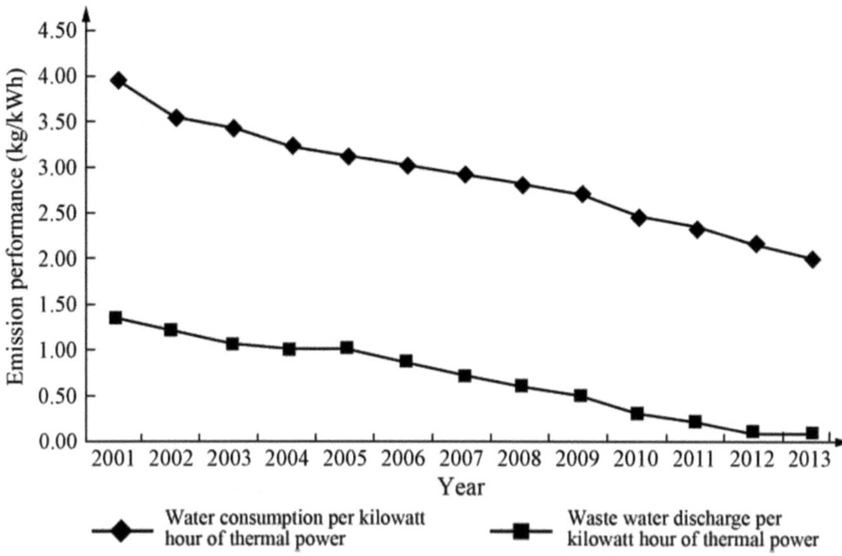


**Fig. 3.8** Water consumption and wastewater discharge per kilowatt hour of electricity of China's thermal power plants between 2001 and 2013

#### 3.2.1.6 Solid Waste

1. Fly ash

Fly ash is the main solid waste discharged from coal-fired power plants and one of the industrial solid wastes with the largest discharge capacity in China. If fly ash is not used properly, it will produce dust and other sources of environmental pollution; if used properly, it is a resource. Therefore, proper disposal and comprehensive utilization of fly ash are one of the important tasks of environmental protection for thermal power plants.

Methods for comprehensive utilization of fly ash mainly include production of building materials (cement, brick, block, ceramicite), construction works (concrete, mortar), road building (embankment, road base, pavement), backfilling (structure backfilling, building backfilling, filling of mines, sea beach), agriculture (soil improvement, compound fertilizer production, land reclamation), fly ash filling materials and recycling useful materials and products in fly ash.

According to the statistics of China Electricity Council, in 2013 China's coal-fired power plants produced about 550 million tons of fly ash, up 20% over 2012; fly ash comprehensive utilization rate was about 69%, up 2% over 2012. See Fig. 3.9 for China's fly ash comprehensive utilization rate between 2005 and 2013.

2. Desulfuration gypsum

More and more China's coal-fired power plants have had flue gas desulfurization units installed, producing a large amount of desulfurization gypsum. Desulfurization

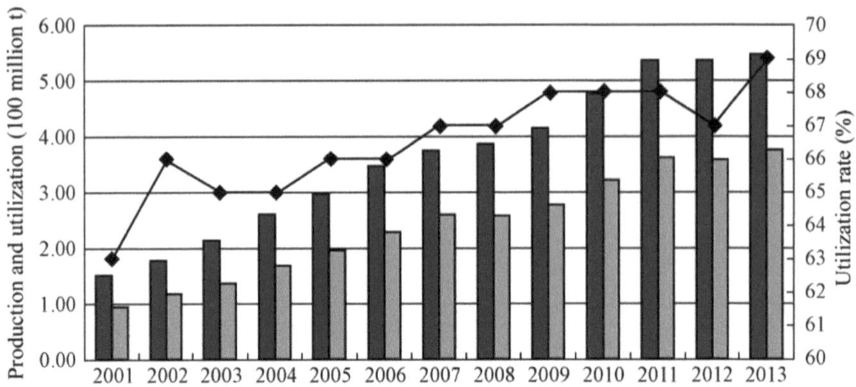


**Fig. 3.9** China's fly ash comprehensive utilization rate between 2005 and 2013

gypsum has a very short history in China, and its comprehensive utilization has just started. China's utilization of desulfurization gypsum can be divided into two categories: ① direct use of desulfurization gypsum, mainly used as cement retarding agent and saline alkali soil improvement. ② Roasting desulfurization gypsum calcination into building gypsum which can be used to produce paperbacked gypsum board, gypsum block, stucco gypsum, gypsum plank, and other gypsum products. At present, desulfurization gypsum produced by China's coal-fired power plants is mainly used in the cement and gypsum board industry; in addition, it is also used in the fields such as stucco gypsum, gypsum powder, gypsum binder, agriculture, mine mortar filling, and highway subgrade materials, but does not form industrial scale. Desulfurization gypsum has been in Europe and Japan for more than 20 years ago. In Japan in 1997, utilization rate of desulfurization gypsum reached 98% and it is mainly used for producing paperbacked gypsum and cement; in Europe, it is mainly used for producing paperbacked gypsum board, gypsum block, and self-leveling floor; in the USA, it is mainly used for producing cement, gypsum plaster for building purposes, and various gypsum boards.

According to statistics, in 2013 China's electric power industry produced about 75.5 million tons of desulfurization gypsum, up 11% over 2012 and the comprehensive utilization rate of desulfurization gypsum was about 72%, equivalent to that in 2012. See Fig. 3.10 for comprehensive utilization of desulfurization gypsum in China between 2005 and 2013.

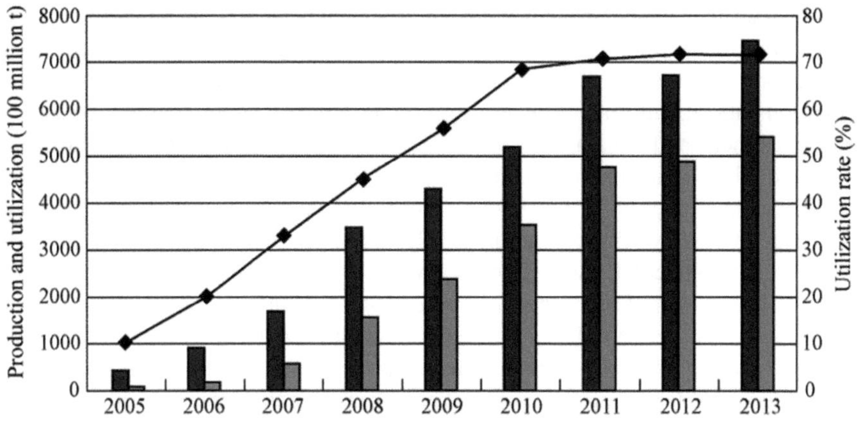


**Fig. 3.10** Comprehensive utilization of desulfurization gypsum in China between 2005 and 2013

## *3.2.2 Analysis of Major Environmental Protection Constraints*

### 3.2.2.1 Analysis of Laws, Regulations, and Policies

1. Requirements of existing laws, regulations, and policies

According to China's current environmental protection regulations and administrative requirements, the emissions of pollutants from electric power (industry) enterprises are measured in three ways. ① Standards for emissions of pollutants from all thermal power plants, for example, *Standard for Emissions of Air Pollutants from Thermal Power Plants* (GB 13223—2011) and *Integrated Wastewater Discharge Standard* (GB8978—2002). ② Requirements for the control of total emissions in a certain region, for example, control of total sulfur dioxide emissions. ③ Environment quality standard for environmental impact assessment of construction projects, for example, *Atmospheric Environment Quality Standard, Environment Quality Standard* (GB 3095—2012) and *Environment Quality Standards for Surface Water* (GB 3838—2002). In addition, long-distance transport of cross-border pollutants (mainly heavy metals discharged from power plants) should conform to relevant international conventions, for example, *Helsinki Convention* and *Rotterdam Convention*.

2. Analysis of laws, regulations, and policies made during the "12th Five-Year Plan" period

In terms of laws, environmental *Protection Law of the People's Republic of China* revised and adopted at the Eighth Session of the Standing Committee of the 12th National People's Congress on April 24, 2014, will come into force on January 1, 2015. The revision of environmental *Protection Law of the People's Republic of*

*China* mainly includes strengthening environmental protection publicity, raising public awareness of environmental protection, drawing the red line of ecological protection, governance of smog and haze pollution, define the legal status of environmental monitoring agencies, improving administrative coercive measures, encouragement and organization of the research into the impact of environment quality on the public health, changing of pollution discharge fee into environmental protection tax, improving the regional restricted approval system and pollution discharge license system, protection of relevant informants, expansion of the main body of environmental public interest lawsuit and increasing environmental violations. The revised *Environmental Protection Law* further defines the responsibilities of the government for environmental protection supervision and management, improves basic environmental protection systems such as the red line of ecological protection, control of total emissions of pollutants, environmental and health monitoring and impact assessment, cross-administrative region joint prevention, strengthens corporate responsibility for pollution prevention and control, increases legal sanctions against environmental violations, and makes systematic regulations on environmental information disclosure by the government and enterprises and the public participation in and supervision of environmental protection. The original 47 provisions are increased to 70, enhancing the law's performability and operability. The revised law is of great significance to protecting and improving environment, preventing pollution and other public hazards, ensuring human health and promoting the construction of ecological civilization and the sustainable development of economy and society. In addition, relevant departments of the Chinese government are arranging to amend the *Law on Prevention and Control of Atmospheric Pollution* and will further increase the strength of pollution prevention and control and strengthen legal liability.

In terms of regulations and policies, the Chinese central government and relevant ministries and commissions have formulated and issued a number of energy conservation and emission reduction plans and notices and put forward specific environmental protection requirements for the electric power industry during the "12th Five-Year Plan" period. For example, the State Council issued the *Action Plan on Prevention and Control of Air Pollution* (Document No. 37, 2013 issued by General Office of the State Council) to carry out comprehensive prevention and control of smog and haze and issued the *12th Five-Year Plan for Energy Development* (Document No. 2, 2013 issued by General Office of the State Council) in which goals for 2015 such as total energy consumption and efficiency, energy production and supply, energy structure optimization, national comprehensive energy base construction, ecological and environmental protection, energy consumption of urban and rural residents and energy system reform and goals for reduction of emissions from electricity production, namely reducing the emissions of sulfur dioxide and nitrogen oxide by 8 million tons and 7.5 million tons, respectively, are set. The Ministry of Environmental Protection issued the *Notice on the Implementation of Special Emission Limits of Air pollutants* (No. 14, 2013) in which the coal-fired generating units in key control regions (47 cities in 19 provinces) are required to, on April 1, 2013, start implementing the special emission limits, namely 20, 50, and 100 mg/m$^3$ for smoke and dust emissions, sulfur dioxide emissions and nitrogen oxide emissions, respectively; for in-service generating units, those in downtown of these key control regions are required to start implementing the special emission limits on July 1, 2014 and those in non-downtown areas of the key control regions are required to implement the special emission limits during the "13th Five-Year Plan" period. In the *12th Five-Year Plan for Energy Conservation and Emission Reduction* (Document No. 40, 2012 issued by General Office of the State Council), the indicators of the goals of reducing emissions from electricity production are also specified. For example, sulfur dioxide emissions should be reduced from 9.56 million tons in 2010 to 8 million tons in 2015, and a decrease of 16% while nitrogen oxide emissions should be reduced from 10.55 tons in 2010 to 7.5 million tons in 2015, a decrease of 29%. In the *Responsibility Agreement on Goals of Reducing Emissions of Major Pollutants during the "12th Five-Year Plan" Period*, the Ministry of Environmental Protection makes public the *Responsibility Agreement on Sulfur Dioxide and Nitrogen Oxide Governance Projects* signed with State Grid Corporation of China, China Huaneng Group, China Datang Corporation, China Huadian Corporation, China Guodian Corporation, State Power Investment Corporation, and other major electric power enterprises. In addition, the Ministry of Environmental Protection requires the governance projects listed in the Responsibility Agreement should be completed by the end of 2012 and strengthens the supervision and inspection.

In terms of standard revision, the more stringent version of *Standard for Emissions of Air Pollutants from Thermal Power Plants* (GB 13223—2011) has been officially implemented since January 1, 2012. The new standard tightens the limits of emissions of air pollutants from thermal power plants and becomes the strictest operation regulation for the thermal power enterprises as it is more stringent than all existing laws, regulations, and administrative requirements including the requirements for the control goal of total sulfur dioxide and nitrogen oxide emissions during the "12th Five-Year Plan" period. The latest revised version of *Environment Quality Standard* (GB 3095—2012) has been implemented since January 1, 2012. Compared with the original one, the revised version is stricter and adds the average concentration limit of PM2.5 and the average 8 h ozone concentration limit. It tightens the annual average PM10 concentration limit of the Class 2 standard to 0.07 mg/m$^3$ and tightens the annual, daily and 1 h average $NO_2$ concentration limits of the Class 2 standard to 0.04, 0.08 and 0.20 mg/m$^3$, respectively. National Energy Administration issued No. 4 Announcement of 2012 and approved the release of the 56 standards of the electric power industry. Among these standards, 20 ones are for thermal power generation, 6 ones, for hydropower generation, 2 ones, for new energy power generation, and 28 ones, for power transmission distribution. The approval of standards for resource conservation and clean production including *Technical Specification for Evaluation of Energy Consumption of Coal-fired Power Plants*, *Guidelines for Online Calculation of Coal Consumption of Thermal Power Generating Units*, *Performance of Flue Gas Denitration Devices Used by Coal-fired Power Plants* and *Guidelines for Energy Efficiency, Technical and Economic Evaluation of Distribution Transformers* has promoted the standardization of environmental protection in China's electric power industry.

The revision and issuing of the above regulations, policies, and standards shows that China will further strengthen the control the power of sulfur dioxide, nitrogen oxide and smoke and dust emissions from electricity production and implement more stringent standards for concentration and concentration control during the "12th Five-Year Plan" period.

(1) Analysis of environmental protection constraints. In terms of the impact on environment quality, with the national and industry environmental protection requirements becoming increasingly stringent, coal-fired power plants have high-level dust removal technologies and the comprehensive dust removal efficiency has reached 99% and even higher. The layout of coal-fired power plants has undergone great changes. Except for heat and power plants, new coal-fired power plants are no longer built in cities. In-service power plants have taken measures such as control of pollutant emissions, shutting down of small thermal power generating units and building of large generating units and wet flue gas desulfurization and dust removal to further improve the smoke and dust emission control level of coal-fired power plants. The causes of fine particulate matters are very complicated. In addition to natural factors, pollutant emissions from coal combustion and motor vehicles and construction dust are also important causes. However, through further analysis we find that, as one of the causes of generating smog and haze, pollutant emissions are not resulted from coal combustion of power plants. Instead, a large number of low and short surface sources in cities including industrial small boilers, domestic stoves and scattered coal burning directly emit pollutants into the atmospheric surface layer, contributing more to the concentrations of PM2.5, TSP, $SO_2$, and $NO_x$. We cannot use the proportion of the smoke and dust emissions to measure its impact on environment quality.

In terms of compliance of emission standards, the emission limits for new generating units specified in the *Standard for Emissions of Air Pollutants from Thermal Power Plants* (GB 13223—2011) are more stringent than the standards of the same class made by other countries and regions, becoming the most stringent standard in the world. See Figs. 3.11 and 3.12 for comparison of smoke and dust emission standard limits for new and in-service generating units between major countries in the world.

Now, we are going to analyze the technical feasibility of smoke and dust emission control. First, from the analysis of coal ash content, we know that in China ash content in coal reaches 28%, one time or two times higher than that in Europe and America, with ash content in coal in Europe averaging 14.4% and ash content in coal in America averaging below 10%. In addition, as in China the calorific value of coal is lower than that in Europe and America, the difference in ash content in coal between China and Europe and America is even larger. Therefore, in order to meet the same emission standard, China's comprehensive dust removal efficiency

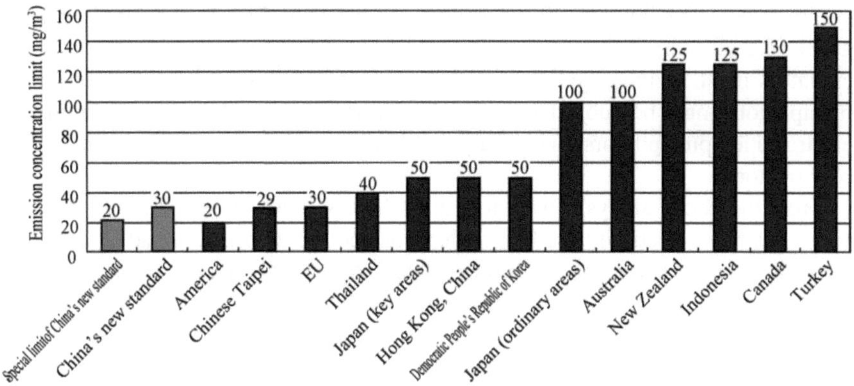


**Fig. 3.11** Comparison of smoke and dust emission concentration limits for new large coal-fired power plants between major countries and regions in the world. *Notes* America's and EU's limits are the daily average over a several-day period

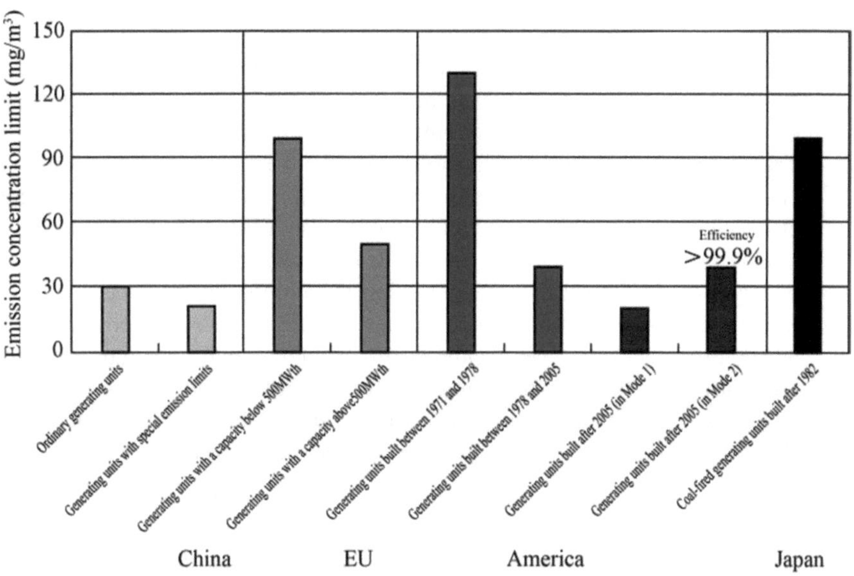


**Fig. 3.12** Comparison of smoke and dust emission standard limits for in-service generating units between major countries in the world. *Notes* America's and EU's limits are the daily average over a several-day period

must be higher than Europe and America, which requires China to pay a higher economic cost. Second, from the analysis of the existing dust removal technology, we know that suppose in China the average ash content in coal is 25%, the dust removal efficiency of precipitators must be higher than 99.9% in order to meet the required emission limit of 20–30 mg/m$^3$. As a result, 73.7% of China's existing generating units must be reconstructed. The capacity of existing electrostatic precipitators must be increased by one to two electric fields or existing electrostatic precipitators must be subject to technology improvement, for example, replacing electrostatic precipitators with fabric filters or electrostatic-fabric integrated precipitator.

From the above analysis, we know that China's control level of smoke and dust emissions from electricity production has reached the international advanced level. What is more, as a result of power plants' optimized layout, smoke and dust emissions from electricity production have made an increasingly small impact on environment and air quality in cities and the smoke and dust emissions from coal-fired power plants are no longer the main cause of environmental pollution. However, as China has made increasingly stringent control requirements for smoke and dust emissions from thermal power plants, most coal-fired power plants have to upgrade and reconstruct precipitators. Specifically, they have to choose from different dust removal technologies, invest heavily in technical reconstruction and make proper arrangements for production and reconstruction in a short period of time.

(2) Analysis of obligatory emission limits in the "12th Five-Year Plan" period. On the one hand, China has tightened requirements for environmental protection. On the other hand, China's dust removal technologies have further developed. Take electrostatic dust removal as an example: The increasingly mature efficient dust removal technologies (system with low-low temperature ESP, electromechanical multi-mode and double-zone ESP, $SO_3$ flue gas conditioning technology, and moving electrode electrostatic precipitator) have been widely used in China; dust condensation technology has become mature in foreign countries, while some countries in China have mastered its core technology; the filter bag technologies, including high pressure spunlace nondestructive filtering technology, new technology of cleaning filtering materials, surface nanocoating technology, and high temperature hot press laminating technology, are being developed and improved.

In China, the ash content in coal is higher than that in other developed countries, averaging 28%. For example, the annual average ash content in coal of a power generation group between 2010 and 2012 was 28.5, 28.4 and 26.5%, respectively. With the changes in the coal market in recent years, the ash content in coal has also reduced. Suppose the average ash content coal is 28% and the precipitator inlet smoke and dust concentration is 30 g/m$^3$, in order to meet the emission limit of 30 g/m$^3$ the comprehensive dust removal efficiency must be higher than 99.9%. However, judging from the actual situation, it is impossible for China's all in-service generating units to completely meet the emission standard. Suppose the comprehensive dust removal efficiency is 99.9%, the smoke and dust emissions from existing generating units will reach 430,000 tons. If the fact that in the key regions the special emission limit of 20 mg/m$^3$ implemented are taken into consideration, the emissions of in-service generating units are even lower, being about 400,000 tons.

It is predicted that in 2015 due to the higher dust removal efficiency of new generating units and reconstructed existing units, the average dust removal efficiency of coal-fired generating units will increase to 99.75–99.8% and smoke and dust emissions will be about 1.1 million tons; in 2020, the average dust removal efficiency of coal-fired generating units will increase to 99.9–99.95% and smoke and dust emissions will be 0.5–0.6 million tons.

2. Sulfur dioxide

(1) Analysis of environmental protection constraints. China has made specific control requirements for sulfur dioxide emissions from electricity production. Among all the requirements, the new emission standard, namely the revised *Standard for Emissions of Air Pollutants from Thermal Power Plants* and the *Notice on the Implementation of Special Emission Limits for Air pollutants,* are the most stringent. As long as the new standard and requirements for special emission limits are met, all other environmental protection requirements will pose no problem. See Figs. 3.13 and 3.14 for comparison of sulfur dioxide emission standard limits for new generating units and in-service generating units between major countries in the world.

Now, we are going to analyze the technical feasibility of sulfur dioxide emission control. First, in China the sulfur content in coal for electricity production is high and unstable. According to the power environment protection statistics over the years, in 2005 the weighted average sulfur content in coal of above-scale power generation groups was about 0.89% and it increased to 1.02% in 2010. The actual average sulfur content in coal was higher than the designed sulfur content, which might affect the stable sulfur dioxide emissions from power plants that meet the standard. Second, as China's emission standard is the most stringent in the world, in terms of desulfurization technology, except for wet limestone–gypsum flue gas desulfurization technology, all other desulfurization technologies (including dry flue gas desulfurization and sea water desulfurization technology) cannot be used to achieve the stable sulfur dioxide emissions from generating units burning high-sulfur coal that meet the standard. According to the experimental results of Electric Power Research Institute (EPRI) and past experience in engineering practice, it is easy to increase the desulfurization efficiency from 85 to 95%, but the increase from 95 to 99% will cost a high price.

(2) Analysis of environmental protection constraints on thermal power installed capacity. The constraints of sulfur dioxide on thermal power installed capacity fundamentally depend on environment quality requirements mainly including the requirement that the concentration of sulfur dioxide in environment should meet the required quality standard and that the acid deposition caused by sulfur dioxide should meet environmental critical load requirement. Therefore, environment quality requirements are the basis for making requirements for total emissions. However, in the actual operation due to the following reasons,

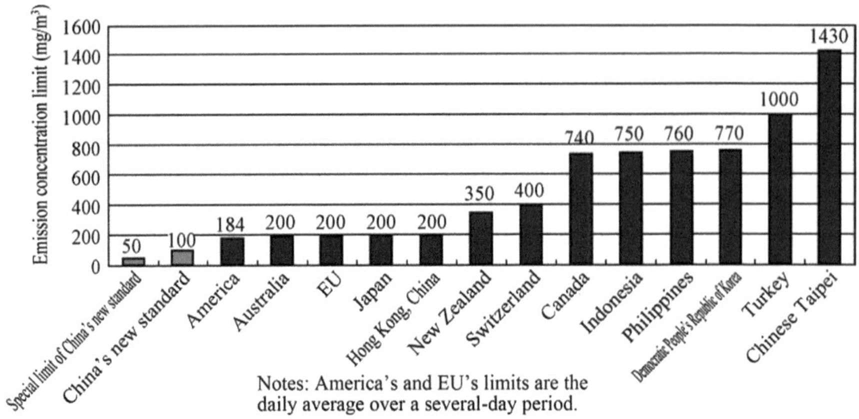


**Fig. 3.13** Comparison of sulfur dioxide emission concentration limits for new large coal-fired power plants between major countries and regions in the world

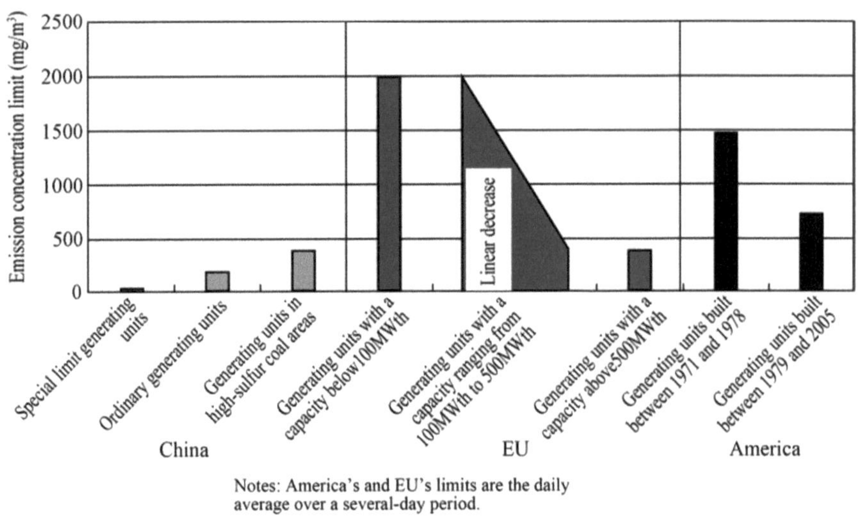


**Fig. 3.14** Comparison of sulfur dioxide emission standard limits for in-service generating units between different countries in the world

it is difficult to make control requirements for sulfur dioxide emissions from electricity production from the perspective of environmental quality. ① The impact of sulfur dioxide on environment quality not only depends on sulfur dioxide emissions, but also various factors such as the layout of emission sources, pollutant emission mode (number and height of stacks and flue gas temperature, etc.), terrain conditions of areas where pollution sources are located and atmospheric physical conditions (such as flow field and dispersion parameters) and atmospheric chemical conditions. The relationship between sulfur dioxide emissions and environmental impact is very complicated non-linear relationship. In other words, with the same environment quality requirement, due to a variety of different conditions, allowed sulfur dioxide emissions might vary. ② The impact of sulfur dioxide on environment quality depends on environmental conditions and other conditions in the affected area. For example, in some areas due to the high background concentration, a small amount of new emissions can be accommodated; some areas are sensitive to acid rain while others are not. ③ Acid deposition mainly results from the combined effect of sulfur dioxide and nitrogen oxides. As a result, the limits of sulfur dioxide emissions cannot be made based on the requirement for acid deposition. What is more, China has not set the legal requirement for critical load. ④ Sulfur dioxide emissions from electricity production account for less than 50% of China's total sulfur dioxide emissions. As sulfur dioxide generated in the electricity production process is mainly emitted outside cities and belongs to elevated emissions, the impact of sulfur dioxide emissions from electricity production on the air and environment quality is smaller than other emission sources. Therefore, when all the above-mentioned boundary conditions are uncertain, it is impossible to set the limit for China's total sulfur dioxide emissions. Even if the boundary conditions are identified, theoretically speaking, we can only roughly predict the impact of sulfur dioxide emissions on environment quality through a large number of complicated forecasting tools. In practical application, such forecasting tools can be used to evaluate the pollution situation and optimize the industrial layout, instead of calculating total permitted emissions. Moreover, from the perspective of environment quality requirements, it is unscientific and unrealistic to set the limit of sulfur dioxide emissions from electricity production without considering the emissions of other sources.

However, it does not mean that China cannot set a limit of sulfur dioxide emissions. In both the *Law on Prevention and Control of Atmospheric Pollution* and *12th Five-Year Plan for National Economic and Social Development of the People's Republic of China*, the sulfur dioxide emission control requirement is made. However, it is an administrative requirement made by the Chinese government or relevant departments based on the overall environmental evaluation as well as technical and economic conditions instead of environment quality. During the "12fth Five-Year Plan" period, sulfur dioxide emissions from electricity production can be controlled through technical measures so as to meet the air and environment quality standard requirement. What is more, sulfur dioxide emissions from non-electric power sectors have a greater impact on environment quality. Therefore, from the "12fth Five-Year Plan" period on control of sulfur dioxide emissions from electricity production will still be based on reducing acid deposition and the "administrative" distribution of sulfur dioxide emissions adopted during the "11fth Five-Year Plan" period will continue to be used. The distribution of total emissions will still be based on technical and economic conditions.

Column Interpretation

Environment Quality, Environmental Critical Load, and Acid Deposition

Environment quality: It generally refers to the suitability of the overall environment within a certain range or some environmental elements for the survival, life, and development of human beings.

Environmental critical load: It is the maximum amount of acid deposition that does not lead to the chemical changes in ecological system that have long-term harmful effects.

Acid deposition: It is the process in which acids in the atmosphere drop to the surfaces in the form of precipitation or under the action of airflow. Acids are deposited via two processes, "wet" and "dry deposition". Wet precipitation generally refers to the process by which acids with a pH normally below 5.6 are removed from the atmosphere in rain, snow, sleet, or hail. Acid rain is the most common form of wet precipitation. Dry deposition is the process by which acids are deposited on the surfaces under the action of airflow. At present, acid rain is most thoroughly studied and acid deposition is usually equated with acid rain.

(3) Analysis of environment protection constraints on the distribution of thermal power installed capacity. Environmental space for the development of coal power in China is affected by many factors. In terms of layout, China's coal power is mainly distributed in economically developed areas in Central and East China. Due to dense population and industrial concentration, concentrated sulfur dioxide emissions (including emissions from electric power sector and other sectors such as industry and transportation), large amount of sulfur deposition and acid rain, these areas usually have serious environmental pollution problems. By contrast, West and North China abound in resources, but due to backward economic development have relatively a small thermal power installed capacity, though thermal power is developing at a fast speed. In addition, in the mode of balancing power development on the spot, more and more coal has been transferred to the resource-deficient areas in Central and East China and shortage of coal power in these areas has been constantly intensified. In the future with the increase of energy consumption and demand in Central and East China as well as external energy dependence, the electric power safety in these areas will become increasingly prominent.

If the local coal resources, regional economic and social development demand, ecological environment carrying capacity, comprehensive energy transportation capacity, and many other factors are taken into consideration, in the "12th Five-Year Plan" period, coal power development should adjust the traditional balance-focused coal power layout mode in order to adapt to the inevitable development trend of "replacing coal and oil with power and long-distance power transmission". The major coal production bases in West China are the key areas for the future new coal power development, while the eastern coastal areas with poor energy resources and limited environmental protection space will be the areas with controlled new coal power development.

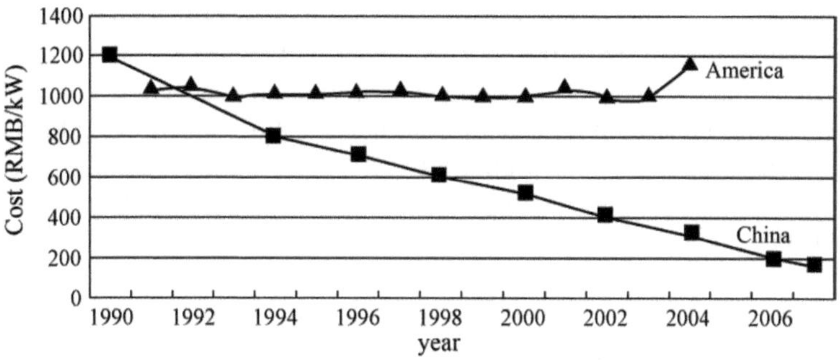


**Fig. 3.15** Comparison of average installation cost of flue gas desulfurization device between China and America

(4) Analysis of economic constraints on desulfurization technology

(1) Cost-level analysis. Judging from the cost of the desulfurization project, as China has vigorously promoted the localization and marketization of flue gas desulfurization equipment, the localization rate of desulfurization equipment and materials has reached 90% and even more than 95%. The cost of the desulfurization project has significantly reduced. The cost of flue gas desulfurization per kW of newly built large coal-fired generating units has reduced from RMB 1200 at the end of the "8th Five-Year Plan" period to RMB 500 at the end of the "9th Five-Year Plan" period and then to the current below RMB 200 which is about one-fifth of the cost of America's flue gas desulfurization device. See Fig. 3.15 for comparison of average installation cost of flue gas desulfurization device between China and America.

(2) Analysis of the sensitivity of the change of sulfur content in coal to the cost of removing one kilogram of sulfur dioxide. When the cost of removing one kilogram of sulfur dioxide is used to measure economic cost, sulfur content in coal is most sensitive to economical efficiency of desulfurization, presenting the negative exponential relationship. See Fig. 3.16 for details. For example, if we take 1% as the benchmark of the sulfur content in coal, when the sulfur content in coal increases by one time, the cost of removing one kilogram of sulfur dioxide will reduce by 42%; when the sulfur content in coal reduces by 50% (namely, when sulfur content in coal is 0.5%), the cost of removing one kilogram of sulfur dioxide will increase by 85%.

(3) Analysis of sensitivity of the change of sulfur content in coal to electricity price increment. When the desulfurization efficiency is certain, with the increase of the sulfur content in coal, total amount of removed sulfur dioxide increases accordingly and so does total cost and electricity price. There is a linear relationship between the increase of the sulfur content in coal and the increase of electricity price. When the sulfur content in coal is 1%, desulfurization will increase the electricity price by RMB 0.0151/kWh. If we take 1% as the benchmark of sulfur content in coal, when the sulfur content in coal

increases to 2%, the electricity price will increase by 13%; when sulfur content in coal reduces to 0.5%, the electricity price will reduce by only 4%. It further demonstrates that when total amount of emission reduction is certain, the generating units burning coal with a higher sulfur content have higher economical efficiency of desulfurization.

(4) Analysis of the impact of cost on desulfurization electricity price. The special studies conducted by China Electricity Council show that in addition to the impact of sulfur content on desulfurization electricity price, the impact of change of other cost elements on desulfurization electricity price is as follows. ① The installation of gas heater (GGH) will increase desulfurization investment and electricity consumption, and the desulfurization electricity price of generating units equipped with GGH will be higher than those not equipped with GGH. For example, the difference in desulfurization electricity price between 300 MW generating units equipped with GGH and those not equipped with GGH is RMB 0.003/kWh and the difference between 600 MW generating units equipped with GGH and those not equipped with GGH is RMB 0.002/kWh. ② The annual utilization hours are most sensitive to desulfurization electricity price, followed by investment, electricity price, and unit price of limestone. The following estimates are made based on different sulfur contents in coal and generating units equipped with or not equipped with GGH: When the annual utilization hours change within a range of 20%, the change of desulfurization electricity price ranges from 8.5 to 15.2%; when the investment changes within a range of 20%, the change of desulfurization electricity price ranges from 6.4 to 11.7%; when the auxiliary electricity price changes within a range of 20%, the change of desulfurization electricity price ranges from 5.4 to 7.5%; when the average load changes within a range of 20%, the change of desulfurization electricity price ranges from 3.3 to 5.6%; when the unit price of limestone changes within a range of 20%, the change of desulfurization electricity price ranges from 1.2 to 5.7%. ③ When power plants' sulfur content in coal as received basis is low, investment, auxiliary electricity price, and repair cost are the three major factors affecting the desulfurization electricity price. With the increase of sulfur content in coal as received basis, the proportion of depreciation cost and desulfurization electricity power cost will reduce while the proportion of cost of limestone and other desulfurizing agents will increase and become the fourth major factor affecting the desulfurization electricity price.

(5) Analysis of obligatory emission limits in the "12th Five-Year Plan" period. At present, the average sulfur content in coal for electricity production in China is 1.0–1.1% (for example, the average sulfur content in coal in environmental accounting of an electric power group between 2010 and 2012 was 1.11, 1.08 and 1.15%, respectively). If the average $SO_2$ concentration at the inlet of desulfurization equipment and the $SO_2$ emissions of in-service generating units are 2200 and 200 mg/m$^3$, respectively, and the average desulfurization efficiency is 92%, especially when the flue gas bypass is removed, the operation rate of desulfurization equipment will reach 100% (without considering

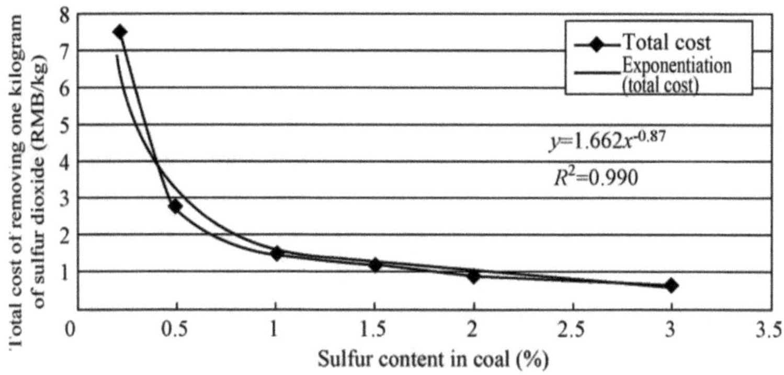


**Fig. 3.16** Analysis of sensitivity of change of sulfur content in coal to cost of removing 1 kg of sulfur dioxide

safety factors). If all the emission standards are met (without considering high-sulfur coal and special emission limits), as the high $SO_2$ emission concentration in high-sulfur coal areas is offset by the low $SO_2$ emission concentration in key areas, the $SO_2$ emissions of in-service generating units that meet the standard will be 3.97 million tons. According to the present total emissions accounting mode and the maximum desulfurization efficiency after removal of bypass being 85%, it is predicted that in 2015 the $SO_2$ emissions from electricity production will be about 6.5 million tons; in 2020, the average desulfurization efficiency will be increased to 90–95% and the $SO_2$ emissions will be 4 million to 4.5 million tons.

3. Nitrogen oxides

(1) Analysis of environment protection constraints. Similar to requirements for $SO_2$ emissions, among all the national requirements for nitrogen oxide emissions from electricity production, the new emission standard, the new emission standard, namely the revised *Standard for Emissions of Air Pollutants from Thermal Power Plants* and the *Notice on the Implementation of Special Emission Limits of Air pollutants,* are the most stringent. See Figs. 3.17 and 3.18 for comparison of nitrogen oxide emission standard limits for new generating units and in-service generating units between major countries and regions in the world.

Now, we are going to analyze the technical feasibility of nitrogen oxide emission control. First, due to poor and variable quality of China's coal as well as unstable coal supply, denitration devices have very bad operating conditions and the parameters of flue gas at the inlet of denitration devices are complicated and variable, which makes a negative impact on the design, construction, and operation of denitration devices. Second, in-service generating units, especially coal-fired generating units built after 2000, are generally equipped with low $NO_x$ burners. However, as the existing low $NO_x$ combustion technology cannot meet the requirement of the new standard, the flue gas denitration technology must be used.

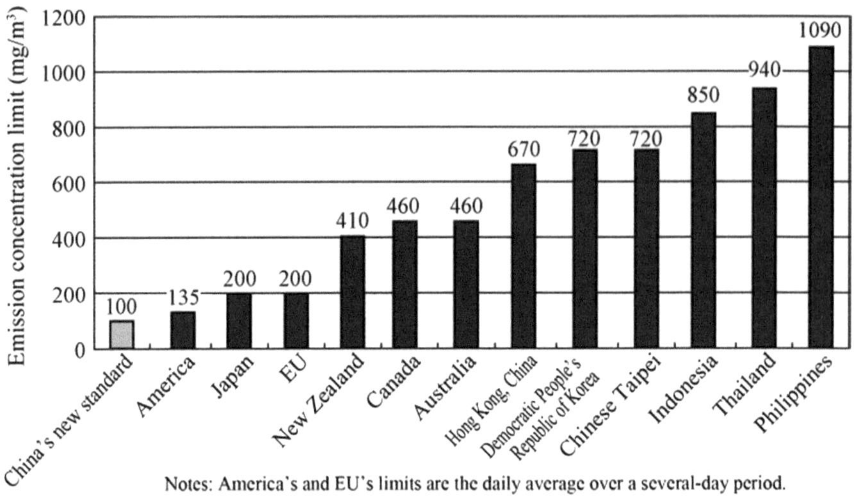


**Fig. 3.17** Comparison of nitrogen oxide emission concentration limits for new large coal-fired power plants between major countries and regions in the world

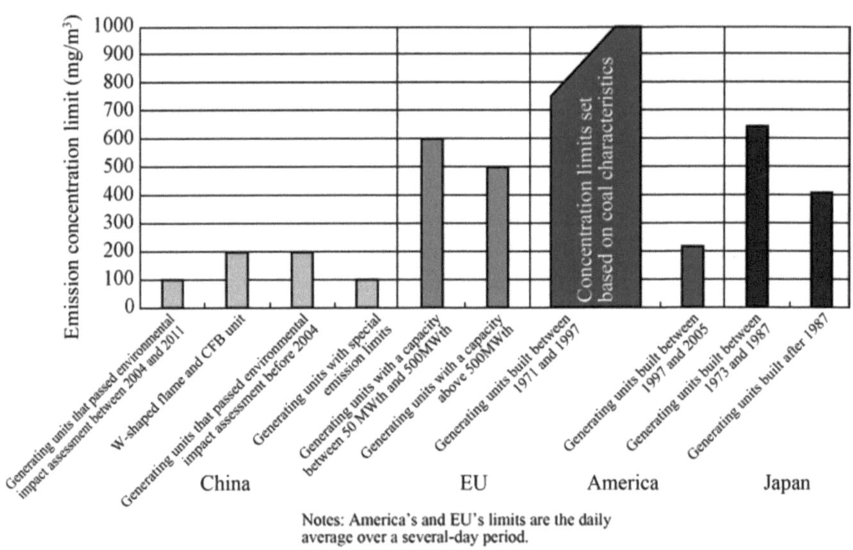


**Fig. 3.18** Comparison of nitrogen oxide emission standard limits for in-service generating units between different countries in the world

As double arch flame furnaces produce a large amount of $NO_x$ which even exceeds the processing limit of SCR technology. When using SCR technology alone cannot make nitrogen oxide emissions meet the standard and technical and economic conditions are feasible, the SNCR+SCR method can be considered. If the nitrogen oxide emissions of existing CFB generating units exceed the emission standard of 200 mg/m$^3$, priority should be given to the SNCR technology.

(2) Analysis of environmental protection indicator limits in the "12th Five-Year Plan" period. In-service generating units, especially coal-fired generating units built after 2000, are generally equipped with low $NO_x$ burners. However, in spite of the use of the low $NO_x$ burner, the concentration of the generated $NO_x$ is closely related to the type of boiler. China's coal-fired power plants mainly use "four-corner tangentially boiler" and "wall-type (cyclone)" boiler and use the W-shaped flame and CFB boiler as supplement. The four-corner tangentially boiler accounts for 50% of generating units of 100,000 kW and above. The $NO_x$ emissions of various boilers are as follows: four-corner tangentially boiler: 250–800 mg/m$^3$; wall-type boiler: 350–1200 mg/m$^3$; W-shaped flame boiler: 800–2000 mg/m$^3$; CFB boiler: <200 mg/m$^3$. For four-corner tangentially boilers and wall-type boilers, very mature low $NO_x$ combustion technologies are available, but as the existing low $NO_x$ combustion technologies cannot meet the requirement of the new standard revised in 2011, the flue gas denitration technology must be used. The W-shaped flame boiler is not applicable to low $NO_x$ combustion technologies, so efficient denitration facilities must be adopted.

According to the current situation of denitration reconstruction, it is expected in 2015, in addition to part of the planned shutdown group, 300 MW (including some 200 MW) or more units to complete the whole denitration. In addition, considering the unit load (denitrification doesn't work under certain unit load) and other factors, is expected to 2015, the power industry emissions of nitrogen oxides fell to about 600 million tons; by 2020, power nitrogen oxide emissions fell to 400 million to 450 million tons.

## 3.2.3 *Solutions and Policy Suggestions*

### 3.2.3.1 Solutions and Policy Suggestions for Fine Particulate Matter

(1) Change the pollution control idea of focusing on major emission sources. The generation of smog and haze and PM2.5 cannot be attributed to major emitters such as electric power enterprises. Instead, it results from the combined effect of hundreds of millions of uncontrolled small pollution sources. Therefore, we should focus on controlling pollution sources that emit a small amount of pollutants, but cover a wide range of areas and change the focus of environmental supervision from the "point sources" of overhead power supply of power plants to crisscrossing "line sources" (motor vehicles) and "surface sources" of numerous small emission sources.

(2) Give full play to the important role of the electric power industry in smog and haze governance from the overall and systematic perspective. With the continuous development of the electric power industry, the electric power industry has made more and more contributions to the improvement of environment quality and the scientific development of the electric power industry will provide powerful material basis and technical means for solving the problem of smog and haze and PM2.5. ① Accelerate the construction and application of smart grids to provide strong basic support for the prevention of smog and haze. ② Increase the proportion of renewable energy power generation, further improve the proportion of electricity in the final energy consumption, actively and effectively promote replacing heat and power plants with small coal-fired heat producers in cities, using electricity instead of coal and development of electric vehicles in cities in a scientific, reasonable, economical, and effective way. ③ In economically feasible areas with adequate gas supply, establish gas-fired power stations; in areas with inadequate gas supply, instead of immediately replacing coal-fired heat and power plants, continue to keep clean and efficient coal-fired heat and power plants in order to supply limited gas to urban or rural residents, and reduce scattered burning of coal and restrain power plants from competing with small and medium energy users. ④ Improve the proportion of converting coal into electricity and the proportion of pollutant emissions from electricity production in total pollutant emissions. Both theory and foreign practice have proved that the higher the proportion of air pollutant emissions from electricity production, the better the energy structure and the less environmental pollution. ⑤ Encourage coal-fired power plants to use and import low sulfur, low ash, and high-quality steam coal for power generation; restrain and even prohibit scattered small combustion equipment, kilns, and residents from using high sulfur and high ash coal inefficiently or without proper pollution control measures; meanwhile if it is difficult to change coal sources and coal quality, convert inferior coal to electricity, strengthen the classification and use of coal for power generation and prevent power plant from scrambling for high-quality coal and imposing inferior coal on residents and small-sized combustion equipment so as to optimize coal utilization at the social level, reduce pollutant emissions and improve the overall environmental and economic benefits. ⑥ For power plants located in areas with large environmental capacity, equip them with proper environmental protection facilities and make reasonable emission standards for them and draw up a reasonable schedule for the construction of pollution control equipment in thermal power plants; avoid constructing environmental protection facilities in large quantities and then reconstructing them before they are put into operation due to the declined quality.

#### 3.2.3.2 Solutions and Policy Suggestions for Sulfur Dioxide

(1) Change the mode of controlling total sulfur dioxide emissions with administrative requirements as the control target; develop the *Standard for Emissions of Air Pollutants from Thermal Power Plants* (GB 13223—2011) in a scientific way with improving environment quality as the goal, the BAT as the means and technical and economic conditions as constraints, and allocate the sulfur dioxide emission indicator to each sector in a scientific way.
(2) Strengthen law-based control, clean up, rectify, and improve the existing sulfur dioxide emission control laws, regulations, and policies to implement sulfur dioxide emission control in compliance with laws and regulations and promote the implementation of emission reduction by enterprises and supervision of it by governments according to law.
(3) Strengthen environmental protection supervision and at the same time ensure the full and timely implementation of the desulfurization electricity price and speed up the introduction of the electricity price formation mechanism with environmental costs taken into consideration.
(4) Find out the problem of poor-quality desulfurization equipment caused by lack of experience and vicious low-price competition in the explosive development of flue gas desulfurization equipment construction in recent years, carry out research on improving the reliability of desulfurization devices, and formulate relevant technical regulations to provide technical basis for law-based administration and technical supervision.
(5) Speed up the development of the resource-saving desulfurization industrialization which can change sulfur dioxide emissions into useful materials, and facilitate the development of circular economy.
(6) Really give enterprises the right to choose pollution control means and manage emission reduction, reduce unnecessary administrative intervention, and at the same time strengthen industrial self-discipline and public supervision.

#### 3.2.3.3 Solutions and Policy Suggestions for Nitrogen Oxides

(1) The Chinese central government should carry out an in-depth investigation into the nitrogen oxide emissions from environmental pollution sources all over the country and strengthen the research into $NO_x$ emissions and its distribution characteristics. ① Study the $NO_x$ emission forecasting and statistical methods. ② Study automatic monitoring-based technologies and methods. ③ Study the industrial and geographical distribution of China's total $NO_x$ emissions, and predict the future $NO_x$ emissions based on it to provide data basis for the control of $NO_x$ emissions.
(2) Revise nitrogen oxide emission standards for thermal power plants in a scientific way. Comprehensively, consider environmental, technical, and

economic conditions, region, time (period of time), characteristics of generating units, and characteristics of fuel. On the principle of distinction, govern the nitrogen oxide emissions from newly established power plants according to more stringent standards than those for the governance of nitrogen oxide emissions from existing power plants and those in-service power plants from which nitrogen oxide emissions have to be governed in a strict and prompt way should be compensated, which is international practice and in line with China's national characteristics and the principle of *Administrative License Law of The People's Republic of China.*

(3) Identify the technical route for control of nitrogen oxide emissions from electricity production.

 (1) The technical route should be determined according to the principle of economy, efficiency, safety, reliability, resource conservation, and comprehensive utilization.
 (2) All power plants including newly established power plants, existing power plants, and power plants using all kinds of coal should give priority to low $NO_x$ combustion technologies to maximally reduce the generation of nitrogen oxides in the combustion process of the boiler.
 (3) If the nitrogen oxide emissions from newly established and in-service generating units burning anthracite and meager (low volatile component) coal cannot meet the emission standard after the low $NO_x$ combustion technologies are used, the SCR denitration technology should be used. If the SCR device cannot be installed due to the constraints of conditions such as space and boiler structure, the SNCR technology or combine the SCR-SNCR combined denitration technology. Small-sized power plants with small changes of load and unstable fuel quality should choose the SNCR technology.
 (4) If the nitrogen oxide emissions from generating units cannot meet the emission standard after the low $NO_x$ combustion technologies are used or generating units located in sensitive areas burn bituminous coal and lignitous coal, the SCR denitration technology should be the first choice.
 (5) Encourage and promote power plants to integrate desulfurization, denitration, and dust removal technologies, and carry out collaborative control research and development and engineering demonstration.
 (6) Identify the reducing agent system suitable for the local technical and economic conditions. In using the liquid ammonia system, security requirements should be fully considered and preventive measures should be taken in accordance with legal provisions for safe production.

(4) Strengthen the nitrogen oxide emission control operation management of coal-fired power plants. Coal-fired power plants should use the low $NO_x$ combustion technologies to give full play to the functions of the low $NO_x$ combustion devices; establish and improve the daily management system of flue gas denitration facilities; set up operation and maintenance procedures, and work out nitrogen oxide emission control operation management measures

such as the ledger system. Environmental protection administrative departments at all levels should strengthen the operation supervision of nitrogen oxide emission control technology and equipment. Nitrogen oxide online continuous monitoring devices that have been constructed should be connected with the management information system of environmental protection administrative departments. Formulation and experimental work of technical specifications for flue gas denitration project construction should be further improved.

(5) Accelerate the flue gas denitration environmental protection industrialization.

  (1) Promote the industrialization of environmental protection with the law as the criterion, the market as the orientation, science and technology as the guide, benefits as the center, and enterprises as the main body.
  (2) Speed up the development of denitration technologies suitable for China's national conditions and with independent intellectual property rights. ① Provide financial support and tax preferences. ② Strictly restrict repeated introduction of denitration technologies. ③ Actively support the construction of demonstration projects to give full play to denitrification technologies with independent intellectual property rights, and continuously improve and optimize denitration technologies so as to make them reach the international advanced level.
  (3) Accelerate the localization of catalysts. Localization of catalysts is the key to flue gas denitration industrialization. Government departments or relevant industrial organizations commissioned by government departments should conduct an in-depth investigation into the development, raw material, production, use, test, recycling, and disposal of flue gas denitration catalysts to collect detailed information of the catalyst production and supply market, grasp the trend of changes, assess the impact of catalyst failure disposal on environment, and put forward relevant standards and management regulations for catalysts as well as suggestions for accelerating localization of catalysts. As there are several catalyst manufacturers in China, government departments should arrange for experts to carry out investigations to find out the situation, release relevant information, and guide the correct development of the industry in order to avoid redundant construction of catalyst manufacturers and reduce blind technical import and blind investment.
  (4) Introduce economic policies to encourage the thermal power plant to implement flue gas denitration. ① We should comprehensively use various economic means to promote the control of $NO_x$ emissions from power plants to achieve maximum environmental benefits at a minimal cost, for example, making the emissions trading policy. ② Reflect environmental protection cost of denitration in the electricity price in a scientific and reasonable way to make the electricity production cost really reflect resource and environmental costs. ③ All the $NO_x$ sewage charges should be used for $NO_x$ emission control, especially the reconstruction of

low $NO_x$ burners and flue gas denitrification technologies of old power plants. ④ For denitrification devices that cannot be localized for the time being, tax exemption or reduction policies should be developed. ⑤ The flue gas denitration device construction of in-service power plants should be subsidized by Central Financial Special Fund of Environmental Protection or investment fund from the central budget.

(5) Strengthen supervision and industrial self-discipline, and standardize and rectify the denitration market. ① Establish the market access system to avoid vicious competition, ensure project quality, and improve the reliability and stability of the operation. ② Establish denitration (desulfurization) facilities operation monitoring and verification system, and strengthen the supervision and inspection of the facilities operation. ③ Establish the directory of denitration technologies and devices encouraged by the Chinese central government to guide the sound development of the denitration industry and technologies. ④ Suggest the national industrialization management department should join hands with relevant government departments to arrange for the research on thermal power plant flue gas denitration standard system, and make relevant standards in accordance with the priorities. ⑤ Improve the communication and cooperation mechanism, and carry out denitration industrialization information statistics, analysis, and release. Organize and carry out the on-the-job training of the denitration operation management personal, and guide the denitration industry to continuously improve quality.

### 3.2.4 *Summary of Environmental Protection Constraints on Coal-Fired Power Generation*

(1) Smoke and dust. As environmental protection requirements have become increasingly stringent, the layout of coal-fired power plants has undergone great changes. Except for heat and power plants, new coal-fired power plants are no longer built in cities. For in-service generating units, measures such as control of pollutant emissions, shutting down of small thermal power generating units, and building of large generating units and wet flue gas desulfurization and dust removal have been taken to further improve the smoke and dust emission control level of coal-fired power plants. The smoke and dust emissions from coal-fired power plants during the "12th Five-Year Plan" period are not the main cause of environmental pollution. However, as China has made increasingly stringent control requirements for smoke and dust emissions from thermal power plants, most coal-fired power plants have to upgrade and reconstruct precipitators. Specifically, they have to choose from different dust removal technologies, invest heavily in technical reconstruction,

and make proper arrangements for production and reconstruction in a short period of time.

(2) Sulfur dioxide. By the end of the "12th Five-Year Plan" period, flue gas desulfurization devices will be installed and operated in all coal-fired generating units. In the 12th Five-Year Plan, it is required that in 2015 sulfur dioxide emissions from electricity production should be controlled within 6.5 million to 7.5 million tons and the emission performance should be 1.4–1.7 g/kWh. During the "12th Five-Year Plan" period, sulfur dioxide emission control will still be one of the major environmental constraints on electric power development. In this book, we mainly put forward the following solutions to it. ① Determine the limits of the air pollutant emissions from thermal power plants during the "12th Five-Year Plan" period with environmental protection requirements (instead of administrative requirements) as the goal, the BAT as the means and technical and economic conditions as constraints. ② Strengthen law-based control, clean up, rectify, and improve the existing sulfur dioxide emission control laws, regulations, and policies, really give enterprises the right to choose pollution control means and manage emission reduction and reduce unnecessary administrative intervention. ③ Strengthen environmental protection supervision, ensure the full and timely implementation of the desulfurization electricity price, and speed up the introduction of the electricity price formation mechanism with environmental costs taken into consideration. ④ Carry out research on improving the reliability of desulfurization devices, and formulate relevant technical regulations to provide technical basis for law-based administration and technical supervision. ⑤ Speed up the development of the resource-saving desulfurization industrialization which can change sulfur dioxide emissions into useful materials, and facilitate the development of circular economy. ⑥ Strengthen industrial self-discipline and public supervision.

(3) Nitrogen oxides. Comprehensively, consider optimal operation, low $NO_x$ combustion technologies, flue gas denitration, and other nitrogen oxide emission control measures and it is predicted that by 2015 $NO_x$ emissions from electricity production will be controlled within 6 million to 7 million tons. The nitrogen oxide emission control will be one of the major environmental constraints on electric power development. In this book, we mainly put forward the following solutions to it. ① The Chinese central government should carry out an in-depth investigation into the nitrogen oxide emissions from environmental pollution sources all over the country and strengthen the research into $NO_x$ emissions and its distribution characteristics. ② Determine limits of nitrogen oxide emissions from thermal power plants based on the comprehensive consideration of environmental, technical, and economic conditions, region, time (period of time), characteristics of generating units, and characteristics of fuel. ③ Priority should be given to low $NO_x$ combustion technologies. If the nitrogen oxide emissions still cannot meet the emission standard after the low $NO_x$ combustion technologies are used, the SCR technology, SNCR technology, or SCR-SNCR combined denitration

technology should be used. Identify the reducing agent system through comparison of technical and economic conditions. ④ Speed up the development of denitration technologies suitable for China's national conditions and with independent intellectual property rights. Accelerate the localization of catalysts. Encourage and promote power plants to integrate desulfurization, denitration, and dust removal technologies, and carry out collaborative control research and development and engineering demonstration. ⑤ Introduce economic policies to encourage the thermal power plant to implement flue gas denitration. ⑥ Strengthen supervision, industrial self-discipline, and standardize and rectify the denitration market.

## 3.3 Hydropower Development

### *3.3.1 Current Situation of Environmental Impacts and Mitigation*

China's hydropower development has gone through four stages, technical constraints, investment constraints, market constraints, and ecological constraints. Nowadays, ecological constraints and the social problems caused by resettlement of affected residents of hydropower projects have become the key constraints on the hydropower development in China.

1. Ecological constraints

China's hydropower potential is mainly concentrated in the western, the upper reaches of China's major rivers. Western China is the ecological screen and "source" of ecological balance of the middle and lower reaches of the rivers. The environmental quality and ecosystem level in the upper region will have a direct impact on these in the middle and lower reaches of the rivers. Western China has an ecological environment characterized by fragility, vulnerability, and difficult recovery because of located in the inland plateau. Under the integration of natural and human factors, ecological and environmental problems such as land desertification, decline of water conservation function of soils, destruction of biodiversity, and frequent disasters have become increasingly prominent, which exacerbated the adverse effects of the development of hydropower resources on the ecological environment and threatened the ecological safety in the middle and lower reaches of the Yangtze River and the Yellow River and other major rivers.

Important ecological and environmental constraints include:

(1) Hydropower construction and operation change the hydrological process and river channel characteristics of the lower reaches, and then impact downstream important aquatic and terrestrial habitats and sensitive protection targets of the reservoir area.

(2) Water temperature stratification in the reservoir area leads to low temperature water released, water eutrophication in reservoirs, and the dam water discharge, which may cause supersaturation of dissolved gas or dissolved oxygen deficiency in water in the downstream channels.
(3) Flooding, impeding, and runoff regulation of a reservoir will destroy river connectivity and impact the quality, quantity, and accessibility of the habitat of aquatic organisms, especially for fishes. Migration and other activities of fishes therefore might be terminated or postponed. Mitigation of ecological and environmental constraints is the major issue of hydropower development, which requires both ecological protection technology and environmental management policies.

2. Social problems caused by resettlement

The social problems caused by resettlement are another important constraint on hydropower development. Although China's current policy on resettlement of water conservancy and hydropower projects has increased the compensation standards, the existing "agriculture-based" resettlement mode and disposable compensation method cannot get rid of chronic poverty or share the benefits of hydropower development. Reservoir inundation and resettlement problems are still very prominent to restrict the hydropower development.

Meanwhile, hydropower bases in Western China are mainly located in regions populated by ethnic minorities. Due to inconvenient transportation and underdeveloped economy in these regions, hydropower enterprises have to shoulder more social responsibilities for promoting the local social and economic development. In addition, resettlement involves the issue of religious beliefs, which will have a far-reaching impact on the future hydropower development.

Column Interpretation

Ecological and environmental impact of Xiluodu Hydropower Station on Jinshajiang River and corresponding mitigation measures during its construction period

Located in the Xiluodu Gorge contiguous to Leibo County, Sichuan Province and Yongshan County, Yunnan Province and as the third cascade in the development planning for the lower reaches of Jinsha River, Xiluodu Hydropower Station has a total installed capacity of 12.6 million kW and an annual generation of 57.12 billion kWh. Ranked the third in the world, it is the second largest hydropower station in China. The Xiluodu Gorge hydropower project focuses on hydropower generation and additional comprehensive benefits such as flood control, sediment retention, and improving downstream shipping conditions. In addition, it can provide cascade compensation for downstream hydropower projects. The concrete double curvature arch dam of the key project is 278 m high and has a normal impounded water level of 600 m, a total reservoir capacity of 11.57 billion $m^3$, and a total investment of RMB 79.234 billion.

After the construction of the Xiluodu hydropower project, the reservoir will inundate the main spawning ground of Chinese paddlefish, coreius guichenoti, and other rare and specific fishes; impeding of the dam will split a complete living environment of aquatics into two sections of the reservoir and downstream, specifically reduce the living environment of fishes; the low temperature water discharged from the reservoir in spring and summer will postpone the spawning period of Chinese paddlefish and other fishes; the changes in hydrological and sediment conditions will change the water level, flow velocity, and channel morphology of natural river.

Main mitigation measures taken during the construction period include: the adjustment scheme of natural reserve area, Chinese paddlefish artificial culture research and artificial propagation, artificial reproduction and releasing of river sturgeon and mullet, coreius guichenoti artificial reproduction research, and artificial reproduction and strengthening monitoring and management.

On the morning of December 7, 2008, China Three Gorges Corporation held the "first releasing of rare and specific fishes in Xiluodu Xiangjiaba Hydropower Station on Jinsha River" and altogether 20,000 rare and specific fishes including river sturgeon, mullet, Procypris rabaudi, Leptobotia elongata, and Megalobrama pellegrini were released to the Yangtze River.

## *3.3.2 Analysis of Major Environmental Constraints*

### 3.3.2.1 Analysis of Laws, Regulations, and Policies

1. Analysis of existing laws, regulations, and policies

Since the founding of People's Republic of China more than 60 years ago, China implements the principle of "actively developing hydropower." As an important electric energy, hydroelectricity has been closely related to the power sector development. The Communist Party of China and the Chinese central government have stuck to the policy of actively developing hydropower resources. In 1982, the editorial entitled *Paying Special Attention to Hydropower Development* published in the *People's Daily* pointed out that hydropower was a renewable energy; the construction of hydropower stations was a kind of project in which the development of primary energy and conversion of secondary energy could be completed at the same time; we should develop and utilize more hydropower to gradually replace some coal-fired power. At the National Electric Power Work Meeting held in 1997, it was proposed the slogans of "vigorously developing hydropower" and "increasing the proportion of hydropower to 30% by 2010." The goal of developing hydropower was further defined in the *11th Five-Year Plan for Renewable Energy Development* and the *Medium and Long-Term Development Plan for Renewable Energy in China*. At the National Electric Power Work Meeting held in August 2009, the role of hydropower in the renewable energy development and its future development goal were reiterated. The goal of "steady and rapid development of hydropower construction" was clearly defined in the *12th Five-Year Plan for*

**Table 3.7** China's current hydropower development promotion policies

| Policy | Time | Main content |
|---|---|---|
| Renewable energy law of the People's Republic of China | 2005 | Hydropower is a renewable energy source. Financial institutions will provide financial discount preferential loans for projects being included in the national renewable energy industry guidance catalogue and meeting credit terms. Meanwhile, the government will provide tax preferences for these hydropower projects |
| 11th Five-Year Plan for national economic and social development of the People's Republic of China (2006–2010) | 2006 | Orderly developing hydropower on the basis of protecting the eco-environment, actively developing renewable energy (including hydropower), implementing preferential fiscal and tax policy and investment policy and mandatory market share policy, encouraging the production and consumption of renewable energy, and increasing the proportion of renewable energy consumption in primary energy consumption |
| 11th Five-Year Plan for renewable energy development | 2006 | Actively developing hydropower on the premise of ensuring environmental protection and resettlement |
| Medium- and long-term development plan for renewable energy in China | 2007 | China has made to international communities to increase its non-fossil fuel energy to 15% of its total primary energy consumption by 2020. Installed capacity of hydropower will reach 190 and 300 GW in 2010 and 2020, respectively |
| 12th Five-Year Plan for renewable energy development<br>12th Five-Year Plan for hydropower development | 2012 | In the period of 12th FTP, new conventional hydropower installed capacity under construction will reach 1.2, and 40 GW for pumped storage hydropower; total hydropower installed capacity will reach 290 GW with an annual generating of 910 billion kWh by 2015; total hydropower installed capacity will reach 420 GW with an annual generation of 1200 billion kWh by 2020 |

*Hydropower Development* released by National Energy Administration in 2012. See Table 3.7 for China's current hydropower development promotion policies.

See Table 3.8 for China's existing laws, regulations, and policies for hydropower development. All of them divide into two kinds of constraints: ① ecological and environmental protection, including conservation of fish-focused aquatics and terrestrials, soil conservation, and protection of important protected areas, such as

**Table 3.8** China's existing laws, regulations, and policies for hydropower development

| Category | Law/regulation/policy | Time | Content |
|---|---|---|---|
| Main laws | *Water Law of the People's Republic of China* | 2002 | It stipulates in Article 26 that the government encourages the development and utilization of water resources. On rivers rich in hydropower, multi-purpose and cascade development shall be promoted in a planned manner. In construction of hydropower stations, attention shall be paid to protection of the ecological environment, and consideration shall, at the same time, be given to the need of flood control, water supply, irrigation, navigation, bamboo and log rafting, fishery, etc. |
| | *Environmental Protection Law of the People's Republic of China* | 1989 | It stipulates in Article 17 that the people's governments at various levels shall take measures to protect regions representing various types of natural ecological systems and regions. Damage to the above shall be strictly forbidden. It stipulates in Article 18 that within the scenic spots or historic sites, nature reserves and other zones that need special protection, as designated by the State Council, the relevant competent department under the State Council, and the people's governments of provinces, autonomous regions and municipalities directly under the Central Government, no industrial production installations that cause environmental pollution shall be built |
| | *Law of Land Administration of the People's Republic of China* | 2004 | It stipulates in Article 31 that the government protects the cultivated land and strictly controls the conversion of cultivated land into non-cultivated land. The states foster the system of compensations to cultivated land to be occupied. Land in rural areas shall be collectively owned by farmers. In addition, the standards for compensation and subsidy and legal procedures for requisition of land are also stipulated |
| International treaties | *World Heritage Convention* | | It gives greater freedom to sovereign nations in the protection and development of resources. The sovereign nations recognize this kind of heritage is part of the world heritage when the property right under the national legislation is not damaged |

(continued)

**Table 3.8** (continued)

| Category | Law/regulation/policy | Time | Content |
|---|---|---|---|
| Special laws and regulations | *Fisheries Law of the People's Republic of China* | 2004 | It stipulates in Article 32 that when building sluices and dams which will have serious effects on fishery resources on the migration routes of fish, shrimp, and crabs, the construction units must build fish passages or adopt other mitigation measures |
| | *Law of the People's Republic of China on the Protection of Wildlife* | 2004 | It stipulates in Article 20 that in nature reserves and areas closed to hunting, and during seasons closed to hunting, the hunting and catching of wildlife and other activities which are harmful to the living and breeding of wildlife shall be prohibited |
| | *Regulations of the People's Republic of China on Wild Plants Protection* | 1997 | It stipulates in Article 9 that the government shall protect wild plants and the environment for their survival. All units and individuals shall be forbidden to illegally collect wild plants or damage the environment for their survival |
| | *Interim Regulations of the People's Republic of China on the National Scenic and Historic Areas* | | It stipulates in Article 26 that activities of destroying landscape, vegetation, and landform shall be prohibited in scenic and historic areas. It stipulates in Article 27 that it is forbidden to build various types of development zones in scenic and historic areas and build other buildings irrelevant to protection of scenic resources in the core scenic areas in violation of the scenic and historic area planning |
| | *Regulations of the People's Republic of China on Nature Reserves* | 1994 | It stipulates in Article 18 that nature reserves may be divided into three parts: the core area, buffer zone, and experimental zone. The integral natural ecosystems and the areas where the rare and endangered animals or plants are concentrated within nature reserves shall be included in the core area into which no units or individuals are allowed to enter. Scientific research activities are generally prohibited in the core area except for those approved according to Article 27 of the Regulations |
| | *Measures for the Administration of National Forest Parks* | 1994 | It stipulates in Article 11 that in the precious scenery, important scenic areas, and core scenic areas, except for necessary protection and ancillary facilities, the construction of hotels, guest houses, nursing homes, and other engineering facilities shall be forbidden |

(continued)

**Table 3.8** (continued)

| Category | Law/regulation/policy | Time | Content |
|---|---|---|---|
| | *Regulations on the Protection and Management of Geological Relics* | 1995 | It stipulates in Article 11 that categorization of protection: Protection of geological relics in protected areas can be categorized into first-grade protection, second-grade protection, and third-grade protection. First-grade protection: First-grade protection shall be provided for geological relics which are extremely rare and of important scientific value in the world or in China and without approval access shall be prohibited. Second-grade protection: Second-grade protection shall be provided for geological relics in the large area—Scientific research, teaching, academic exchange, and appropriate tourism activities can be organized in these geological relics. Third-grade protection: Third-grade protection shall be provided for geological relics which are of a certain value—tourism activities can be organized in these geological relics |
| | *Law of the People's Republic of China on Land Contract in Rural Areas* | | Land can be contracted to individual villagers for operation and management, and villagers' land contract rights and related circulation rights are protected by law |
| | *Regulations on the Protection of Basic Farmland* | 1998 | It stipulates in Article 15 that once a protected area of basic farmland is designated, no plant/institution or individual may alter or occupy the said land without authorization. When selecting sites for important national construction projects related to energy, communications and water conservancy, military facilities, etc., and there is no way to avoid in fact a need to occupy cultivated land in protected areas of basic farmland, it shall be approved by the State Council |
| | *Regulations on Land Requisition Compensation and Resettlement for Construction of Large and Medium-sized Water Conservancy and Hydropower Projects* | 2006 | Applicable in the whole China, it is made based on the characteristics of hydropower engineering and summary of the historical experience and lessons learned in resettlement. It makes corresponding provisions for standards for land requisition compensation and principle of resettlement. These provisions are one of the major legal bases for implementation of resettlement in recent years |

**Table 3.8** (continued)

| Category | Law/regulation/policy | Time | Content |
|---|---|---|---|
| Special plans | Comprehensive utilization plan for water resources in China and sub-basins, national water resources protection plan, river basin comprehensive utilization plan, river hydropower development plan, national and regional water function zoning, etc. | | Take river basin comprehensive utilization plan as an example: Revise and implement water resources comprehensive utilization plan, combine hydropower development and comprehensive utilization of water resources, ecological engineering construction, and regional economy organically, realize the comprehensive development of water resources in the river basin, and give full play to the guidance role of water resources comprehensive utilization in hydropower design, construction, and management |
| | Outline of National Ecological Environmental Protection of the People's Republic of China, national and regional ecological function zoning, etc. | | Take the Outline of National Ecological Environmental Protection of the People's Republic of China as an example: All development activities that lead to the continuous degeneration of ecological functions and other human sabotage activities shall be stopped; the construction of all engineering projects that produce serious environmental pollution shall be stopped |

natural reserves, scenic spots, forest parks, geological relics, and world natural heritage. ② Resettlement.

2. Trends of regulations and policies

The *Energy Law of the People's Republic of China* and *Renewable Energy Law of the People's Republic of China* are being revised, and it is expected that the role of renewable energy of large hydropower will be further defined. It is clearly put forward by the State Council in approving *Notice on Opinions of Deepening Economic Restructuring in 2009* made by the National Development and Reform Commission (NDRC) for release that we should vigorously promote the resource product price and energy conservation and environmental protection system reform. On November 20, 2009, the NDRC published the on-grid tariff adjustment. Up-regulation of hydropower on-grid price was larger than that of benchmark thermal coal-fired power price. The gap between hydropower on-grid price and thermal coal-fired power on-grid price was further narrowed. Benchmark hydropower price, i.e., the hydropower on-grid pricing mechanism, was suspended temporarily. Instead, the interim hydropower price was implemented. In recent years, the "same grid and same price" has been vigorously proposed. In addition, in order to promote economic and social development of the hydropower reservoir area and resettlement area, and to solve the problem of lack of reconstruction funds, the NDRC will further raise the hydropower on-grid price during the "12th Five-Year Plan" period and make relevant policies to promote the ecological protection and resettlement of affected residents in hydropower development.

At the end of 2009, the revised or newly revised comprehensive plannings for all major river basins were released. The revised or newly revised comprehensive plans for all major river basins put more emphasis on multi-purpose development of resources and strengthened the ecological protection of river basins. In addition, the hydropower development plan was revised, laying a foundation for the orderly development of hydropower.

#### 3.3.2.2 Analysis of Major Ecological and Environmental Constraints

1. Rare and specific fishes

(1) The upper reaches of the Yangtze River. The upper reaches of the Yangtze River are a concentrated distribution area of rare fishes including Chinese paddlefish, river sturgeon, mullet, and 117 fishes specific to the upper reaches of the Yangtze River. In April 2005, general office of the State Council approved the plan for changing *Rare Fish Species National Nature Reserve at Hejiang-Leibo Section of Yangtze River* to *Rare and Specific Fish Species National Nature Reserve in the Upper Reaches of Yangtze River* proposed by the Ministry of Agriculture. The scope of the reserve changed from the 420 km Leibo-Hejiang river section to the 1162.6 km river section which starts from the estuary of the Hejiang River, 1.8 km down Xiangjiaba Dam to

Masang Stream, Chongqing, including the main stream and tributaries of Chishui River, lower reaches of Minjiang River, tributaries of Yuexi River and estuaries of Nanguang River, Changning River, Tuojiang River, and Yongning River. The adjusted reserve crosses Sichuan, Guizhou, Yunnan, and Chongqing and has total river area of the 331.7 km$^2$. The following three kinds of fishes are protected: ① rare and endangered fish species in the upper reaches of the Yangtze River such as national first-grade and second-grade key protected aquatic wildlife including river sturgeon, Chinese paddlefish, and mullet; ② 117 fish species specific to the upper reaches of the Yangtze River; ③ the aquatic environment rare and endangered species and specific fishes in the upper reaches of the Yangtze River rely on for survival.

(2) Yalong River. The Yalong River is the largest first-grade tributary on the left bank of Jinsha River. According to the relevant river basin comprehensive planning results, in the middle reaches of the river there are no national protected rare fishes, but there are two Sichuan provincial key protected fishes *Schizothorax dolichonema* Herzenstein and Euchiloglanis david and one specific fish Diptychus kaznakovi. All the above-mentioned three fishes are fishes living in running water. *Diptychus kaznakovi* only exists in the upper reaches of Jinsha River and Yalong River and has been listed in *China Red Data Book of Endangered Animals: Fishes*.

(3) Jinsha River. Hutiao Gorge is the natural divide between fish faunas in the Jinsha River. Living in the channel segment up the entrance to the gorge are mainly cold water fishes typical of the Qinghai–Tibet Plateau including about 20 Schizothoracinae species and 10 loach species. Down the entrance to the gorge is the lower reaches of the Jinsha River (transition section between the lower reaches of the Jinsha River and the upper reaches of the Yangtze River) which is the transition distribution waters between river plain fishes and Qinghai–Tibet Plateau fishes and the staggered distribution area of 160 fish species, mainly river plain fishes, of which 133 ones are the same as those in the middle reaches of the Jinsha River and 151 ones are the same as those in the Rare Fish Species National Nature Reserve in the Upper Reaches of Yangtze River.

According to the research findings of the *Research Report on the Environmental Impacts of Cascade Hydropower Development in the Lower Reaches of the Jinsha River and Countermeasures*, in the lower reaches of the Jinsha River there are three national key protected rare fish species, Chinese paddlefish, river sturgeon, and mullet (Chinese paddlefish and river sturgeon are the first-grade national key protected rare fish species, while mullet is the national second-grade key protected rare fish species); there are 56 fishes specific to the upper reaches of the Yangtze River of which two ones, *Anabarilius songmingensis* and *Anabarilius brevianalis*, are specific to the lower reaches of the Jinsha River; there are 54 other specific fishes of which 16 ones including *Schizothorax prenanti*, *Schizothorax wangchiachii*, *Schizothorax dolichonema Herzenstein*, *Schizothorax kozlovi Nikolsky*,

*Schizothorax chongi*, *P. anteanalis*, *Onychostoma angustistomata*, *Leptobotia elongata*, *Coreius guichenoti*, *Rhinogobio ventralis*, *Xenophysogobio nudicorpa*, *Percocypris pingi,* and *Procypris rabaudi* should be taken as national key protected species according to the characteristics of the distribution of fishes in this section and the environmental impacts of cascade hydropower development.

(4) Dadu River. According to the research on the special investigation of the aquatic organisms in Dadu River and the literature, in the Dadu River basin there are 111 fish species of which 2 ones, Sichuan faimen and mullet, are the national second-grade protected fishes, 11 ones, provincial protected fishes, and 38 ones, specific to the upper reaches of the Yangtze River. These fishes are basically living in running water, for example, *Sichuan faimen*, *Varicorhinus daduensis*, *Percocypris pingi*, *Procypris rabaudi*, *Schizothorax longibarbus*, *Schizothorax davidi*, *Schizothorax prenanti,* and Euchiloglanis david.

(5) Lancang River. According to relevant river basin comprehensive planning results, in the Lancang River the fish faunas mainly include 105 species, 67 genera, and 3 families of the Cypriniformes. The largest genus of the indigenous fishes is Schistura, of which there are eight species. At present no long-distance migratory fishes that migrate between oceans and rivers are found, but most fishes have the habit of spawning migration. There are four typical long-distance spawning migratory fishes, namely giant pangasius, *Pangasius micronemus*, Pseudopangasius, and *Pangasius beani.*

No fishes in Lancang River are listed in the national first-grade or second-grade protected fishes or in the Appendix of CITES. Listed in the Yunnan provincial second-grade protected fishes are biforated carp, *Macrochirichthys macrochiriu,* and giant pangasius which are mainly distributed in the lower reaches of the Lancang River. With the three provincial protected fishes included, a total of ten fishes in the Lancang River basin are listed in *China Red Data Book of Endangered Animals: Fishes*. The other seven ones are grooved isthmus barbell, *Cosmochilus cardinalis*, *Puntioplites proctozysron*, *Kryptopterus moorei*, *Akysis brachybarbatus*, yellow sisorid-catfish, and *Trichogaster trichopterus*. All these ten fishes are distributed in the middle and lower reaches of Lancang River.

(6) Nujiang River. At present, there are 7 kinds of 72 fish species, including 58 indigenous fish species which belong to 32 genera, 8 families, and 5 orders. Among the 58 indigenous species, there are 40 rare fishes, accounting for 68.97% of the indigenous fish species, 6 endangered species, accounting for 10.34%, and one national second-grade protected fish species. The *Anguilla nebulosa*, *Epalzeorhynchos bicornis*, *Schizothorax griseus Pellegrin*, *Schizothorax lantsangensis*, *P. myzostoma,* and *P. gracilicaudata* are listed in *China Red Data Book of Endangered Animals: Fishes. Anguilla nebulosa* (commonly known as Monopterus alba) is one of the national second-grade protected fish species. *Epalzeorhynchos bicornis* (commonly known as

bihorned barb) is a specific fish species. *Placocheilus cryptonemus* is a rare fish species and listed in the Vulnerable Animal Catalogue. *Bagarius yarrelli* is a provincial protected fish species and distributed in Gaoligongshan Mountains National Nature Reserve. Acrossocheilus (Lissochilichthys) hemispinus is a rare fish species and only found in underground swallet stream flowing into Nujiang River near Liuku Town and is not found outside Gaoligongshan Mountains National Nature Reserve.

(7) Wujiang River. Wujiang River is a first-grade tributary on the right bank in the upper reaches of the Yangtze River. A total of 123 fish species are distributed in it including 9 Chongqing municipal key protected aquatic wild fishes and more than 30 fishes specific to the upper reaches of the Yangtze River. According to *Overall Planning for Wujiang River-Changxi River Fish Nature Reserve*, there are 12 rare fish species in Wujiang River basin mainly including Chinese paddlefish, river sturgeon, mullet, *Leptobotia elongata*, *Leptobotia rubrilabris*, *Luciobrama microcephalus*, *Ochetobius elongatus* Kner, *Percocypris pingi*, *Procypris rabaudi*, *Jinshaia sinensis*, *Sinogastromyzon szechuanensis* and *Metahomaloptera omeiensis*.
Among all the above-mentioned protected fish species, there have been no exact distribution records for Chinese paddlefish and river sturgeon for years. As the two fish species are mainly distributed in the main stream of the Yangtze River, in the newly built "Rare and Specific Fish Species National Nature Reserve in the Upper Reaches of the Yangtze River" they can be well protected. Similarly, Leptobotia elongata, *Leptobotia rubrilabris*, *Luciobrama microcephalus*, *Ochetobius elongatus* Kner, *Percocypris pingi*, *Procypris rabaudi,* and *Jinshaia sinensis* are also widely distributed in this reserve. In addition, the amphibious animal *Andrias davidianus* appearing in the river basin is a second-grade national protected animal and should also be protected. Therefore, the core protected rare and specific fishes in the Wujiang River section include two second-grade national key protected animals, mullet and *Andrias davidianus* and *Sinogastromyzon szechuanensis* and *Metahomaloptera omeiensis*. Particularly, mullet is mainly distributed in the lower reaches of the Wujiang River.

(8) The upper reaches of the Yellow River. Yellow River is relatively poor in aquatic species. The biomass is low, and there are only a small number of fish species. Due to the special geographical environment and climate characteristics in the upper reaches of the Yellow River, there have formed the indigenous fishes specific to the Yellow River that adapt to the plateau ecological environment. These fishes are concentrated in the channel segment up Longyang Gorge, and the fish fauna in the upper reaches mainly includes central Asia mountain complex fishes. According to relevant river basin comprehensive planning results, in the upper reaches of the Yellow River, three national aquatic germplasm resources conservation areas including Zhaling Lake and Eling Lake Gymnocypris eckloni Herzensten and Platypharodon extremus National Aquatic Germplasm Resources

Conservation Area, Specific Fish Species National Aquatic Germplasm Resources Conservation Area at Upper Reaches of Yellow River and *Silurus lanzhouensis* National Aquatic Germplasm Resources Conservation Area at Liujia Gorge at Yellow River. Key protected fish species include plateau cold water fish such as *Triplophysa siluroides*, *Chuanchia labiosa*, *Platypharodon extremus*, *Gymnocypris eckloni*, *Schizopygopsis pylzovi* and *Triplophysa pappenheimi* and *Silurus lanzhouensis* and *Cyprinus carpio*.

According to the survey of the former State Bureau of Aquatic Products, at present in the upper reaches of the Yellow River there are a total of 47 fish species including 3 endangered fish species and 15 indigenous fish species, but no long-distance migratory fishes or national key protected fishes. *Triplophysa siluroides*, *Chuanchia labiosa*, and *Platypharodon extremus* are listed in *China Red Data Book of Endangered Animals: Fishes*. Compared with the 1980s, the fish species in the Yellow River have declined sharply and the rare endangered and indigenous fish species have reduced. In the channel segments where hydropower resources are developed in a concentrated way, the fish habitat has undergone great changes and indigenous fish species resources have declined dramatically.

(9) North main stream of the Yellow River. The channel segment between Ketou Town in the middle reaches of the Yellow River and Yumenkou is known as the north main stream of the Yellow River and the longest continuous canyon in the main stream of the Yellow River. The fish fauna in the north main stream of the Yellow River mainly includes a great variety of river plain complex fishes. Most of these fishes are widely distributed. In order to protect aquatic germplasm resources and their living environment, the Ministry of Agriculture has built three national aquatic germplasm resources conservation areas including Silurus Lanzhouensis National Aquatic Germplasm Resources Conservation Area at Weining Section of Yellow River, Rhinogobio Nasutus National Aquatic Germplasm Resources Conservation Area at Qingshi Section of Yellow River, and Yellow River Catfish National Aquatic Germplasm Resources Conservation Area at Ordos Section of Yellow River in the channel segment between Liujia Gorge and Toudaoguai. Key protected fish species include *Silurus lanzhouensis*, *Cyprinus carpio*, *Rhinogobio nasutus,* and *Coreius septentrionalis*. At present in the north main stream of the Yellow River, there are no long-distance migratory fishes or national key protected fishes. The *Coreius septentrionalis* distributed in the channel segment between Qingtong Gorge and Shizuishan is first-grade endangered animal and listed in *China Red Data Book of Endangered Animals: Fishes*.

(10) Nanpanjiang River–Hongshui River. According to relevant river basin comprehensive utilization planning results, the channel segment between the middle and lower reaches of Nanpanjiang River and the upper reaches of Hongshui River are concentrated spawning and reproduction areas of fishes. In this channel segment, there are 90 indigenous fish species which belong to 65 genera, 16 families, and 6 orders and most of which are cypriniformes. In this

**Table 3.9** Standard for fish constraint indicator evaluation in important hydropower bases

| Score | Ecological implications |
|---|---|
| 5 | National protected long-distance migratory fishes |
| 4 | First-grade national protected fishes |
| 3 | Second-grade national protected fishes |
| 2 | Short-distance migratory fishes and China Red Data Book of Endangered Animals: Fishes |
| 1 | Provincial protected fishes, specific fishes, national aquatic germplasm resources conservation areas |
| 0 | None |

channel segment, there are a total of 11 specific species including *Acrossocheilus iridescens*, *Paraspinibarbus hekouensis*, Sparulo justo quatuor requiruntur caldwelli, *Abbottina yunnanensis*, *Schistura macrotaenia*, *Glyptothorax honghensis*, *Microphysogobio yunnanensis*, *Triplophysa gejiuensis*, *Ctenogobius hongheensis*, *Schistura callichromus*, and *Schistura laterivittata*. Listed in the *China Red Data Book of Endangered Animals: Fishes* are *Anguilla marmorata*, *Paraspinibarbus alloiopleurus*, *Semilabeo obscurus* Lin, *Triplophysa gejiuensis*, and *Parazacco spilurus* of which the migratory fish *Anguilla marmorata* is second-grade national protected fish species.

(11) Fish constraint indicators evaluation standards and results. Dam building will change the original water habitat of the river course, while fish channel impeding will have an adverse impact on the fishes living in running water, especially for migratory fishes. According to the national requirements for biodiversity and fish resources protection, the fish constraint indicator evaluation standard (see Table 3.9) is defined based on the indigenous significance, peculiarity, economic value, and endangered degree of rare and specific fishes in the hydropower bases combined with the results of the present situation survey. At the same level, the river basins with a large number of fish species are further categorized. See Table 3.10 for constraint indicatory evaluation results.

2. Biodiversity Conservation

National ecological function zoning is to analyze the spatial distribution regularities and determine the ecological functions of different regions based on ecological status investigation, ecological sensitivity, and ecological service function evaluation. The national terrestrial ecosystem is divided into three ecological areas: eastern monsoon ecological region, western arid ecological region, and Qinghai–Tibet alpine ecological region, and the national ecologically functional zones are divided into the following three grades:

**Table 3.10** Fish constraint level in China's main hydropower stations

| River basin | Score | Notes |
|---|---|---|
| Upper reaches of Yangtze River | 5 | Rare and Specific Fish Species National Nature Reserve in the upper reaches of the Yangtze River; 3 first- and second-grade national key protected fish species (Chinese paddlefish, river sturgeon, mullet); 117 fish species specific to the upper reaches of Yangtze River |
| Main stream of Jinsha River | 5 | Three first- and second-grade national key protected fish species (Chinese paddlefish, river sturgeon, mullet); 4 provincial key protected fish species; 54 fish species specific to the upper reaches of Yangtze River; 2 fish species specific to the lower reaches of Jinsha River |
| Main stream of Yalong River | 2 | No long-distance migratory fish species or national key protected fish species; provincial key protected fish species; one specific fish species; one fish species listed in *China Red Data Book of Endangered Animals: Fishes* |
| Main stream of Lancang River | 2.5 | No long-distance migratory fish species or national key protected fish species; there exist anadromous and migratory fish species in Mekong River; three provincial key protected fish species; ten fish species listed in *China Red Data Book of Endangered Animals: Fishes* |
| Main stream of Dadu River | 5 | Second-grade national key protected fish species (Sichuan faimen and mullet); 11 provincial key protected fish species; 38 fish species specific to the upper reaches of the Yangtze River |
| Main stream of Nujiang River | 5 | 1 s grade national key protected fish species (Anguilla nebulosa); 40 specific fish species; 6 fish species listed in *China Red Data Book of Endangered Animals: Fishes* |
| Nanpanjiang River-Hongshui River | 5 | First-grade national key protected fish species (Anguilla marmorata); 11 specific fish species; five fish species listed in *China Red Data Book of Endangered Animals: Fishes* |
| Wujiang River | 5 | Second-grade national key protected fish species (mullet and Andrias davidianus); 9 Chongqing municipal key protected fish species; 30 fish species specific to the upper reaches of the Yangtze River |
| Upper reaches of Yellow River | 2.3 | Two national aquatic germplasm resources conservation areas; no long-distance migratory fish species or national key protected fish species; three fish species listed in *China Red Data Book of Endangered Animals: Fishes* |
| North main stream of Yellow River | 2.1 | Three national aquatic germplasm resources conservation areas; no long-distance migratory fish species or national key protected fish species; one fish species listed in *China Red Data Book of Endangered Animals: Fishes* |

(1) According to the natural properties of the ecosystem and dominant service function types, China is divided into three first-grade ecologically functional zones, ecological regulation, product supply, and human habitat security.

(2) The first-grade ecologically functional zones are further divided into second-grade ecologically functional zones based on the importance of ecological functions. The ecological regulation functions include water conservation, soil conservation, windbreak and sand fixation, biodiversity conservation, flood regulation and storage, and other functions; product supply functions include agricultural products, livestock products, aquatic products, and forest products; human habitat security functions include densely populated and economy-intensive metropolitan groups and key town groups. The value of protecting the biodiversity of different regions depends on the distribution of endangered rare animals and plants as well as typical ecological system distribution.
(3) The second-grade ecologically functional zones are divided into the third-grade ecologically functional zones according to the ecological system, spatial differentiation characteristics of ecological functions, topographic differences, and land use.

Based on the plan for dividing ecologically functional zones into three grades, this book uses the ratio of the sum of biodiversity conservation areas to the total area of hydropower development base, to measure the constraint of biodiversity conservation on the hydropower base development. The results show that the Dadu River hydropower base and Wujiang River hydropower base are 100% within the third-grade biodiversity conservation zones, while Jinsha River hydropower base and Lancang River main stream hydropower base occupy a relatively smaller area in the third-grade biodiversity conservation zones.

3. Ecological conservation priority area

According to relevant river basin comprehensive planning reports, environmental impact assessment documents, *Overall Construction Planning for Rare Fish Species National Nature Reserve at Hejiang-Leibo Section of Yangtze River*, *Overall Planning for Wujiang River-Changxi River Fish Species Nature Reserve* and literature review (Zhou Jianping, and others 2011), we review the world natural and cultural heritage, national scenic and historic interest areas, national nature reserves, and national forest parks that might get involved in cascade development of all river basins and hydropower bases. See Table 3.11 for details. Our obtained results indicate that it needs to further demonstrate dam site selection of Hutiao Gorge on Jinsha River, identify the relationship between ecological protection priority areas and important hydropower stations in the upper reaches of Yalong River, Dadu River, and Yellow River, recalculate the hydropower station size, and further increase environmental protection investment.

4. Technical and economic analysis

In addition to the benefits brought by power generation, development of hydropower resources has comprehensive utilization benefits such as flood control, irrigation, water supply, shipping, and tourism. The technical and economic analysis we make in this book of major hydropower bases to be promoted during the "12th

Table 3.11 Important ecological and environmental protection priority areas of China's major hydropower bases

| Hydropower base | Cascade hydropower station | Protection scope | Name of reserve |
|---|---|---|---|
| Jinsha River | Longpan Hydropower Station | From Judian to Daju | Landscape resources such as first bend of the Yangtze River and Tiger Leaping Gorge Grand Canyon; Yulong Snow Mountain and Haba Snow Mountain Provincial Nature Reserve |
| | Liangjiaren Hydropower Station; Liyuan Hydropower Station | Four areas: Hongshan Mountain, Haba Snow Mountain, Laojun Mountain, Baima-Meri Snow Mountain | World Heritage Sites located in these eight areas; Three Parallel Rivers of Yunnan National Key Scenic and Historic Interest Area |
| | Xiangjiaba Hydropower Station | The channel segment from 1.8 km down the dam axis of Xiangjiaba Dam to Yibin City | Rare and Specific Fish Species National Nature Reserve in the Upper Reaches of Yangtze River |
| Yangtze River | Xiaonanhai Hydropower Station, etc. | From Yibin City to Nanxi Town | River sturgeon, Chinese paddlefish, mullet, and other rare and specific fish resources in the Yangtze River basin and their living environment; typical river wetland; Rare and Specific Fish Species National Nature Reserve in the upper reaches of Yangtze River |
| | | From Nanxi Town to Shatuozi | |
| | | From Tuojiang River mouth to Mituo Town | |
| | | From Mituo Town to Songji Town | |
| | | From Songji Town to Luohuang Town | |
| Lancang River | Huangdeng Hydropower Station, Tuoba Hydropower Station, Lidi Hydropower Station, Wunonglong Hydropower Station, Guonian Hydropower Station, Gushui Hydropower Station | Eight areas: Gaoligong Mountain (North), Baima-Meri Snow Mountain, Laowo Mountain, Yunling Mountain, Julong Lake, Laojun Mountain, Haba Snow Mountain, Qianhu Mountain, and Hongshan Mountain | World Heritage Sites located in these eight areas; Three Parallel Rivers of Yunnan National Key Scenic and Historic Interest Area |
| | Jinghong Hydropower Station, Ganlanba Hydropower Station, Mengsong Hydropower Station | From Jinghong to Mengla | Xishuangbanna National Scenic and Historic Interest Area and National Nature Reserve |
| Dadu River | Shenxigou Hydropower Station, Zhentouba Hydropower Station | Northwestern area of Jinkouhe District, Leshan City, eastern area of Hanyuan County, Ya'an City, northern area of | Dadu River Grand Canyon National Geopark |

(continued)

**Table 3.11** (continued)

| Hydropower base | Cascade hydropower station | Protection scope | Name of reserve |
|---|---|---|---|
| | | Ganluo County, Liangshan Yi Autonomous Prefecture | |
| | Luding Hydropower Station, Yinliangbao Hydropower Station, Dagangshan Hydropower Station, Longshitou Hydropower Station, Laoyingyan Hydropower Station | Kangding County, Luding County, Jiulong County, and Daofu County, Ganzi Tibetan Autonomous Prefecture and Shimian County, Ya'an City | Gongga Mountain National Scenic and Historic Interest Area |
| Yalong River | Lianghekou Hydropower Station | 26 km north of Yajiang County town | Qingdagou Second-Grade Forest Park |
| Nanpanjiang River; Hongshui River | Longtan Hydropower Station | Leye County in Northwest Guangxi Province | Guangxi Yalong National Nature Reserve |
| Wujiang River | Pengshui Hydropower Station | Areas along the rivers in Guizhou Province and Wuchuan County | Mayang River National Nature Reserve |
| Nujiang River | Liuku Hydropower Station, Hushui Hydropower Station, Yabiluo Hydropower Station, Bijiang Hydropower Station, Fugong Hydropower Station, Lumadeng Hydropower Station, Maji Hydropower Station, Bingzhongluo Hydropower Station, Songta Hydropower Station | Gaogongli Mountain area | Three Parallel Rivers of Yunnan National Key Scenic and Historic Interest Area |
| Upper reaches of Yellow River | Guancang Hydropower Station, Mentang Hydropower Station, Tajike Hydropower Station | From Xiarihu Temple to Mentang | Sanjiangyuan National Nature Reserve |
| | Xiaoguanyin Hydropower Station | Jingtai County, Gansu Province | Yellow River Stone Forest National Geopark |

**Table 3.12** Technical and economic analysis of hydropower resources development of China's major hydropower bases

| River basin | Installed capacity (10,000 kW) | Annual electricity generation (100 million kWh) | Operation scale in 2010 (10,000 kW) | New construction scale during "12th Five-Year Plan" period | New operation scale during "12th Five-Year Plan" period | Flood control capacity | Total static investment (RMB 100 million) |
|---|---|---|---|---|---|---|---|
| Upper reaches of Yangtze River | 3400 | 1280 | 2431 | 170 | 420 | 231.5 | 1602 |
| Jinsha River | 7700 | 2908 | 0 | 4640 | 2390 | 221.4 | 2775 |
| Lancang River | 3140 | 1400 | 885 | 2030 | 750 | 32 | 1350 |
| Yalong River | 2600 | 1360 | 330 | 850 | 1080 | 40 | 1228 |
| Dadu River | 2640 | 1010 | 641 | 1180 | 540 | 11.7 | 1390 |
| Wujiang River | 1140 | 418 | 933 | 40 | 170 | 11.7 | 526 |
| Upper reaches of Yellow River | 2530 | 1275 | 1457 | 650 | 40 | 116.7 | 689 |
| Nanpanjiang River; Hongshui River | 1570 | 504 | 1208 | 140 | 60 | 100.2 | 766 |
| Nujiang River | 3600 | 1029 | 0 | 1120 | 0 | 52.3 | 897 |
| Total | 28,320 | 11,184 | 7885 | 10,820 | 5450 | 817.5 | 11,223 |

*Notes* The data of annual generating capacity are based on *Planning for China's 13 Hydropower Bases*; the data of flood control capacity and total static investment are as of 2010; other data are based on the *12th Five-Year Plan for Hydropower Development*

Five-Year Plan" period is mainly based on the construction scale in 2010, started construction scale and new production scale during the 12th FYP period, total static investment and flood control capacity. See Table 3.12 for details.

## 3.3.3 *Solutions and Policy Suggestions*

### 3.3.3.1 Solutions

1. Coordinate overall planning goals and promote hydropower development

China developed the *Planning for National Principal Functional Regions* and officially released in June 2011. It is a strategy and the basis of the overall plan for national economic and social development, regional planning, and urban planning. In addition, we give more consideration to comprehensive river basin requirements. At the end of 2009, China completed the revision or formulation of comprehensive plans for all major river basins, which will make arrangements for river basin development planning and become important basis for river basin development, utilization, water resources conservation and prevention, and control of water disasters. Hydropower development planning should follow the overall goals and principles of watershed planning to accelerate the rapid and good hydropower development.

2. Integrating ecological protection into the process management of hydropower development

The issue of ecological protection in hydropower development has received extensive attention. In reviewing China's hydropower development progress since reform and opening up 30 years ago, we find we have made major progresses in environmental protection for a hydropower plants in the planning and construction stage. However, the ecological environmental constraint to hydropower development is still a hard nut to crack. First, it should optimize hydropower layout and scale and balance development and ecological protection in the planning period. Hydropower development should meet the overall demands of the comprehensive watershed planning. Second, the mitigation measures should be implemented in the hydropower project design and construction period. Hydropower project design should focus on more environmental protection technologies, put forward countermeasures for preventing or mitigating the adverse effects on the environment to ensure that the ecological environment protection measures construct earlier than the construction of hydropower projects. Finally, the ecological operational scheme should be identified in hydropower project operation. Hydropower project operational dispatch management will become the core of ecological protection for the hydropower sector in China. Dispatch scheme of a hydropower should meet the requirements of regional ecological protection goals. We suggest that it should improve the dispatch scheme of a reservoir by learning from the international idea and mechanism of "adaptive management."

3. Establish social wealth sharing mechanism

Hydropower development is a process of converting natural resources into social wealth. China's central government, local governments, enterprises, and citizens should establish the idea of development resources and sharing wealth. Most hydropower bases are located in poor and undeveloped areas. Without the support of hydropower enterprises, hydropower might never become social wealth only by local finance and techniques. Practice experiences shows that we should establish a wealth sharing mechanism to achieve the regional social and economic benefits driving by hydropower development, and thus promote the common prosperity of the whole society; we should recognize the rights of the affected groups to use resources and improve the current mode that local residents and the governments in the reservoir areas fail to share social wealth in a reasonable way; we should actively explore the innovative resettlement mode, reform the existing the distribution mode of benefits and profit from resources, raise the funds for resettlement, and define resettlement compensation standards according to law.

4. Establish guarantee institutions

Reservoir migrants have a characteristic of involuntary. In China, involuntary migrants caused by project construction is not limited to the water conservancy and hydropower sectors. The compensation standard for migrants varies among different sectors and areas. Therefore, we suggest resettlement compensation should be legally standardized to ensure the stable implementation of resettlement.

We suggest "equal electricity with equal price in a grid." It should establish a long-term institution to ensure put funds in resettlement compensation and ecological protection in the hydropower sector projects. We also suggest establishing the watershed compensation mechanism as soon as possible. It should reasonably evaluate the ecological loss and benefits of a hydropower project and establish a cost sharing mechanism to promote the implementation of watershed ecological compensation.

#### 3.3.3.2 Policy Suggestions

1. Improve medium- and long-term hydropower development idea and strategy

According to China's energy strategy goals and GHG emission reduction goals in 2020, we suggest establishing China's overall hydropower development strategy, identifying the development goals and adjusting current hydropower station layout, scale, development mode, and time arrangement in comprehensive watershed planning.

The development degree of China's hydropower bases such as Jinsha River hydropower base, Yalong River hydropower base, Lancang River hydropower base, and Dadu River hydropower base will reach 70% and above by 2020. Then, China will still have a surplus of economically exploitable hydropower installed capacity of 50 GW mainly concentrated in river sources and Tibet. Hydroelectricity transmission from Tibet to the whole country will play an important supporting role in achieving China's energy strategy and promoting Tibet's social and economic development. Meanwhile, China will face ecological protection, migrant resettlement, protection of minority culture, and other problems. We suggest making preparations for developing hydropower resources in Tibet as soon as possible.

China has rich transboundary rivers, but the development degree of these international rivers is very low. In which, Nujiang River and Yarlung Zangbo River are basically not developed. Downstream countries of these transboundary rivers are very complicated. We suggest starting the research on establishing the transboundary river water right allocation mechanism, transboundary water pollution conflict coordination mechanism, biodiversity conservation negotiation mechanism, major emergency warning mechanism, and river development rights coordination mechanism. The results can provide useful information for decision-makers and promote the early utilization of water resources and hydropower potential in transboundary rivers.

We should consider how to use cascade reservoir dispatch to improve the quantity and quality of hydroelectricity generation. Meanwhile, it will a long-term task to obtain the sustainable development of river basins, including water supply, flood control, and eco-environment protection goals. We suggest comprehensively considering the relationship between upstream and downstream cascade power stations in basins. It should improve the operational and scheduling level of the cascade hydropower plants by using advanced decision-making theories and methods to achieve efficient utilization of river hydropower potential.

Construction of pumped storage hydropower is an important solution to the grid peak regulation problem with the rapid development of new energy. We suggest as soon as possible to make pumped storage hydropower planning and construction during the 12th FYP period for the grid security.

2. Improve institutions to establish a policy system for promoting hydropower development

   (1) Tax reduction policy. Current hydropower taxes are too high in China. Taxation basically remains unchanged after tax reform in other sectors in China. However, hydropower taxation has been increased by several times in recent years. Take the value-added tax as an example: Before the tax reform, the value-added tax rate in the electric sector was about 10%; while after the tax reform, value-added tax rate for hydropower sector is 17%, about two times that of coal-fired fuel power. We suggest carrying out tax reform in the hydropower industry to reduce its taxation.
   (2) Price policy. Hydropower price is lower than its value. China's electricity price is controlled and approved by the government. China's electricity price system may be the most complex in the world. Although the Electricity Law of the People's Republic of China required "equal electricity with equal price in a grid," however, it does not implement until now. The on-grid hydropower price in China is the lowest in all electricity. We

suggest implementing the policy of "equal electricity with equal price in a grid" and establishing relevant policies to reflect resettlement and ecological protection cost in hydropower projects.

(3) Current submerged compensation policy does not benefit to hydropower development. Since the promulgation and implementation of the revised Land Administration Law of the People's Republic of China and Forestry Law of the People's Republic of China, the submerged compensation policy has failed to reflect the nature of hydropower projects occupying land. All land is regarded as forest land and farmland, which results in the overestimated value of land in remote areas and the failure to reflect the significance of hydropower development. We suggest to further subdivide land types in reservoir inundation for compensation policy-making.

(4) Environment policies. We suggest carrying out strategic environmental impact assessment for energy development projects with the consideration of national energy security and GHG emission reduction goal. It should weight the GHG emissions and the ecosystem protection to identify the short-, medium-, and long-term environmental protection strategy for national energy development.

(5) Hydropower project construction management policy. We suggest improving hydropower project construction management procedures and measures, identifying the requirements for the prophase, approval, construction, and operation, establishing the relative simple approval system according to the investment system reform requirements and actual situation of hydropower construction, establishing the new hydropower preparatory work fund and operation system according to the market rules, and emphasizing the investment in the preparatory work.

3. Increase technical investment to promote innovations in the science and technology system

We find that each leap in hydropower construction sourced from the breakthrough in the key technologies during China's hydropower development history. China's dam construction technology has reached the world leading level, but it does not mean that the technology problem has been solved completely. After the reform of the national electric power system, the investment in the systematic research on the hydropower development science and technology is obviously insufficient. The ecological protection practice for the hydropower sector lacks basic and systematic research and investigation, which results in the absence of the integrated ecological protection goal for many rivers. As an interdisciplinary, the theory and methodologies of ecological protection for the hydropower sector should be better. We suggest increasing investment in hydropower development technology, innovating the industry-academia-research system and promoting application of interdisciplinary approaches to solve key technical problems that affect the dam security and restrict hydropower development.

4. Suggestions for revision of laws and regulations

We suggest that it should revise the *Renewable Energy Law of the People's Republic of China* to define hydropower as a renewable energy, establishing and improving the hydropower development regulations, making Regulations on Administration of Hydropower Development to propose relevant requirements for hydropower development.

We suggest developing a *Project Construction Resettlement Law* to ensure the rights and interests of reservoir migrants, standardize the migrant resettlement in different hydropower projects, and solve the problem of unrealistic comparison of resettlement standards among different sectors.

5. Suggestions for hydropower development and management

China implements multiple administration by different government sections for hydropower development. The NDRC is responsible for the management of large hydropower, the Ministry of Water Resources for the management of reservoirs and small hydropower focusing on public welfare, and transportation departments for the management of navigation hydropower junction. This situation is seriously harmful to watershed integrated management and water resource utilization. We suggest establishing an integrated organization to implement watershed hydropower management.

# Chapter 4
# Water Resource Constraints for Energy Development

**Dongsheng Wang and Jing Yin**

There is a certain water resource consumption more or less in each kind of energy production, particularly in coal mining or thermal power production. In coal mining, water is used as a medium for mining and processing and the surrounding groundwater is unwatered, which damages the groundwater aquifers of mining areas; in thermal power production, water is used as a medium for cooling, power and hydraulic conveying. There is difference of water consumption per production capacity among enterprises because of different production technology, all enterprises treat water as a basic element in coal mining or thermal power production. Furthermore, China faces water resource shortage because of a per capita water resource less than one-third of the world average level and its water resources are unevenly distributed. What's worse is that from spatial distribution, the coal resources in China are not matched with the water resources in China: most of the coal resources are in north China with serious water shortage; in particular, in the Yellow River basin, the Haihe River basin, and the Songhua River–Liaohe River basin, water resources have seriously curbed coal base construction and such provinces with abundant coal as Shanxi, Inner Mongolia, and Ningxia have serious water shortage. Additionally, in the development of related industries, the damage to groundwater systems in coal mining and the impact of energy production on the ecological environmental quality must be taken into full consideration and argumentation. Whether water resources can support growth of the energy industry or not is an unavoidable question for the energy and water resources administrative authorities.

D. Wang (✉)
China Renewable Energy Engineering Institute, Beijing
People's Republic of China
e-mail: 1340037360@qq.com

J. Yin
China Institute of Water Resources and Hydropower Research,
Beijing, People's Republic of China

 
Z. Wang (ed.), *Constraints and Solutions for Energy and Electricity Development*,
https://doi.org/10.1007/978-981-10-8304-4_4

## 4.1 Constraints of Water Resources for Coal Development

In China, coal is the main energy source and the foundation of energy safety. Since the founding of new China, coal has been accounting for about 70% of the primary energy source consumption structure. According to the forecast by China Electricity Council in accordance with the current energy resource situation and the energy demand tendency, the total coal demand in China will tend to keep increasing before the year of 2030 and the increase rate will be increasingly quicker between 2010 and 2015. In future, the tendency of quick increase of the coal output in China will surely increase the constraints of water resources for coal development. Therefore, studying the constraints of water resources for the coal development in China and improving the support of water resources for coal development are practically significant to the energy safety of China.

### *4.1.1 Relation Between Coal and Water Resources*

Water resources are a basic element in existence and development of the coal industry; poor protection of water resources in coal mining and processing influences the quantities and quality of regional surface water and groundwater significantly and results in water ecology and water environment-related problems such as exhaustion and quality deterioration of surface water, the damage to groundwater circulation systems and reduction of groundwater reserves.

#### 4.1.1.1 Water Demand Characteristics in Coal Production

The operation of coal mines needs water, so it is necessary to find water sources in coal mine searching. Water is necessary in each link of coal mining and processing, from the mud used in drilling during exploration to the heating and ventilation in coal mines in winter, from the dust reduction of working surfaces of mines by means of watering to the washing and processing of coal products, and from the auxiliary production water for the grounds of coal mines to the domestic water of coal enterprises (see Table 4.1).

According to statistics, in the coal production of the coal mines for coal washing and processing, the water use of each ton of coal (including production water and domestic water) is about 1 $m^3$ or more. In both key state-owned coal mines and local state-owned coal mines in the whole coal industry, the average water use by each ton of coal is 2.3 $m^3$ and the average water drainage by each ton of coal is about 2 $m^3$. Both total water use and total drainage quantity of the coal industry are among the top five of all industries in China.

**Table 4.1** Main water use links in coal production

| Coal mining in mines | Coal mining in open-pit coal mine | Coal dressing and processing | Coal processing and conversion |
|---|---|---|---|
| Hydraulic coal mining | Dust reduction of working surfaces of coal mines | Dust reduction by crushing | Evaporative cooling in the coal chemical industry |
| Hydraulic hoisting | Watering for roads for trucks | Gravity separation | Coke quenching in coking |
| Dust reduction by means of watering | Punching, blasting and drilling | Coal flotation | Boilers and steam turbine units in power plants |
| Mechanical coal mining | Land reclamation of waste dumps | Circulating cooling of vacuum pumps | Circulating cooling of power plants |
| Water injection softening for hard roofs | | Circulating cooling of air compressors | Hydraulic sluicing in power plants |
| Hydraulic filling and underground grouting | | | Dust removal with a wet method |
| Fire-proof spraying and pre-filling of coal gangue piles | | | Coal-water slurry preparation |
| Blasting and drilling | | | Briquette suppression |
| Heating and ventilation boilers | | | |
| Cooling of mechanical equipment | | | |

#### 4.1.1.2 Impacts of Coal Production on Water Resource Systems

1. Impacts on water resources

After the coal mining area reaches a certain range, there will be more and bigger vertical fractures in the coal measure stratum and the overlying loose rock series, forming surface deformation and subsidence, whereas the water in the coal measure stratum and the overlying loose rock series seep down quickly and enter shafts, roadways, and goafs or drain outward, forming water level decline caused goaf-centered depression cones. Consequently, there will be water resource-related problems such as regional groundwater level decline, seepage of surface water into underground or pits, dramatic decrease of surface runoffs, spring cut-off, and well drying-up, so that the sustainable utilization of the local water resources will be seriously affected. Furthermore, there will be a direct decrease of surface runoffs because atmospheric precipitations in rainstorms do not form effective surface runoffs to increase the surface water but seep into mines heavily along surface subsidence zones and water-flowing fractured zones when overland flows meet rivers.

Take the Gucheng River to the east of Huangfeng Coal Mine in Shandong as an example. Its average runoff in the year of 1900 was 582 $m^3$/s and kept decreasing after that because of coal mining. In 2001, it had an average runoff of about 1 $m^3$/s and even dried up in the dry season. This was caused by many factors such as precipitation decrease and overwater intaking of both surface water and groundwater. However, it is undeniable that coal mining was a main one among the factors.

2. Impacts on groundwater

Groundwater is the main water source for coal mining, while the coal mining influences regional groundwater systems inevitably. The imbalance between water supply, runoffs, and water drainage of coal measure aquifers is broken by massive drainage from pits, changing the natural flow fields of groundwater, and resulting in exhaustion of regional water resources.

(1) Impacts of pit drainage on groundwater circulation: Coal is a sedimentary rock type underground mineral resource. Viewed from the geological structure, coal measure stratums, aquifers, and watertight stratums are distributed alternately and there is always one or multiple aquifers above a coal measure stratum. Thus, coal mining influences natural groundwater systems inevitably. According to incomplete statistics, the national annual drainage quantity of mine water for coal mines is 2.1–2.4 billion $m^3$. However, the drainage quantities of mine water of specific coal mines in different regions differ greatly, depending on the geological conditions, the climates, the mining methods, etc.: For the coal mines in regions with a humid climate and abundant groundwater, their drainage quantity for each ton of coal may be higher than 80 $m^3$; for the coal mines in regions with a dry climate and insufficient atmospheric precipitation or surface water for groundwater, their drainage quantity for each ton of coal is always lower than 1.6 $m^3$.

Coal and water resources coexist in a geological body. In natural conditions, precipitation, surface water, and groundwater present a certain relation of recharge and discharge and respective occurrence conditions and change laws. The drainage in coal mining breaks the original natural balance of groundwater, results in mine-centered depression cones, partially damages the relation of water supply, runoffs, and water drainage of coal measure aquifers and changes the natural flow fields of groundwater. As a result, precipitation, surface water and groundwater convert and coal measure stratums, aquifers and watertight stratums are connected, making surface water convert into groundwater, seep into pits and discharge into lower reaches. Surface water and groundwater convert and recharge mutually, affecting their quality and resulting in both water resource waste and economic losses. For example, there are one hundred coal mines in an area of 192 $km^2$ in western Xiyang County in Shanxi Province. In 1988, their pits had a total drainage quantity of 7,630,000 $m^3$. Due to the mining area centered depression cones caused by drainage of the coal mines, the horizontal movement of coal measure groundwater has changed into vertical flowing into the pits and 28 (or 72%) of the 39 wells in scope of influence of the depression cones have been significantly influenced.

(2) Impacts of coal mining on groundwater reserves: In the second water resource evaluation of Shanxi titled 'Study on Impacts of the Coal Mining in Shanxi on Water Resources,' the degrees of damage to coal mining of the coal fields in the province to groundwater resources were calculated on the basis of the relation between the coal measure stratums and the aquifers as well as exploration, production, and study results of the coal fields in the past few decades. According to the calculation results, the static reserve of groundwater damaged by the coal mining in Shanxi was 713,880,000 $m^3$ as shown in Table 4.2.

(3) Impacts of coal mining on the water environment: Coal mining pollutes the water environment seriously. In the drainage in coal mining, the direct drainage of the untreated mine water pollutes surface water and results in regional precipitation infiltration pollution through karst fissures, thus turning quality water sources into polluted mine water. The main pollutants in mine water are suspended solids (SS) which mainly include coal dust and rock dust and are mainly in gray–black. Furthermore, mine water has a high COD and the mine water in some regions in China is highly mineralized. According to statistics, the highly mineralized mine water in China has a salinity of above 1000 mg/L to 3000 mg/L or even above 4000 mg/L. Due to coal mining, the mine water of this sort has a high turbidity because of a high content of SS including coal dust, rock dust, etc. According to statistics, in 2004, the coal mines in Shanxi had a water inflow of 202,405,000 and 111,356,000 $m^3$ of the mine water was drained directly without treatment and utilization. According to current data, there are multiple pollutants in the karst water in Shanxi and some of them are in content much higher than the standard value. Besides that, a lot of coal gangue is generated in coal mining and washing, and the leaching water of coal gangue piles usually contains acid and harmful heavy metal elements which contaminate soil and water close to the dumps.

To sum up, the harmony between coal production and regional water resources is a precondition for healthy and sustainable development of the coal industry. Thus, in this book, a suitability study about three main constraints including regional water resources, regional water environment, and regional groundwater is done for the big coal bases in China and these constraints are evaluated as well.

## *4.1.2 Constraint Analysis*

### 4.1.2.1 Layout and Water Demand Forecast of the Coal Bases in China

1. Layout and hydrological regions of thirteen big coal bases in China

**Table 4.2** Calculation result of the degrees of damage of coal mining of the coal fields in Shanxi to static reserves of groundwater

| Coal field | Sandstone | | | Limestone | | | Total |
|---|---|---|---|---|---|---|---|
| | Thickness (m) | Area (10,000 $m^2$) | Groundwater damaged (10,000 $m^3$) | Thickness (m) | Area (10,000 $m^2$) | Groundwater damaged (10,000 $m^3$) | |
| Datong coal field | 95.15 | 55,190 | 6144 | | | | 6144 |
| Ningwu coal field | 50.67 | 36,950 | 2190 | 5.78 | 12,316.6 | 114 | 2304 |
| Xishan coal field | 51.80 | 51,250 | 3106 | 9.88 | 35,576.07 | 562 | 3668 |
| Huoxi coal field | 46.25 | 151,900 | 8219 | 16.25 | 110,041.33 | 2861 | 11,080 |
| Hedong coal field | 59.76 | 141,600 | 9900 | 13.69 | 102,247.29 | 2240 | 12,140 |
| Qinshui coal field | 63.75 | 385,000 | 28,716 | 17.84 | 24,875.36 | 7100 | 35,816 |
| Other coal fields | 30.73 | 5048 | 181 | 7.18 | 4800.09 | 55 | 236 |
| Total | | | 58,456 | | | 12,932 | 71,388 |

In February 2003, the State Council made a decision that 'gives priority to key support to the construction of big coal bases, promotes coal and power pool, and forms several 100-million-ton key coal enterprises' in a bid to solve the main problems in coal production and supply. On the 60th executive meeting of the State Council on August 11, 2004, it was decided to 'strengthen the exploration and support capacity of coal resources, push forward the construction, reconstruction, and expansion of some modern big coal mines, encourage coal and power pool and speed up the construction of big coal bases.' As a response to the decision, 13 big coal bases in all were planned for west China, northeast China, north China, and southwest China.

The 13 big coal bases include Shendong coal base, northern Shanxi coal base, eastern Shanxi coal base, eastern Inner Mongolia (northeast) coal base, Yunnan–Guizhou coal base, Henan coal base, western Shandong coal base, central Shanxi coal base, Huainan–Huaibei coal base, Huanglong coal base, central Hebei coal base, eastern Ningxia coal base, and northern Shaanxi coal base. These big coal bases are distributed in 14 provinces (including autonomous regions) and have 98 mining areas in all, a total area of 2,547,000 $km^2$, a forecast resource reserve of 781.9 billion ton at a 1500 m shallow range. The Shendong coal base, the Yunnan–Guizhou coal base and the central Shanxi coal base have the largest reserves all above 100 billion ton as shown in Table 4.3.

For the sake of easy allocation and management of water resources by basin and region, the whole country is divided into ten first-class hydrological regions of water resources in accordance with characteristics of basins and regional water resources. Each first-class hydrological region is further divided into several second-class hydrological regions. The thirteen big coal bases in China situate in 21 second-class hydrological regions of first-class hydrological regions (Songhua River basin, Liaohe River basin, Haihe River basin, Yellow River basin, Huaihe River basin, and Yangtze River basin) in 14 provincial administrative regions. See Table 4.4.

2. Water demand forecast of the thirteen big coal bases in China.

Water demands of coal mining and dressing of the thirteen coal bases in the years of 2010, 2015, and 2020 are forecast in accordance with related regulations of some provinces and cities about the water use quotas for coal mining and dressing in the years from 2000 to 2004 (Table 4.5), and the coal outputs of these coal bases in the same three years are forecast in the 'Report on Overall Evaluation of Coal Resources, Water Resources and Ecological Environments of Big Coal Bases' project of China National Administration of Coal Geology. The forecast results are shown in Tables 4.5 and 4.6.

According to water use quotas for coal mining and dressing issued by the provinces and autonomous regions in Table 4.6, the total water demands in coal mining and dressing of the thirteen big coal bases will be 3.152 and 3.457 billion $m^3$, respectively, in 2015 and 2020.

**Table 4.3** Basic information of big coal bases in China

| Coal base name | Base area ($km^2$) | Geographical location | Mining area name | Main coal type | Forecast resources | | <1000 m | | 1000–1500 m | |
|---|---|---|---|---|---|---|---|---|---|---|
| | | | | | Resources ($10^8$ ton) | Percent in total resources of coal base (%) | Resources ($10^8$ ton) | Percent (%) | Resources ($10^8$ tons) | Percent (%) |
| Shendong coal base | 9000 | Northern Shaanxi and western Inner Mongolia autonomous Region | Shendong, Wanli, Zhunge'er, Wuhai, Baotou and Fugu | Fat coal, bituminous coal and gas coal | 1723.59 | 22.04 | 763.92 | 44.32 | 959.67 | 55.68 |
| Northern Shanxi coal base | 6700 | Northern Shanxi | Datong, Pingshuo, Shuonan, Xuangang, Lan County and Hequ-Baode-Pianguan | Gas coal and fat coal | 262.89 | 3.36 | 182.07 | 69.26 | 80.82 | 30.74 |
| Eastern Shanxi coal base | 15,060 | Eastern Shanxi | Yangquan, Wuxia, Lu'an and Jincheng | Low-sulphur coal and low-and-medium-sulphur coal | 532.79 | 6.80 | 231.35 | 43.42 | 301.44 | 56.58 |
| Eastern Inner Mongolia coal base | 27,264.7 | Eastern and western (southeastern) Inner Mongolia, Central Liaoning and Eastern Heilongjiang | Zhalainuo'er, Baori Xile, Yimin, Dayan, Huolinhe, Pingzhuang, Baiyinhua, Shengli, Fuxin, Tiefa, Shenyang, Fushun, Jixi, Qitaihe, Shuangyashan and Hegang | Fat coal, gas coal, and lignite | 274.67 | 3.51 | 193.70 | 70.52 | 80.97 | 29.48 |
| Yunnan–Guizhou coal base | 46,872 | Northwestern Guizhou, eastern Yunnan and Southeastern Sichuan | Junlian, Guxu, Pan County, Puxing, Shuicheng, Liuzhi, Zhina, northern Guizhou, Laochang, Xiaolongtan, Zhaotong and Zhenxiong and Enhong | Anthracite, gas coal, coking coal, lignite, and lean coal | 1615.43 | 20.66 | 879.82 | 54.50 | 735.61 | 45.50 |

(continued)

**Table 4.3** (continued)

| Coal base name | Base area ($km^2$) | Geographical location | Mining area name | Main coal type | Forecast resources | | <1000 m | | 1000–1500 m | |
|---|---|---|---|---|---|---|---|---|---|---|
| | | | | | Resources ($10^8$ ton) | Percent in total resources of coal base (%) | Resources ($10^8$ ton) | Percent (%) | Resources ($10^8$ tons) | Percent (%) |
| Henan coal base | 10,849 | Western and eastern (central) Henan | Hebi, Jiaozuo, Yima, Zhengzhou, Pingdingshan and Yongxia | Anthracite, lean coal, fat coal, lean coal, and coking coal | 265.89 | 3.40 | 76.39 | 29.06 | 186.50 | 70.94 |
| Western Shandong coal base | 55,962.82 | Southwestern Shandong | Yanzhou, Jining, Xinwen, Zaoteng, Longkou, Zibo, Feicheng, Linyi, Juye and Huanghebei | Gas coal, fat coal, gas fat coal, lignite, coking coal, and anthracite | 179.42 | 2.29 | 53.71 | 29.94 | 125.71 | 70.26 |
| Central Shanxi coal base | 27,475.06 | Central Shanxi | Xishan, Dongshan, Gaoliu, Fenxi, Huozhou, Xiangning and Huodong Shixi | Fat coal, coking coal, lean coal, and lean coal | 1098.08 | 14.04 | 689.58 | 62.80 | 408.50 | 37.20 |
| Huainan–Huaibei coal base | 19,000 | Central and northern parts of Anhui | Huaibei and Huainan | Fat coal, lean coal, coking coal, and gas coal | 291.05 | 3.72 | 0 | 0 | 291.05 | 100 |
| Huanglong coal base | 11,084 | Central Shaanxi and Eastern Gansu | Binchang, Huangling, Xunyao, Tongchuan, Pubai, Chenghe and Hancheng Huating | Gas coal, lean coal, coking coal, anthracite, and fat coal | 247.97 | 3.17 | 127.22 | 51.30 | 120.75 | 48.70 |
| Central Hebei coal base | 5000 | Southern, central and northwestern parts of Hebei | Fengfeng, Handan, Xingtai, Jingxing, Kailuan, Yu County, Xuanhua-Xiahuayuan, Zhangjiakou and Pingyuan big coal field | Anthracite, gas coal, fat coal, coking coal, and fat coal | 201.75 | 2.58 | 34.63 | 17.16 | 167.12 | 82.84 |
| Eastern Ningxia coal base | 2105 | Eastern Ningxia | Shizuishan, Shitanjing, Lingwu, Shigouyi, Yuanyanghu, Hengcheng, Weizhou and Majimeng | Coking coal, fat coal, lean coal, fat coal, and non-caking coal | 377.70 | 4.83 | 123.21 | 32.62 | 254.49 | 67.38 |

(continued)

**Table 4.3** (continued)

| Coal base name | Base area ($km^2$) | Geographical location | Mining area name | Main coal type | Forecast resources | | <1000 m | | 1000–1500 m | |
|---|---|---|---|---|---|---|---|---|---|---|
| | | | | | Resources ($10^8$ ton) | Percent in total resources of coal base (%) | Resources ($10^8$ ton) | Percent (%) | Resources ($10^8$ tons) | Percent (%) |
| Northern Shaanxi coal base | 18,365.7 | Northern Shaanxi | Yushen and Yuheng | Non-caking coal and fat coal | 751.17 | 9.60 | 751.17 | 100 | 0 | 0 |
| Total | 254,738.3 | – | – | | 7819.40 | 100 | 4106.77 | – | 3712.63 | – |

**Table 4.4** Hydrological regions of thirteen big coal bases in China

| Coal base name | Administrative region | Hydrological region | |
|---|---|---|---|
| | | Second-class hydrological region | First-class hydrological region |
| Shendong coal base | Inner Mongolia and Shaanxi | Lanzhou-Hekou Town | Yellow River basin |
| Eastern Ningxia coal base | Ningxia | | |
| Northern Shaanxi coal base | Shaanxi | Hekou Town-Longmen | |
| Central Shanxi coal base | Shanxi | | |
| Northern Shanxi coal base | Shanxi | | |
| Huanglong coal base | Shaanxi and Gansu | Longmen-Sanmenxia | |
| Henan coal base | Henan | Sanmenxia–Huayuankou | |
| Central Hebei coal base | Hebei | Luanhe River and the rivers in the coastal areas of eastern Hebei | Haihe River basin |
| | | Northern drainage of Haihe River | |
| | | Southern drainage of Haihe River | |
| | | Tuhai River–Majia River | |
| Eastern Shanxi coal base | Shanxi | Southern drainage of Haihe River | |
| Western Shandong coal base | Shandong | Yishusi River | Huaihe River basin |
| | | Shandong peninsula | |
| Huainan–Huaibei coal base | Anhui | Middle reaches of Huaihe River | |
| Yunnan–Guizhou coal base | Yunnan, Guizhou, and Sichuan | Lower reaches of Jinsha River from Shigu | Yangtze River basin |
| | | Wujiang River | |
| | | Yibin-Yichang mainstream | |
| Eastern Inner Mongolia coal base | Inner Mongolia, Liaoning and Heilongjiang | Argun River | Songhua River basin |
| | | Songhua River (down from the Sancha River estuary) | |
| | | Wusuli River | |
| | | West Liaohe River | Liaohe River basin |
| | | Main stream of the Liaohe River | |
| | | Huntai River | |
| | | Rivers close to Yellow Sea and Bohai Sea in northeastern China | |

**Table 4.5** Water use quotas for the coal production in some provinces and autonomous regions in China

| Provincial administrative region | Quota unit | Coal mining | Coal dressing | Data source |
|---|---|---|---|---|
| Inner Mongolia | $m^3/t$ | 0.9 | 2.5 | DB15/T 385—2003 |
| Liaoning | $m^3/t$ | 0.06–0.07 | 0.15 | DB21/T 1237—2003 |
| Heilongjiang | $m^3/t$ | 0.9 | | Department of Water Resources of Heilongjiang Province, 2000; |
| Hebei | $m^3/t$ | 1.0 | 3.0 | Department of Water Resources of Hebei Province, 2002 |
| Henan | $m^3/t$ | 0.5 | 0.2 | DB41/T 500—2004 |
| Shanxi | $m^3/t$ | 0.02 (Pingshuo open-pit coal mine) | 0.23 | Department of Water Resources of Shanxi Province, 2003 |
| | | 0.7 | | |
| Gansu | $m^3/t$ | 0.4 | – | Department of Water Resources of Gansu Province, 2004 |
| Sichuan | $m^3/t$ | 1.6 | – | Department of Water Resources of Sichuan Province, 2002 |
| Ningxia | $m^3/t$ | 0.7 | 1.0 | Department of Water Resources of Ningxia Hui Autonomous Region, 2004 |
| Shaanxi | $m^3/t$ | 1.0 Mining and washing of bituminous coal and anthracite | | Department of Water Resources of Shaanxi Province |
| | | 0.8 Mining and dressing of lignite | | |
| Yunnan | $m^3/t$ | 1.0 Mining and washing of bituminous coal and anthracite | | DB53/T 168—2006 |
| | | 0.5–0.7 Mining and dressing of lignite | | |

The period from 2010 to 2020 will be a key to the economy of China. In the period, all basins in China will face an increasingly larger water demand pressure due to economic development. In consideration of this, changes of the water demands in production of the coal bases in different hydrological regions in the Twelfth Five-Year Plan period and the Thirteenth Five-Year Plan period are compared in Table 4.7.

According to Table 4.7, all the coal bases in China, except the western Shandong, central Hebei, and northern Shaanxi coal bases, will witness a large increase rate of water demand in the Twelfth Five-Year Plan period, and all of them except both northern Shaanxi and central Hebei coal bases will witness a smaller

**Table 4.6** Water demand forecasts for the coal mining and dressing of big coal bases in China in different target years

| Coal base name | Name/Region of the mining area | Planned output of the mining area ① (10,000 ton) | | | Province where the mining area is | Quota of coal mining ($m^3/t$) | Quota of coal dressing ($m^3/t$) | Water demand of the mining area (10,000 $m^3$) | | |
|---|---|---|---|---|---|---|---|---|---|---|
| | | 2010 | 2015 | 2020 | | | | 2010 | 2015 | 2020 |
| Shendong coal base | Shendong | 16,720 | 19,120 | 20,260 | Inner Mongolia | 0.9 | 2.5 | 56,848 | 65,008 | 68,884 |
| | Wanli | 2100 | 2800 | 4000 | Inner Mongolia | 0.9 | 2.5 | 7140 | 9520 | 13,600 |
| | Zhunge'er | 6300 | 7400 | 9100 | Inner Mongolia | 0.9 | 2.5 | 21,420 | 25,160 | 30,940 |
| | Wuhai | 2150 | 2150 | 2300 | Inner Mongolia | 0.9 | 2.5 | 7310 | 7310 | 7820 |
| | Baotou | 540 | 540 | 540 | Inner Mongolia | 0.9 | 2.5 | 1836 | 1836 | 1836 |
| | Fugu | 1200 | 2000 | 2300 | Shaanxi | 0.25 | 0.1 | 420 | 700 | 805 |
| Northern Shanxi coal base | Datong | 13,530 | 13,685 | 13,275 | Shanxi | 0.7 | 0.23 | 12,583 | 12,727 | 12,346 |
| | Pingshuo | 7900 | 10,330 | 10,100 | Shanxi | 0.02 | 0.23 | 1975 | 2583 | 2525 |
| | Shuonan | 400 | 1400 | 3200 | Shanxi | 0.7 | 0.23 | 372 | 1302 | 2976 |
| | Xuangang | 1450 | 1335 | 1675 | Shanxi | 0.7 | 0.23 | 1349 | 1242 | 1558 |
| | Lan County | 1120 | 1100 | 1100 | Shanxi | 0.7 | 0.23 | 1042 | 1023 | 1023 |
| | Hequ-Baode-Pianguan | 3600 | 3650 | 3650 | Shanxi | 0.7 | 0.23 | 3348 | 3395 | 3395 |
| Eastern Shanxi coal base | Yangquan | 4600 | 5900 | 7400 | Shanxi | 0.7 | 0.23 | 4278 | 5487 | 6882 |
| | Wuxia | 500 | 700 | 1500 | Shanxi | 0.7 | 0.23 | 465 | 651 | 1395 |
| | Lu'an | 6100 | 7400 | 7600 | Shanxi | 0.7 | 0.23 | 5673 | 6882 | 7068 |
| | Jincheng | 7800 | 10,000 | 11,500 | Shanxi | 0.7 | 0.23 | 7254 | 9300 | 10,695 |
| Eastern Inner Mongolia coal base | Yimin, and so on | 10,320 | 11,370 | 12,170 | Inner Mongolia | 0.9 | 2.5 | 35,088 | 38,658 | 41,378 |
| | Fuxin, and so on | 4580 | 4337 | 4092 | Liaoning | 0.07 | 0.15 | 1008 | 954 | 900 |
| | Hegang, and so on | 8100 | 7793 | 7738 | Heilongjiang | 0.9 | 1.3 | 17,820 | 17,145 | 17,024 |
| Yunnan–Guizhou coal base ② | Junlian | 825 | 1245 | 1800 | Sichuan | 1.6 | – | 1320 | 1992 | 2880 |
| | Guxu | 775 | 1155 | 1500 | Sichuan | 1.6 | – | 1240 | 1848 | 2400 |
| | Pan County | 1960 | 2680 | 2680 | Guizhou | 1 | | 1960 | 2680 | 2680 |

(continued)

**Table 4.6** (continued)

| Coal base name | Name/Region of the mining area | Planned output of the mining area ① (10,000 ton) | | | Province where the mining area is | Quota of coal mining ($m^3/t$) | Quota of coal dressing ($m^3/t$) | Water demand of the mining area (10,000 $m^3$) | | |
|---|---|---|---|---|---|---|---|---|---|---|
| | | 2010 | 2015 | 2020 | | | | 2010 | 2015 | 2020 |
| | Puxing | 780 | 780 | 900 | Guizhou | 1 | | 780 | 780 | 900 |
| | Shuicheng | 2030 | 2330 | 2330 | Guizhou | 1 | | 2030 | 2330 | 2330 |
| | Liuzhi | 800 | 780 | 780 | Guizhou | 1 | | 800 | 780 | 780 |
| | Zhina | 1330 | 1840 | 2320 | Guizhou | 1 | | 1330 | 1840 | 2320 |
| | Northern Guizhou | 2790 | 3390 | 3690 | Guizhou | 1 | | 2790 | 3390 | 3690 |
| | Laochang | 1100 | 1600 | 1600 | Yunnan | 1 | | 1100 | 1600 | 1600 |
| | Xiaolongtan | 1100 | 1280 | 1500 | Yunnan | 0.5 | | 550 | 640 | 750 |
| | Zhaotong | 500 | 500 | 600 | Yunnan | 0.5 | | 250 | 250 | 300 |
| | Zhenxiong | 500 | 560 | 600 | Yunnan | 1.1 | | 550 | 616 | 660 |
| | Enhong | 600 | 660 | 700 | Yunnan | 1 | | 600 | 660 | 700 |
| Henan | Hebi | 1165 | 1120 | 1035 | Henan | 0.5 | 0.2 | 816 | 784 | 725 |
| | Jiaozuo | 1000 | 1050 | 1245 | Henan | 0.5 | 0.2 | 700 | 735 | 872 |
| | Yima | 1665 | 1600 | 1540 | Henan | 0.5 | 0.2 | 1166 | 1120 | 1078 |
| | Zhengzhou | 3500 | 3500 | 3620 | Henan | 0.5 | 0.2 | 2450 | 2450 | 2534 |
| | Pingdingshan | 5800 | 5800 | 5630 | Henan | 0.5 | 0.2 | 4060 | 4060 | 3941 |
| | Yongxia | 1370 | 1430 | 1430 | Henan | 0.5 | 0.2 | 959 | 1001 | 1001 |
| Western Shandong coal base ③ | Yanzhou | 3139 | 2346 | 2190 | Shandong | 0.5 | 0.2 | 2197 | 1642 | 1533 |
| | Jining | 3505 | 3305 | 3100 | Shandong | 0.5 | 0.2 | 2454 | 2314 | 2170 |
| | Xinwen | 1548 | 1281 | 1222 | Shandong | 0.5 | 0.2 | 1084 | 897 | 855 |
| | Zaoteng | 3107 | 3074 | 2857 | Shandong | 0.5 | 0.2 | 2175 | 2152 | 2000 |
| | Longkou | 588 | 588 | 378 | Shandong | 0.5 | 0.2 | 412 | 412 | 265 |

(continued)

**Table 4.6** (continued)

| Coal base name | Name/Region of the mining area | Planned output of the mining area ① (10,000 ton) | | | Province where the mining area is | Quota of coal mining ($m^3/t$) | Quota of coal dressing ($m^3/t$) | Water demand of the mining area (10,000 $m^3$) | | |
|---|---|---|---|---|---|---|---|---|---|---|
| | | 2010 | 2015 | 2020 | | | | 2010 | 2015 | 2020 |
| | Zibo | 387 | 277 | 243 | Shandong | 0.5 | 0.2 | 271 | 194 | 170 |
| | Feicheng | 537 | 30 | | Shandong | 0.5 | 0.2 | 376 | 21 | 0 |
| | Linyi | 84 | 69 | 60 | Shandong | 0.5 | 0.2 | 59 | 48 | 42 |
| | Juye | 1820 | 3115 | 3205 | Shandong | 0.5 | 0.2 | 1274 | 2181 | 2244 |
| | Huanghebei | 285 | 415 | 745 | Shandong | 0.5 | 0.2 | 200 | 291 | 522 |
| Western Shandong coal base Central Shanxi coal base | Xishan | 2900 | 3500 | 3500 | Shanxi | 0.7 | 0.23 | 2697 | 3255 | 3255 |
| | Dongshan | 200 | 200 | 200 | Shanxi | 0.7 | 0.23 | 186 | 186 | 186 |
| | Gaoliu | 2450 | 2850 | 3350 | Shanxi | 0.7 | 0.23 | 2279 | 2651 | 3116 |
| | Fenxi | 1660 | 1940 | 2100 | Shanxi | 0.7 | 0.23 | 1544 | 1804 | 1953 |
| | Huozhou | 1890 | 1410 | 1250 | Shanxi | 0.7 | 0.23 | 1758 | 1311 | 1163 |
| | Xiangning | 1700 | 1900 | 1900 | Shanxi | 0.7 | 0.23 | 1581 | 1767 | 1767 |
| | Huodong | 600 | 600 | 600 | Shanxi | 0.7 | 0.23 | 558 | 558 | 558 |
| | Shixi | 100 | 100 | 100 | Shanxi | 0.7 | 0.23 | 93 | 93 | 93 |
| Huainan–Huaibei coal base ④ | Huaibei | 4347 | 4291 | 4357 | Anhui | 0.5 | 0.2 | 3043 | 3004 | 3050 |
| | Huainan | 6153 | 7209 | 8143 | Anhui | 0.5 | 0.2 | 4307 | 5046 | 5700 |
| Huanglong coal base | Binchang | 2370 | 3070 | 3570 | Gansu | 0.4 | – | 948 | 1228 | 1428 |
| | Huangling | 1040 | 1010 | 1010 | Gansu | 0.4 | – | 416 | 404 | 404 |
| | Xunyao | 80 | 80 | 80 | Gansu | 0.4 | – | 32 | 32 | 32 |
| | Tongchuan | 865 | 705 | 705 | Gansu | 0.4 | – | 346 | 282 | 282 |
| | Pubai | 400 | 360 | 360 | Gansu | 0.4 | – | 160 | 144 | 144 |

(continued)

**Table 4.6** (continued)

| Coal base name | Name/Region of the mining area | Planned output of the mining area ① (10,000 ton) | | | Province where the mining area is | Quota of coal mining ($m^3/t$) | Quota of coal dressing ($m^3/t$) | Water demand of the mining area (10,000 $m^3$) | | |
|---|---|---|---|---|---|---|---|---|---|---|
| | | 2010 | 2015 | 2020 | | | | 2010 | 2015 | 2020 |
| | Chenghe | 655 | 655 | 955 | Gansu | 0.4 | – | 262 | 262 | 382 |
| | Hancheng | 960 | 920 | 920 | Gansu | 0.4 | – | 384 | 368 | 368 |
| | Huating | 1530 | 1900 | 1900 | Gansu | 0.4 | – | 612 | 760 | 760 |
| Central Hebei coal base | Fengfeng | 1160 | 1320 | 1280 | Hebei | 1 | 3 | 4640 | 5280 | 5120 |
| | Handan | 855 | 825 | 780 | Hebei | 1 | 3 | 3420 | 3300 | 3120 |
| | Xingtai | 995 | 920 | 905 | Hebei | 1 | 3 | 3980 | 3680 | 3620 |
| | Jingxing | 60 | 60 | 105 | Hebei | 1 | 3 | 240 | 240 | 420 |
| | Kailuan | 2560 | 2640 | 2610 | Hebei | 1 | 3 | 10,240 | 10,560 | 10,440 |
| | Yu County | 1040 | 825 | 815 | Hebei | 1 | 3 | 4160 | 3300 | 3260 |
| | Xuanhua-Xiahuayuan | 210 | 210 | 210 | Hebei | 1 | 3 | 840 | 840 | 840 |
| | Northern Zhangjiakou | 512 | 402 | 402 | Hebei | 1 | 3 | 2048 | 1608 | 1608 |
| | Pingyuan big coal field | 30 | 90 | 320 | Hebei | 1 | 3 | 120 | 360 | 1280 |
| Eastern Ningxia coal base | Shizuishan | 250 | 450 | 450 | Ningxia | 0.7 | 1 | 425 | 765 | 765 |
| | Shitanjing | 592 | 592 | 592 | Ningxia | 0.7 | 1 | 1006 | 1006 | 1006 |
| | Lingwu | 1513 | 1480 | 1780 | Ningxia | 0.7 | 1 | 2571 | 2516 | 3026 |
| | Shigouyi | 100 | 100 | 100 | Ningxia | 0.7 | 1 | 170 | 170 | 170 |
| | Yuanyanghu | 1000 | 2000 | 2800 | Ningxia | 0.7 | 1 | 1700 | 3400 | 4760 |
| | Hengcheng | 0 | 0 | 240 | Ningxia | 0.7 | 1 | 0 | 0 | 408 |
| | Weizhou | 45 | 165 | 255 | Ningxia | 0.7 | 1 | 77 | 281 | 434 |
| | Majimeng | 15 | 15 | 15 | Ningxia | 0.7 | 1 | 26 | 26 | 26 |

(continued)

**Table 4.6** (continued)

| Coal base name | Name/Region of the mining area | Planned output of the mining area ① (10,000 ton) | | | Province where the mining area is | Quota of coal mining ($m^3/t$) | Quota of coal dressing ($m^3/t$) | Water demand of the mining area (10,000 $m^3$) | | |
|---|---|---|---|---|---|---|---|---|---|---|
| | | 2010 | 2015 | 2020 | | | | 2010 | 2015 | 2020 |
| Northern Shaanxi coal base | Yushen | 3000 | 3690 | 5400 | Shaanxi | 1 | – | 3000 | 3690 | 5400 |
| | Yuheng | 390 | 400 | 400 | Shaanxi | 1 | – | 390 | 400 | 400 |
| 合计 | | 187,827 | 209,694 | 227,459 | – | – | – | 283,441 | 315,168 | 345,736 |

① The output data of the thirteen coal bases in the years of 2010, 2020, and 2030 were extracted from Report on Overall Evaluation of Coal Resources, Water Resources, and Ecological Environments of Big Coal Bases
② The water use quotas for coal mining and dressing in Guizhou refer to the water use quotas for coal mining and dressing in Yunnan
③ The water use quotas for coal mining and dressing in Shandong
④ Anhui refer to the water use quotas for coal mining and dressing in Henan

**Table 4.7** Changes of the water demands in production of the coal bases in different hydrological regions

| Hydrological region | Coal base name | 2010 | 2015 | 2020 | Increase rate during 2010–2015 (%) | Increase rate during 2015–2020 (%) |
|---|---|---|---|---|---|---|
| Yellow River basin | Shendong coal base | 94,974 | 109,534 | 123,885 | 15.3 | 13.1 |
| | Eastern Ningxia coal base | 5975 | 8164 | 10,595 | 36.6 | 29.8 |
| | Northern Shaanxi coal base | 3390 | 4090 | 5800 | 20.6 | 41.8 |
| | Central Shanxi coal base | 10,696 | 11,625 | 12,091 | 8.7 | 4.0 |
| | Northern Shanxi coal base | 20,669 | 22,272 | 23,823 | 7.8 | 7.0 |
| | Huanglong coal base | 3160 | 3480 | 3800 | 10.1 | 9.2 |
| | Henan coal base | 10,151 | 10,150 | 10,151 | 0 | 0 |
| | Average | | | | 14.2 | 15.0 |
| Haihe River basin | Central Hebei coal base | 29,688 | 29,168 | 29,708 | -1.8 | 1.9 |
| | Eastern Shanxi coal base | 17,670 | 22,320 | 26,040 | 26.3 | 16.7 |
| | Average | | | | 12.3 | 9.25 |
| Songhua River–Liaohe River basin | Eastern Inner Mongolia coal base | 53,916 | 56,757 | 59,302 | 5.3 | 4.5 |
| Huaihe River basin | Western Shandong coal base | 10,502 | 10,152 | 9801 | −3.3 | −3.5 |
| | Huaibei–Huainan coal base | 7350 | 8050 | 8750 | 9.5 | 8.7 |
| | Average | | | | 3.1 | 2.6 |
| Yangtze River basin | Yunnan–Guizhou coal base | 15,300 | 19,406 | 21,990 | 26.8 | 13.3 |

increase rate of water demand after the Twelfth Five-Year Plan period. During the two periods, the western Shandong coal base will witness an increasingly smaller water demand. Viewed from hydrological regions of the coal bases, the coal bases in the Yangtze River basin, the Yellow River basin, and the Haihe River basin will witness a larger increase rate of water demand in the Twelfth Five-Year Plan period.

3. Evaluation of reuse of mine drainage

There are mine water inflows in coal mining. For water resource saving and higher water resource utilization rates, the mine drainage reuse technology has been used in most of the mining areas in China. Table 4.8 shows the annual average drainage quantity and reuse rate of mine water in the coal industry in China in the Eleventh Five-Year Plan period provided by China Coal Processing and Utilization Association.

According to the data in the table above, the reuse rates of mine water of the thirteen big coal bases in China in 2015 and 2020 are forecast below, and water supplies and water demands of these coal bases are forecast on the basis of three schemes (high, moderate, and low). See Table 4.9.

Based on considerations of reuse of mine water, water demands of the thirteen big coal bases in China will decrease sharply in 2015 and 2020, and water supplies of these coal bases will therefore increase to some extent. However, mine water is actually groundwater resources damaged in coal mining, so its damage to groundwater circulation system is huge and irreversible. Furthermore, the utilization rates of mine water are actually estimated optimistically for future, but the fact is that the mine water in some mining areas will be difficult to be utilized because of

**Table 4.8** Annual average drainage quantity and reuse rate of mine water in the coal industry in China in the Eleventh Five-Year Plan period (unit: $10^8$ ton)

| Year | Coal output | Drainage quantity of mine water | Utilization rate of mine water | Reuse rate of mine water (%) |
|---|---|---|---|---|
| 2006 | 23.73 | 46 | 16 | 35 |
| 2007 | 25.36 | 48 | 21.5 | 45 |
| 2008 | 27.88 | 52 | 27 | 52 |

**Table 4.9** Analysis of balance between water supply and demand of the coal bases in China in future in different schemes (0.1 billion t)

| Year | Coal output of the coal base | Scheme | Reuse rate of mine water (%) | Mine water used | Production water demand of the coal base | External water supply necessary for the production of coal bases |
|---|---|---|---|---|---|---|
| 2015 | 20.97 | High | 70 | 27.89 | 31.52 | 3.63 |
| | | Moderate | 60 | 23.91 | | 7.61 |
| | | Low | 50 | 19.92 | | 11.6 |
| 2020 | 22.75 | High | 70 | 30.26 | 34.57 | 4.31 |
| | | Moderate | 60 | 25.94 | | 8.63 |
| | | Low | 50 | 21.61 | | 12.96 |

great differences in the drainage quantities of mine water in the mining areas in different regions in China. Amount of the mine water reused in future is highly uncertain. Thus, it is not proper to use the reclaimed water of mines as the water supply sources of mining areas but make efforts to improve coal mining technologies to minimize mine water inflows.

#### 4.1.2.2 Spatial Suitability of Coal Bases and Water Resources

Regional water resource protection, water environment protection, and groundwater protection are the most important three water resource constraints for development of the coal industry. Suitability of the thirteen big coal bases in China and the constraints are analyzed below.

1. Suitability of coal bases and regional water resources

(1) Water resource conditions of coal bases: China covers a vast geographic area, so that its different regions have quite different natural conditions, economic and social development and water resource conditions. Table 4.10 shows the water resources of the 21 second-class hydrological regions of the thirteen big coal bases.

Viewed from water resource amounts of the second-class hydrological regions in the coal bases, both central Hebei and eastern Shanxi coal bases in the Haihe River basin had the minimum water resource amounts. In other words, both had a poor water resource condition, followed by the coal bases of Shendong, eastern Ningxia, northern Shaanxi, central Shanxi, northern Shanxi, Huanglong, and Henan in the Yellow River basin. The Yunnan–Guizhou coal base in the Yangtze River basin had the largest water resource amount. In other words, it had a good water resource condition. The coal bases in the Huaihe River basin and those in the Huaihe River basin had a moderate water resource condition.

Viewed from per capita water resources of the second-class hydrological regions of the coal bases, the per capita water resources of the Yunnan–Guizhou coal base in the Yangtze River basin and that of the eastern Inner Mongolia coal base in the Songhua River basin were both higher than the national average, i.e., 2114 $m^3$; the per capita water resources of the coal bases in Haihe River basin, Yellow River basin, and Huaihe River basin were all lower than the national average.

Thus, it can be seen that suitability of the big coal bases in China and regional water resources was poor. Viewed from the six hydrological regions of thirteen big coal bases in China, only parts of the Yunnan–Guizhou coal base in the Yangtze River basin and the eastern Inner Mongolia coal base in the Songhua River basin had abundant regional water resources and the remaining eleven coal bases in the Yellow River basin, the Haihe River basin, the Huaihe River basin, and the Liaohe River basin had water resource shortage.

(2) Suitability of coal bases and regional water resource supply capacities: With the quick economic and social development of China, all the basins in China now

**Table 4.10** Water resources of the basins of the thirteen big coal bases in China in the year of 2006

| Name of coal base | Hydrological region | | Annual average water resource amount ($10^8$ $m^3$) | Per capita water resources ($m^3$) |
|---|---|---|---|---|
| | First-class hydrological region | Second-class hydrological region | | |
| Shendong coal base | Yellow River basin | Lanzhou—Hekou Town | 40.1 | 248 |
| Eastern Ningxia coal base | | | | |
| Northern Shaanxi coal base | | Hekou Town—Longmen | 62.8 | 719 |
| Central Shanxi coal base | | | | |
| Northern Shanxi coal base | | | | |
| Huanglong coal base | | Longmen—Sanmenxia | 160.3 | 311 |
| Henan coal base | | Sanmenxia—Huayuankou | 63.1 | 463 |
| Central Hebei coal base and eastern Shanxi coal base | Haihe River basin | Luanhe River and the rivers in the coastal areas of eastern Hebei | 35.1 | 325 |
| | | Northern drainage of Haihe River | 78.2 | 252 |
| | | Southern drainage of Haihe River | 166.5 | 216 |
| | | Tuhai River-Majia River | 28.4 | 164 |
| Western Shandong coal base | Huaihe River basin | Yishusi River | 211.0 | 401 |
| | | Shandong peninsula | 117.0 | 328 |
| Huainan–Huaibei coal base | | Middle reaches of Huaihe River | 371.0 | 446 |
| Yunnan–Guizhou coal base | Yangtze River basin | Lower reaches of Jinsha River from Shigu | 1149.1 | 5477 |
| | | Wujiang River | 551.1 | 2488 |
| | | Yibin-Yichang mainstream | 634.8 | 2013 |
| Eastern Inner Mongolia coal base | Songhua River basin | Argun River | 135.8 | 9836 |
| | Songhua River basin | Songhua River (down from the Sancha River estuary) | 411.6 | 1825 |

(continued)

**Table 4.10** (continued)

| Name of coal base | Hydrological region | | Annual average water resource amount ($10^8$ $m^3$) | Per capita water resources ($m^3$) |
|---|---|---|---|---|
| | First-class hydrological region | Second-class hydrological region | | |
| | | Wusuli River | 101.5 | 2196 |
| | Liaohe River basin | West Liaohe River | 70.2 | 1094 |
| | | Mainstream of the Liaohe River | 69.9 | 750 |
| | | Huntai River | 69.0 | 465 |
| | | Rivers close to Yellow Sea and Bohai Sea in northeastern China | 121.1 | 702 |

face a supply–demand imbalance of water resources more or less. In order to meet the water demands of social and economic development of all the basins and improve the regional water resource supply capacities, all the basins have planned to take effective actions of new resource increasing and consumption reduction such as construction of new water supply projects, water saving, use of unconventional water, and inter-basin water diversion in accordance with their regional water resource characteristics. Information about the supplies and demands of water resources of the second-class hydrological regions of the big coal bases in China in different target years in future can be found in official documents including the *Comprehensive Planning of Water Resources of the Yellow River Basin*, the *Comprehensive Planning of Water Resources of the Haihe River Basin*, the *Comprehensive Planning of Water Resources of the Huaihe River Basin and the Shandong Peninsula*, the *Comprehensive Planning of Water Resources of the Yangtze River Basin and the Rivers in Southwest China (to the West of Lancang River)*, the *Comprehensive Planning of Water Resources of the Songhua River Basin*, the *Comprehensive Planning of Water Resources of the Liaohe River Basin*, etc. Suitability of coal bases and regional water resource supply capacities is analyzed below in accordance with comprehensive planning of water resources of the basins and water demand changes in the thirteen big coal bases in China in different target years, as shown in Tables 4.11 and 4.12.

According to the analysis above, the supplies of water resources of the hydrological regions all increased thanks to actions such as construction of new water supply projects, water saving, use of unconventional water and inter-basin water diversion. Thus, the supply–demand imbalance of water resources in the Twelfth Five-Year Plan period was less serious more or less in all the hydrological regions. If water demands of the coal bases remain unchanged in future, the suitability of the demand and supply of regional water resources will differ too because of large differences in water resource conditions of the hydrological regions.

(1) The Shendong coal base and the eastern Ningxia coal bases both in second-class hydrological regions of the Yellow River basin with a poor water resource condition, as well as parts of the eastern Shanxi coal base of the Haihe River basin tend to have larger water demands while their corresponding regional water resource supply capacities tend to decrease. So the development of the coal bases will make further the supply–demand imbalance of the regional water resources worse. In the year of 2020, these coal bases will still have large supply–demand gaps of water resources, and suitability between suitability of the water demands of these coal bases and the corresponding regional water resource supply capacities will become worst, even with the actions of new resource increasing and consumption reduction above.
(2) The coal bases of northern Shaanxi, central Shanxi, northern Shanxi, Huanglong, and Henan in the second-class hydrological regions of the Yellow River basin and the Huainan–Huaibei coal base in a second-class hydrological

**Table 4.11** Comparison between water demands of coal bases and regional water supplies (unit: $10^8$ $m^3$)

| Coal base name | Hydrological region | | Water supply of hydrological region | | | Water demand of coal base | | |
|---|---|---|---|---|---|---|---|---|
| | First-class hydrological region | Second-class hydrological region | 2010 | 2015 | Increase rate (%) | 2010 | 2015 | Increase rate (%) |
| Shendong coal base | Yellow River basin | Lanzhou—Hekou Town | 160.07 | 158.01 | −2.06 | 10.10 | 11.77 | 1.67 |
| Eastern Ningxia coal base | | | | | | | | |
| Northern Shaanxi coal base | | Hekou Town—Longmen | 19.21 | 21.64 | 2.43 | 3.48 | 3.80 | 0.32 |
| Central Shanxi coal base | | | | | | | | |
| Northern Shanxi coal base | | | | | | | | |
| Huanglong coal base | | Longmen—Sanmenxia | 116.96 | 123.92 | 6.96 | 0.32 | 0.35 | 0.03 |
| Henan coal base | | Sanmenxia—Huayuankou | 28.63 | 31.84 | 3.22 | 1.02 | 1.02 | 0 |
| Central Hebei coal base | Haihe River basin | Luanhe River and the rivers in the coastal areas of eastern Hebei | 208.24 | 222.46 | 14.23 | 2.97 | 2.92 | −0.05 |
| | | Northern drainage of Haihe River | | | | | | |
| | | Southern drainage of Haihe River | | | | | | |
| | | Tuhai River-Majia River | | | | | | |
| Eastern Shanxi coal base | | Southern drainage of Haihe River | 22.02 | 19.46 | −2.56 | 1.77 | 2.23 | 0.46 |
| Western Shandong coal base | Huaihe River basin | Yishusi River | 186.61 | 183.78 | −2.83 | 1.05 | 1.02 | −0.03 |
| | | Shandong peninsula | 88.21 | 93.64 | 5.43 | | | |

(continued)

**Table 4.11** (continued)

| Coal base name | Hydrological region | | Water supply of hydrological region | | | Water demand of coal base | | |
|---|---|---|---|---|---|---|---|---|
| | First-class hydrological region | Second-class hydrological region | 2010 | 2015 | Increase rate (%) | 2010 | 2015 | Increase rate (%) |
| Huainan–Huaibei coal base | | Middle reaches of Huaihe River | 223.58 | 236.99 | 13.41 | 0.74 | 0.81 | 0.07 |
| Yunnan–Guizhou coal base | Yangtze River basin | Lower reaches of Jinsha River from Shigu | 91.05 | 100.85 | 9.80 | 1.53 | 1.94 | 0.41 |
| | | Wujiang River | 62.01 | 67.42 | 5.41 | | | |
| | | Yibin-Yichang mainstream | 97.57 | 103.46 | 5.89 | | | |
| Eastern Inner Mongolia coal base | Songhua River basin | Argun River | 3.74 | 7.26 | 3.52 | 5.40 | 5.68 | 0.28 |
| | | Songhua River (down from the Sancha River estuary) | 133.86 | 141.98 | 8.12 | | | |
| | | Wusuli River | 64.40 | 74.03 | 9.63 | | | |
| | Liaohe River basin | West Liaohe River | 48.12 | 50.50 | 2.38 | | | |
| | | mainstream of the Liaohe River | 41.00 | 42.95 | 1.95 | | | |
| | | Huntai River | 62.12 | 67.11 | 4.99 | | | |
| | | Rivers close to Yellow Sea and Bohai Sea in northeastern China | 39.93 | 41.39 | 1.46 | | | |

**Table 4.12** Supplies and demands of water resources of the thirteen big coal bases in the year 2020

| Coal base name | Hydrological region | | Water demand | Water supply | Water deficiency ratio (%) | Remarks |
|---|---|---|---|---|---|---|
| | First-class hydrological region | Second-class hydrological region | | | | |
| Shendong coal base | Yellow River basin | Lanzhou—Hekou Town | 197.26 | 155.98 | 20.9 | Actions such as water saving of basins, use of unconventional water and construction of new water supply projects are fully considered |
| Eastern Ningxia coal base | | | | | | |
| Northern Shaanxi coal base | | Hekou Town—Longmen | 26.20 | 24.37 | 7.0 | |
| Central Shanxi coal base | | | | | | |
| Northern Shanxi coal base | | | | | | |
| Huanglong coal base | | Longmen—Sanmenxia | 150.94 | 131.29 | 13.0 | |
| Henan coal base | | Sanmenxia—Huayuankou | 37.72 | 35.42 | 6.1 | |
| Central Hebei coal base | Haihe River basin | Luanhe River and the rivers in the coastal areas of eastern Hebei | 237.9 | 221.3 | 7.0 | Actions including water saving of basins and use of unconventional water (seawater, brackish water and reclaimed water) are fully considered |
| | | Northern drainage of Haihe River | | | | |
| | | Southern drainage of Haihe River | | | | |
| | | Tuhai River-Majia River | | | | |
| Eastern Shanxi coal base | | Southern drainage of Haihe River | 17.6 | 17.2 | 2.3 | |
| Western Shandong coal base | Huaihe River basin | Yishusi River | 293.2 | 280.4 | 4.3 | Actions including water saving of basins, improvement of water resource supply capacities and water diversion from other basins are fully considered |
| | | Shandong peninsula | | | | |
| Huainan–Huaibei coal base | | middle reaches of Huaihe River | 263.9 | 251.2 | 4.8 | |

(continued)

**Table 4.12** (continued)

| Coal base name | Hydrological region | | Water demand | Water supply | Water deficiency ratio (%) | Remarks |
|---|---|---|---|---|---|---|
| | First-class hydrological region | Second-class hydrological region | | | | |
| Yunnan–Guizhou coal base | Yangtze River basin | Lower reaches of Jinsha River from Shigu | 298 | 294.7 | 1.1 | Actions including water saving of basins, improvement of water resource supply capacities, better development of water resources, and higher utilization rate of water resources are fully considered |
| | | Wujiang River | | | | |
| | | Yibin-Yichang mainstream | | | | |
| Eastern Inner Mongolia coal base | Songhua River basin | Argun River | 471.8 | 463.2 | 1.8 | Actions including water saving of basins, use of unconventional water, improvement of water resource supply capacities and water diversion between first-class hydrological regions and second-class hydrological regions are fully considered |
| | | Songhua River (down from the Sancha River estuary) | | | | |
| | | Wusuli River | | | | |
| | Liaohe River basin | West Liaohe River | | | | |
| | | Mainstream of the Liaohe River | | | | |
| | | Huntai River | | | | |
| | | Rivers close to Yellow Sea and Bohai Sea in northeastern China | | | | |

region of the Huaihe River basin tend to have larger water demands and higher regional water resource supply capacities because of actions of new resource increasing and consumption reduction, but these coal bases will still have large supply–demand gaps and regional coal development will not be effectively supported. Thus, the suitability between these coal bases has too poor suitability of demand and supply of the corresponding regional water resources. The central Hebei coal base in the second-class hydrological region of the Haihe River basin will have a smaller water demand, but the suitability of demand and supply of the corresponding regional water resources will still be poor because of the serious regional supply–demand imbalance of water resources.

(3) The eastern Inner Mongolia coal base in a second-class hydrological region of the Songhua River–Liaohe River basin and the Yunnan–Guizhou coal base in a second-class hydrological region of the Yangtze River basin have a good water resource condition, and the corresponding regional water resource supply capacities will be improved greatly because of actions such as new resource increasing and consumption reduction. Thus, in spite of the tendency of larger water demands, these coal bases will have much-improved supply–demand imbalance of regional water resources. As a result, these coal bases will have the better suitability of supply and demand of the corresponding regional water resources in future.

Obviously, the serious supply–demand imbalance of water resources in the Twelfth Five-Year Plan period will curb the development of the bases of Shendong, eastern Ningxia, northern Shaanxi, central Shanxi, Huanglong, Henan, central Hebei, northern Shanxi, and eastern Shanxi. In spite of supply–demand imbalance of water resources, both eastern Inner Mongolia and Yunnan–Guizhou coal bases, respectively, in a second-class hydrological region of the Songhua River–Liaohe River basin and the Yangtze River basin both with a good water resource condition will witness continuous improvement of the corresponding regional water resource supply capacities because of actions of new resource increasing and consumption reduction. Both will present a better suitability of supply and demand of regional water resources. So the water demands in their development can be met.

2. Suitability of groundwater amount in coal bases and regions

(1) Groundwater of the Yellow River basin (Table 4.13): According to Table 4.13, the exploitation degrees of regional groundwater in the coal bases differ a lot. From 2006 to 2030, the groundwater in both Huanglong and Henan coal bases was or will be in overwater intaking (their overwater intaking rates in the year of 2006 were 12 and 105%, respectively). The Shendong, eastern Ningxia, northern Shaanxi, central Shanxi, and northern Shanxi coal bases have a certain exploitation potential of groundwater, so they will enjoy a larger planned groundwater yield in the future.

(2) Groundwater of the Haihe River basin: The Haihe River basin has the most serious overdraft of groundwater in China. In the year of 2006, groundwater in its second-class hydrological regions had a total overwater intaking of 8.1 billion $m^3$ including 4.3 billion $m^3$ of shallow groundwater and 3.7 billion $m^3$ of deep confined water. Hydrological regions of the central Hebei coal base and the eastern Shanxi coal base had overdraft of groundwater; the southern drainage of Haihe River basin had the largest overdraft of groundwater (77% of the total regional overwater intaking) in the whole basin. See Table 4.14.

According to the data, the central Hebei coal base shows the largest overdraft of groundwater which has caused some serious environmental geological problems.

**Table 4.13** Exploitation of groundwater of the coal bases in the Yellow River basin (unit: $10^8$ $m^3$)

| Coal base name | Second-class hydrological region | Groundwater reserves | Exploitable yield | Water supplies in different target years | | |
|---|---|---|---|---|---|---|
| | | | | 2006 | 2020 | 2030 |
| Shendong coal base | Lanzhou—Hekou Town | 50.60 | 38.52 | 16.84 | 24.40 | 26.38 |
| Eastern Ningxia coal base | | | | | | |
| Northern Shaanxi coal base | Hekou Town—Longmen | 17.50 | 12.78 | 4.55 | 7.48 | 8.62 |
| Central Shanxi coal base | | | | | | |
| Northern Shanxi coal base | | | | | | |
| Huanglong coal base | Longmen—Sanmenxia | 52.28 | 41.87 | 47.27 | 47.00 | 46.77 |
| Henan coal base | Sanmenxia—Huayuankou | 7.63 | 6.68 | 13.73 | 13.76 | 13.57 |

The Haihe River plain has a shallow overdraft of groundwater area of 60,000 $km^2$, resulting in eleven groundwater depression cones of 18,200 $km^2$ in places including Tangshan, Shunyi-Tongzhou, Fangshan, Baoding, Shijiazhuang, Xingtai, Gaoyi, Handan, Suning, Anyang-Hebi-Puyang, Shen County-Xiajin. The depression cone in southern Shijiazhuang has a maximum length of 52 m. The Haihe River plain had a deep confined water overwater intaking area of 56,000 $km^2$, resulting in seven big deep water depression cones in places including Tangshan, Tianjin, Langfang, Jizhou-Zaoqiang-Taocheng, Xingtai Junxin, Cangzhou, and Dezhou. The depression cones in places such as Jizhou-Zaoqiang-Taocheng, Cangzhou, Dezhou and Xingtai Junxin have formed huge combined depression cone groups. The mining areas in such places in the central Hebei coal base as Handan, Xingtai, and northern Zhangjiakou are distributed in these shallow or deep depression cones.

The overdraft of groundwater in the southern Haihe River plain in the eastern Shanxi coal base caused such eco-geological problems as decrease of spring area, decline of surface water level, groundwater depression cone rivers and land subsidence as well as the social problem of impacting the regional drinking water. Current outflows of big springs in Shanxi such as Niangziguan Spring, Shentou Spring, Pingshang Spring, and Xin'an Spring are only half of their corresponding amounts in the 1950s. Datong Basin and Xinding Basin in Shanxi have a overdraft of groundwater area of 550 $km^2$, resulting in land subsidence. In many places in

**Table 4.14** Exploitation of groundwater of the coal bases in the Haihe River basin in 2006 (unit: $10^8$ $m^3$)

| Coal base name | Second-class hydrological region | Shallow yield | Deep yield | Total yield | Exploitable yield | Shallow overwater intaking | Total overwater intaking |
|---|---|---|---|---|---|---|---|
| Central Hebei | Luanhe River and the rivers in the coastal areas of eastern Hebei | 14.7 | 2.6 | 17.3 | 9.2 | 5.5 | 8.1 |
| | Northern drainage of Haihe River | 32.0 | 4.3 | 36.3 | 28.2 | 3.8 | 8.1 |
| | Southern drainage of Haihe River | 105.6 | 28.9 | 134.5 | 71.8 | 33.8 | 62.7 |
| | Tuhai River–Majia River | 25.2 | 2.4 | 27.6 | 26.3 | 0 | 2.4 |
| Eastern Shanxi coal base | Southern drainage of Haihe River | 105.6 | 28.9 | 134.5 | 71.8 | 33.8 | 62.7 |
| Total ① | | 177.5 | 38.2 | 215.7 | 135.5 | 43.1 | 81.3 |

① The values in the 'Total' line are based on once calculation of the southern drainage of the Haihe River basin

Shanxi, such as Datong and Yangquan, the safety of water use in neighboring urban and rural areas was affected because of exhaustion of the groundwater due to coal mining.

In consideration of the increasingly serious overdraft of groundwater in the Haihe River plain, it is planned in the *Comprehensive Planning of Water Resources of the Haihe River Basin* to basically maintain groundwater supply at 18.5 billion $m^3$ from 2020 to 2030 and decrease the proportion of groundwater supply in the total regional water supply gradually. Obviously, the central Hebei coal base and the eastern Shanxi coal base, particularly the latter, will have a large regional groundwater constraint in their coal mining in future.

(3) Groundwater of the Huaihe River Basin: The Huaihe River basin reveals overdraft of groundwater too. According to related investigation and statistics, the Huaihe River plain now has 85 overdraft of groundwater areas in all with a total overdraft of groundwater area of 42,800 $km^2$ (22 of them have serious overdraft of groundwater and a total underground overwater intaking area of 6979 $km^2$), resulting in 2720 $km^2$ of land subsidence and 509 $km^2$ of surface subsidence. Shandong Peninsula has a seawater intrusion area of 1079 $km^2$. See Table 4.15 for groundwater exploitation of the coal bases of the Huaihe River basin in different target years.

According to Table 4.15, Shandong Peninsula where the western Shandong coal base is had been in overdraft of groundwater and the Yishusi River basin had a

**Table 4.15** Groundwater exploitation of the coal bases of the Huaihe River basin in different target years (unit: $10^8$ $m^3$)

| Coal base name | Second-class hydrological region | Groundwater reserves | Exploitable yield of groundwater | Groundwater supplies in different target years | | |
|---|---|---|---|---|---|---|
| | | | | 2006 | 2020 | 2030 |
| Western Shandong | Yishusi River | 99.9 | 67.7 | 48.9 | 36.6 | 30.3 |
| | Shandong peninsula | 58.9 | 39.7 | 45 | 36.6 | 36.8 |
| Huainan–Huaibei | Middle reaches of Huaihe River | 168 | 79.5 | 75.6 | 66.8 | 72.4 |

certain exploitable yield in 2006; groundwater exploitation in the middle reaches of Huaihe River basin where the Huainan–Huaibei coal base is had been close to the upper limit of the exploitable yield. In the *Comprehensive Planning of Water Resources of the Huaihe River Basin* and Shandong Peninsula, the groundwater yields of the Yishusi River basin, Shandong Peninsula, and the middle reaches of the Huaihe River basin in 2020 and 2030 are required to decrease to be within the exploitable yield. Thus, the coal mining in both western Shandong and Huainan–Huaibei coal bases will be confined by groundwater in the future.

(4) Groundwater of the Songhua River–Liaohe River basin: In the Liaohe River basin, where the eastern Inner Mongolia coal base is, groundwater has been well developed and utilized; shallow groundwater supply has accounted for 44.30% of the total water supply; middle and large-sized cities have suffered serious overdraft of groundwater. For example, Shenyang and Liaoyang in Liaoning Province, Siping in Jilin Province, and Tongliao in Inner Mongolia Autonomous Region have formed extensive groundwater depression cones with a total area of 1102 $km^2$, endangering the local ecological environments as shown in Table 4.16.

According to Table 4.16, most of the hydrological regions in the eastern Inner Mongolia coal base, particularly the rivers close to Yellow Sea and Bohai Sea in northeastern China, had suffered overdraft of groundwater in 2006. In both official documents, the *Comprehensive Planning of Water Resources of the Songhua River Basin* and the *Comprehensive Planning of Water Resources of the Liaohe River Basin*, groundwater yields of the seven hydrological regions where the eastern Inner Mongolia coal base is (including the hydrological regions without overdraft of groundwater at present) will decrease to protect groundwater resources in 2020 and 2030. Thus, the coal mining in the eastern Inner Mongolia coal base will be confined by groundwater amount in the future.

(5) Groundwater of the Yangtze River basin: The Yangtze River basin has an annual average of groundwater resources of 249.2 billion $m^3$, accounting up

**Table 4.16** Groundwater exploitation of the eastern Inner Mongolia coal base in different target years (unit: $10^8$ $m^3$)

| Hydrological region | | Groundwater reserves | Exploitable yield | Groundwater supplies in different target years | | |
|---|---|---|---|---|---|---|
| First-class hydrological region | Second-class hydrological region | | | 2006 | 2020 | 2030 |
| Songhua River basin | Argun River | 43.27 | 5.84 | 2.06 | 1.30 | 1.50 |
| | Songhua River (down from the Sancha River estuary) | 135.82 | 71.29 | 51.45 | 49.10 | 49.27 |
| | Wusuli River | 44.15 | 31.05 | 35.29 | 21.65 | 24.14 |
| Liaohe River basin | West Liaohe River | 53.75 | 39.42 | 41.38 | 30.00 | 32.01 |
| | Mainstream of the Liaohe River | 44.17 | 33.65 | 24.53 | 22.87 | 23.39 |
| | Huntai River | 34.77 | 26.32 | 27.28 | 19.20 | 18.85 |
| | Rivers close to Yellow Sea and Bohai Sea in northeastern China | 33.95 | 13.71 | 17.63 | 11.77 | 11.85 |

30.3% of the national groundwater resources, suggesting it enjoys an abundant groundwater reserve. Therefore, the hydrological regions where the Yunnan–Guizhou coal base is now constrained by groundwater.

According to the analyses above, only the Yunnan–Guizhou coal base in some second-class hydrological regions of the Yangtze River basin, and the bases of Shendong, eastern Ningxia, northern Shaanxi, central Shanxi, and northern Shanxi all in some second-class hydrological regions of the Yellow River basin still have a potential of groundwater exploitation increase and are suitable for the corresponding regional groundwater supply capacities to some extent. The hydrological regions where other coal bases have overdraft of groundwater more or less, and their future groundwater yields will decrease according to the comprehensive planning of water resources of corresponding basins. Accordingly, the related coal bases will have a regional groundwater constraint in their coal production and be less suitable for corresponding regional groundwater supply capacities.

3. Suitability of coal bases and functional areas of groundwater
(1) As a part of water resources, groundwater is not only an important natural resource supporting economic and social development but also an important

part of ecology and environment. It can serve economic and social development and maintain the geological environment balance and the surface ecosystem. In China, particularly on the plains in north China, groundwater is important for drinking water, farm irrigation, industrial production, urban development, and ecology and environment maintenance. Unfortunately, due to overdraft of groundwater, some regions have suffered continuous decline of groundwater level and partial exhaustion of groundwater aquifers, resulting in ecological and environmental problems such as land subsidence, seawater intrusion, and land desertification. Due to excessive drainage of sewage and wastewater and continuous worsening of non-point source pollution, some regions have suffered groundwater quality degradation, which influences the ecosystem and the sustainable utilization of groundwater adversely and curbs regional economic and social development.

In order to give full play to the functions of groundwater, develop, use, and protect groundwater resources rationally and strengthen groundwater management, China Renewable Energy Engineering Institute formulated the *Technical Outline for Comprehensive Planning of Water Resources in China—Planning for Development, Utilization and Protection of Groundwater* in accordance with the natural resource attribute, ecological and environmental attributes, and economic and social attributes of regional groundwater, demand of water resource allocation in planning periods for groundwater development and utilization, and objectives of ecological and environmental protection. In this official document, the functional area of groundwater is divided into two classes. The first-class functional area of groundwater is further divided into development and utilization zone, conservation zone, and reservation zone. The functional area of groundwater in the first class are mainly used for coordinating the relation of the water for economic and social development and ecological and environmental protection, and it suggests the state's general plan for rational development, utilization, and protection of groundwater resources. Each first-class functional area of groundwater is further divided into eight second-class functional areas of groundwater according to the leading role of groundwater. The second-class functional areas of groundwater are mainly used for coordinating the relation of different regions, water users or groundwater functions as shown in Table 4.17.

(2) Suitability of the big coal bases in China and corresponding functional areas of groundwater:

Table 4.18 shows areas of the functional areas of groundwater in the thirteen big coal bases in China.

Groundwater source conservation areas, including ecologically vulnerable areas, geological-disaster-prone areas, and groundwater water source conservation areas, are important conservation areas of groundwater and have the largest water constraint for coal bases. Groundwater development and utilization areas, including water source areas with centralized water supply and decentralized development and utilization areas, allow certain water-related activities and have a relatively

**Table 4.17** Functional area system of groundwater

| First-class functional area | Second-class functional area | Protection objective |
|---|---|---|
| Development and utilization area | Water source area with centralized water supply | (1) Water quality standard: For a water source area with a domestic water supply and centralized water supply, its water quality standard should be up to Class–III Quality Standard for Groundwater (GB/T 14848—1993) at least or, when the current water quality is superior to the class III water standard, the current water quality is used as the control standard; for an area with industrial and centralized water supply, its water quality standard should be the current water quality<br>(2) Water quantity standard: An annual average yield not higher than the exploitable yield<br>(3) Water-level standard: No continuous decline of groundwater levels during groundwater exploitation |
| | Area with decentralized development and utilization | (1) Water quality standard: For an area with domestic water supply, its water quality standard should be up to class III Water Quality standard for Groundwater (GB/T 14848—1993) at least or, when the current water quality is superior to the class III water quality standard, the current water quality is used control standard; for an area with industrial water supply, its water quality standard should be up to Class IV Water Quality Standard for Groundwater (GB/T 14848—1993) at least or, when the current water quality is superior to the Class IV Water Quality Standard, the current water quality is used as the control standard; for an area using groundwater for farm irrigation only, its current water quality or water quality after treatment should comply with the water quality standard for farm irrigation and the current water quality should be the protection objective when it is superior to the class-V water standard<br>(2) Water quantity standard: An annual average yield not higher than the exploitable yield<br>(3) Water-level standard: No continuous decline of groundwater levels, or degradation of groundwater systems or surface ecosystems or environmental geological disasters during groundwater exploitation |
| Conservation area | Ecologically vulnerable area | (1) Water quality standard: For an area with good water quality, maintain its current water quality; for a polluted area, take the natural water quality before the water is polluted in the area as a protection objective<br>(2) Water quantity standard: Control the exploitation intensities in development and utilization and keep the groundwater level high enough to avoid wetland degradation and oasis desertification |

(continued)

**Table 4.17** (continued)

| First-class functional area | Second-class functional area | Protection objective |
|---|---|---|
| | | (3) Water-level standard: Maintain a rational ecological water level to avoid wetland degradation and oasis desertification |
| | Geological-disaster-prone area | (1) Water quality standard: For an area with good water quality, maintain its current water quality; for a polluted area, take the natural water quality before the water is polluted in the area as a protection objective<br>(2) Water quantity standard: Control the exploitation intensities in development and utilization and keep the groundwater level high enough to avoid seawater intrusion, saltwater intrusion, surface subsidence, groundwater pollution and other disasters<br>(3) Water-level standard: Maintain a rational ecological water level to avoid seawater intrusion, saltwater intrusion, surface subsidence, groundwater pollution and other disasters |
| | Groundwater source conservation area | (1) Water quality standard: For an area with good water quality, maintain its current water quality; take the natural water quality before the water is polluted in the area as a protection objective<br>(2) Water quantity standard: Limit groundwater exploitation and keep certain outflows in the spring outcrop areas or ecological basic flows for rivers<br>(3) Water-level standard: During groundwater development, maintain a high groundwater level and keep certain outflows in the spring outcrop areas or ecological basic flows for rivers |
| Reservation area | Area unsuitable for exploitation | Maintain the current situation of groundwater |
| | Reserve area | Maintain the current situation of groundwater |
| | Emergency water source area | Usually, stick to exploitation forbidding and strict protection |

smaller constraint for coal bases. Among the thirteen big coal bases in China, the both Shendong coal base and the eastern Inner Mongolia coal bases have the largest groundwater protection objective constraint because areas of the groundwater source conservation areas in them take a largest proportion in their total areas of the functional zones. The western Shandong coal base, the Huainan–Huaibei coal base, the northern Shaanxi coal base, and the Yunnan–Guizhou coal base have a relatively smaller groundwater protection objective because the groundwater development and utilization areas take a largest proportion in their total area of the functional zones.

**Table 4.18** Areas of the functional areas of groundwater in the thirteen big coal bases in China

| Coal base name | Development and utilization area | | Conservation area | | Reservation area | | Total area of functional areas in coal base (km) |
|---|---|---|---|---|---|---|---|
| | Area ($km^2$) | Proportion in total area (%) | Area ($km^2$) | Proportion in total area (%) | Area ($km^2$) | Proportion in total area (%) | |
| Shendong coal base | 1917.1 | 25.2 | 5374.1 | 70.7 | 315.00 | 4.1 | 7606.2 |
| Eastern Ningxia coal base | 37.5 | 2.4 | 216.9 | 14.1 | 1285.4 | 83.5 | 1539.8 |
| Northern Shaanxi coal base | 9078.3 | 60.5 | 5923.8 | 39.5 | 0 | 0.0 | 15,002.1 |
| Central Shanxi coal base | 6548.1 | 20.9 | 1900.2 | 6.1 | 22,813.4 | 73.0 | 31,261.7 |
| Northern Shanxi coal base | 111.7 | 10.4 | 135.6 | 12.6 | 828.6 | 77.0 | 1075.9 |
| Huanglong coal base | 7413.2 | 40.9 | 10,726.1 | 59.1 | 0 | 0.0 | 18,139.3 |
| Henan coal base | 4903.4 | 48.8 | 4588.1 | 45.6 | 565.1 | 5.6 | 10,056.6 |
| Central Hebei coal base | 20,216.7 | 41.5 | 14,839.1 | 30.5 | 13,625.5 | 28.0 | 48,681.3 |
| Eastern Shanxi coal base | 2743.7 | 20.2 | 0.5 | 0.0 | 10,866.3 | 79.8 | 13,610.5 |
| Eastern Inner Mongolia coal base | 1460.4 | 23.5 | 4627.5 | 74.5 | 122.8 | 2.0 | 6210.7 |
| Western Shandong coal base | 16,610.0 | 83.0 | 2843.9 | 14.2 | 555.6 | 2.8 | 20,009.5 |
| Huainan–Huaibei coal base | 14,303.2 | 60.2 | 9475.8 | 39.8 | 0 | 0.0 | 23,779 |
| Yunnan–Guizhou coal base | 80,743.6 | 59.0 | 43,510.6 | 31.8 | 12,619.4 | 9.2 | 136,873.6 |

4. Suitability of coal bases and water environment

(1) Water functional areas: In order to provide a scientific basis for the development, utilization, protection, and management of waters to realize sustainable utilization of water resources, China Renewable Energy Engineering Institute worked out a water system of the first-class functional area and the second-class functional area in accordance with the natural attributes of corresponding waters, social demand, and the whole-part relationship. The first-class water functional area is divided into conservation area, reservation area, buffer area, and development and utilization area and are mainly used for coordinating the water use relation between different regions to handle the development, utilization, and protection of water resources at macro-level to realize sustainable development; the second-class water functional area based on the development and utilization areas among the first-class functional areas of water include drinking water source area, industrial water areas, agricultural water areas, fishery water areas, landscape and entertainment water areas, transition areas, and pollution control areas and are mainly used for coordinating the water use relation between different water users. Figure 4.1 shows the system about classification system of water functional areas of water and Table 4.19 for the classification standards.

(2) Standards of the water functional areas in the regions where coal bases are located see Table 4.20 for up-to-standard information of the water functional areas in the regions where the thirteen big coal bases in China are located on the basis of official documents such as the *Comprehensive Planning of Water Resources of the Yellow River Basin*, the *Comprehensive Planning of Water Resources of the Huaihe River Basin and the Shandong Peninsula*, the *Comprehensive Planning of Water Resources of the Yangtze River Basin and the Rivers in Southwest China (to the West of Lancang River)*, the

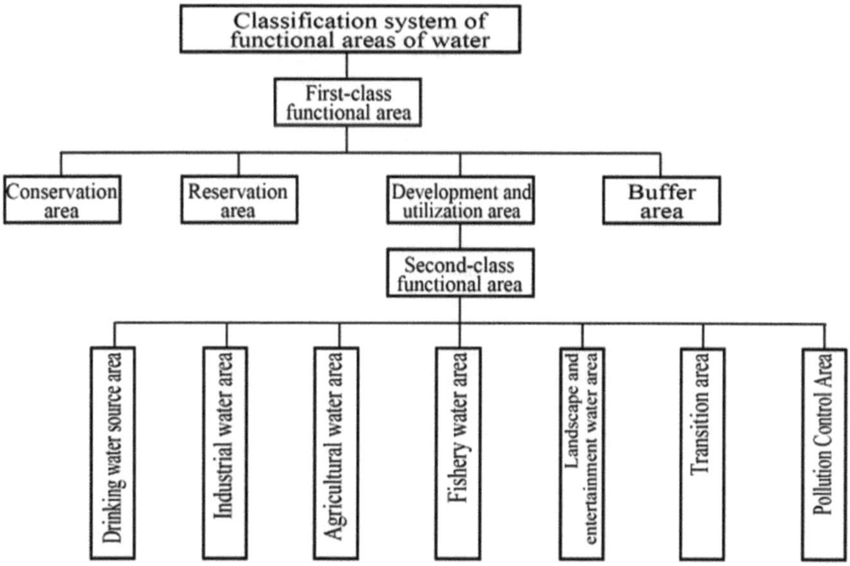


**Fig. 4.1** Classification system of water functional areas

**Table 4.19** Classification standards for functional areas of water

| Functional area of water | Definition | Classification condition | Classification standard | Water quality standard |
|---|---|---|---|---|
| Conservation area | Waters significant for the protection of water resources, natural ecosystems and rare and endangered species | (1) Waters in national or provincial nature reserves or waters in natural habitats and typically significant for ecological protection (2) Water source areas of big inter-basin, inter-provincial, and intra-provincial water diversion projects (3) Source water source conservation areas specially divided from the source reaches of important rivers to converse and protect water sources to protect water resources | Catchment area, water yield, water diversion quantity, protection level, etc. | Refer to the quality standard for water of class I or II or maintain the current water quality. See Environmental quality standards for surface water (GB 3838—2002) |
| | | | Catchment area, water yield, water diversion quantity, protection level, etc. | Refer to the quality standard for water of class I or II or maintain the current water quality. See Environmental quality standards for surface water (GB 3838—2002) |
| Buffer area | Waters for the purposes of coordinating the water use relation between different regions (provinces) with serious water pollution and meeting requirements for the water quality of functional areas | (1) Rivers connecting different provinces, autonomous region or municipalities and the boundary waters of lakes (2) Inter-regional waters with a serious water use contradiction | Waters at the sections of provincial boundaries and waters with a serious water use contradiction | Refer to the related water quality standard or maintain the current situation |

(continued)

**Table 4.19** (continued)

| Functional area of water | Definition | Classification condition | Classification standard | Water quality standard |
|---|---|---|---|---|
| Development and utilization area | Waters capable to meet multiple water use demands such as industrial and agricultural production, urban living, fishery and entertainment | Waters with centralized water-intaking quantity and large water-intaking quantity (such as basins' reaches in important cities) | Waters' water quality, output value, population, water use, etc. | Refer to the related water quality standard for second-class hydrological regions |
| Reservation area | Waters with a currently low degree of development and utilization but reserved for future development and utilization and expected to maintain its current situation | (1) Waters less influenced by human activities and with a lower degree of development and utilization of water resources (2) Waters not suitable for development at present (3) Water resource areas reserved for the purpose of sustainable development | Waters' water quality, output value, population, water use, etc. | Maintain the current water quality |

*Comprehensive Planning of Water Resources of the Songhua River Basin*, and the *Comprehensive Planning of Water Resources of the Liaohe River Basin*. According to the comparison above, among the second-class hydrological regions where coal bases are located, the functional areas of water, rivers, lakes, and reservoirs in the Yellow River basin and the Yangtze River basin have up-to-standard rates higher than those of the Haihe River basin, the Huaihe River basin, and the Songhua River–Liaohe River basin. Thus, the nine coal bases of Shendong, eastern Ningxia, northern Shaanxi, central Shanxi, northern Shanxi, Huanglong, Henan, and Yunnan–Guizhou are more suitable for the corresponding regional water environments than the coal bases of central Hebei, eastern Shanxi, western Shandong, Huainan–Huaibei, and eastern Inner Mongolia.

**Table 4.20** Comparison of up-to-standard of water functions of hydrological regions of coal bases of the current year

| Coal base name | Hydrological region | | Amount (number) | Standard hydrological regions (number) | Standard rate (%) | River length (km) | | | Lake area ($km^2$) | | | Reservoir | |
|---|---|---|---|---|---|---|---|---|---|---|---|---|---|
| | First-class hydrological region | Second-class hydrological region | | | | River length | Standard river length | Standard rate (%) | Lake area | Standard area | Standard rate (%) | Storage capacity ($10^8$ $m^3$) | Up-to-standard rate (%) |
| Shendong coal base | Yellow River basin | Lanzhou—Hekou Town | 84 | 35 | 41.7 | 4955.6 | 2434.6 | 49.1 | 301.2 | 293 | 97.3 | | |
| Ningdong coal base | | | | | | | | | | | | | |
| Northern Shaanxi coal base | | Hekou Town—Longmen | 110 | 49 | 44.5 | 4366.9 | 1961.9 | 44.9 | | | | | |
| Central Shanxi coal base | | | | | | | | | | | | | |
| Northern Shanxi coal base | | | | | | | | | | | | | |
| Huanglong coal base | | Longmen—Sanmenxia | 193 | 83 | 43.0 | 7664.2 | 2894.1 | 37.8 | | | | | |
| Henan coal base | | Sanmenxia—Huayuankou | 74 | 39 | 52.7 | 1940.7 | 1007.6 | 51.9 | | | | | |
| Central Hebei coal base | Haihe River basin | Luanhe River and coastal area of eastern Hebei | 473 | 121 | 26 | | | 25 | | | | | |
| | | Northern drainage of Haihe River | | | | | | | | | | | |
| Eastern Shanxi coal base | | Southern drainage of Haihe River | | | | | | | | | | | |
| | | Tuhai River-Majia River | | | | | | | | | | | |
| | | | 129 | 33 | 25.6 | 3652 | 766.9 | 21 | 1641 | 375.8 | 22.9 | 15.9 | 59.7 |

**Table 4.20** (continued)

| Coal base name | Hydrological region | | Amount (number) | Standard hydrological regions (number) | Standard rate (%) | River length (km) | | | Lake area (km$^2$) | | | Reservoir | |
|---|---|---|---|---|---|---|---|---|---|---|---|---|---|
| | First-class hydrological region | Second-class hydrological region | | | | River length | Standard river length | Standard rate (%) | Lake area | Standard area | Standard rate (%) | Storage capacity ($10^8$ m$^3$) | Up-to-standard rate (%) |
| Western Shandong coal base | Huaihe River basin | Yishusi River basin | | | | | | | | | | | |
| | | Shandong Peninsula | 48 | 14 | 29.2 | 1730 | 423.9 | 24.5 | | | | 22.4 | 69.6 |
| Huainan–Huaibei coal base | | Middle reaches of Huaihe River | 202 | 45 | 22.3 | 5604 | 930.3 | 16.6 | | | | 55 | 24.7 |
| Yunnan–Guizhou coal base | Yangtze River basin | – | 3071 | 2061 | 67.1 | | | | | | | | |
| Eastern Inner Mongolia coal base | Songhua River basin | Argun River | 37 | 5 | 13.5 | 3502.8 | 498.3 | 14.2 | 2171.5 | 2171.5 | 100 | | |
| | | Songhua River (down from the estuary of Sancha River) | 150 | 39 | 26.0 | 7612.4 | 2043.6 | 26.8 | | | | 1.27 | 49.6 |
| | | Wusuli River | 32 | 4 | 12.5 | 2234.4 | 238 | 10.7 | | | | | |
| | Liaohe River basin | Xiliao River | 75 | 26 | 34.7 | 4491.5 | 1655 | 36.8 | | | | | |
| | | Mainstream of Liaohe River | 66 | 18 | 27.3 | 2498.7 | 738.6 | 29.6 | | | | 6.292 | 96.5 |
| | | Huntai River | 89 | 33 | 37.1 | 2153 | 863 | 40.1 | | | | 21.4 | 100.0 |
| | | Rivers close to Yellow Sea and Bohai Sea in northeastern China | 96 | 38 | 39.6 | 2173 | 1104 | 50.8 | | | | 8.01 | 58.6 |

(3) Distribution of the functional areas of water in coal bases: See Table 4.21 for quantities of the first-class hydrological regions in the thirteen big coal bases in China.

According to Table 4.21, both central Hebei and western Shandong coal bases in thirteen big coal bases in China have 60 and 32 first-class water functional areas, respectively, and both bases have 5 and 36 conservation areas with the highest water quality requirement among the first-class water functional areas, respectively, accounting for one-fifth and one-seventh, respectively, of the total conservation areas of the thirteen coal bases. So these coal bases are under the largest constraint by water source conservation areas; the northern Shaanxi and the central Shanxi coal base have two water source conservation areas, respectively, and the eastern Inner Mongolia coal base has one water source conservation area. So these coal bases are under a moderate constraint by water source conservation areas; the Shendong, northern Shanxi, Huanglong, and eastern Shanxi coal bases have no water source conservation areas but a few water functional areas, so these coal bases are under a small constraint by water source conservation areas; the eastern Ningxia coal base has no functional areas of water, so it is under the smallest constraint by water source conservation areas, as shown in Table 4.22.

#### 4.1.2.3 Evaluation of Constraints of Water Resources for Coal Development

1. Constraint level

The constraints of regional water resources for big coal bases mainly lie in water resources protection, groundwater protection, and water environment protection. Thus, water deficiency ratio, groundwater functional areas, and water functional areas are used as three indexes for evaluating the constraints of regional water resources for coal bases.

Weights are respectively given to conservation areas, reservation areas, and development and utilization areas of the groundwater functional areas in accordance with the groundwater utilization objectives of different groundwater functional areas; different weights are respectively given to conservation areas, reservation areas, buffer areas, and development and utilization areas of the water functional areas in accordance with the water environment protection objective of the water functional areas. During constraint evaluation, the three indexes above are given a weight of 33.3%, respectively, in accordance with their degrees of importance for coal bases. See Table 4.23 for assignment standards of the three indexes and influence weights of the constraints. According to expert opinions and existing research achievements, the constraints of water resources for coal bases are divided into three levels in accordance with their intensities. Table 4.24 shows the ranges and characteristics of the three levels and Table 4.25 for the result of overall evaluations of constraints of the coal bases.

**Table 4.21** Quantities of the first-class hydrological regions in the thirteen big coal bases in China

| Hydrological region | Coal base | Conservation area | | Reservation area | | Buffer area | | Development and utilization area | | Sub-total |
|---|---|---|---|---|---|---|---|---|---|---|
| | | Quantity | Proportion in the amount (%) | Quantity | Proportion in the amount (%) | Quantity | Proportion in the amount (%) | Quantity | Proportion in the amount (%) | |
| Yellow River basin | Shendong coal base | 0 | 0 | 1 | 12.5 | 1 | 12.5 | 6 | 75 | 8 |
| | Eastern Ningxia coal base | 0 | 0 | 0 | 0 | 0 | 0 | 0 | 0 | 0 |
| | Northern Shaanxi coal base | 2 | 25 | 0 | 0 | 2 | 25 | 4 | 50 | 8 |
| | Central Shanxi coal base | 2 | 12.5 | 2 | 12.5 | 2 | 12.5 | 10 | 62.5 | 16 |
| | Northern Shanxi coal base | 0 | 0 | 0 | 0 | 2 | 50 | 2 | 50 | 4 |
| | Huanglong coal base | 0 | 0 | 1 | 25 | 1 | 25 | 2 | 50 | 4 |
| | Henan coal base | 1 | 20 | 0 | 0 | 0 | 0 | 4 | 80 | 5 |
| | Sub-total | 5 | 11.1 | 4 | 8.9 | 8 | 17.8 | 28 | 62.2 | 45 |
| Haihe River basin | Central Hebei coal base | 5 | 8.3 | 1 | 1.7 | 12 | 20 | 42 | 70 | 60 |
| | eastern Shanxi | 0 | 0 | 1 | 25 | 1 | 25 | 2 | 50 | 4 |
| | Sub-total | 5 | 7.8 | 2 | 3.0 | 13 | 20.2 | 44 | 69 | 64 |
| Songhua River-Liaohe River basin | Eastern Inner Mongolia coal base | 1 | 33.3 | 0 | 0 | 0 | 0 | 2 | 66.7 | 3 |

(continued)

**Table 4.21** (continued)

| Hydrological region | Coal base | Conservation area | | Reservation area | | Buffer area | | Development and utilization area | | Sub-total |
|---|---|---|---|---|---|---|---|---|---|---|
| | | Quantity | Proportion in the amount (%) | Quantity | Proportion in the amount (%) | Quantity | Proportion in the amount (%) | Quantity | Proportion in the amount (%) | |
| Huaihe River basin | Western Shandong coal base | 6 | 18.8 | 0 | 0 | 2 | 6.2 | 24 | 75 | 32 |
| | Huainan–Huaibei coal base | 1 | 2.8 | 0 | 0 | 4 | 11.2 | 31 | 86 | 36 |
| | Sub-total | 7 | 10.3 | 0 | 0 | 6 | 8.8 | 55 | 80.9 | 68 |
| Yangtze River basin | Yunnan–Guizhou coal base | 0 | 0 | 4 | 57.1 | 0 | 0 | 3 | 42.9 | 7 |
| Total | | 35 | 9.6 | 16 | 4.4 | 54 | 14.8 | 259 | 71.2 | 364 |

**Table 4.22** Water source conservation areas in the coal bases in China

| Hydrological region | Coal base | Name of water source conservation area |
|---|---|---|
| Yellow River basin | Northern Shaanxi coal base | Wuqi water source conservation area of Wuding River |
| | | An'sai water source conservation area of Yanhe River |
| | Central Shanxi coal base | Jingle water source conservation area of Fenhe River |
| | | Qinyuan water source conservation area of Qinhe River |
| | Henan coal base | Jin-Yu nature reserve of Qinhe River |
| Haihe River basin | Central Hebei coal base | Water delivery conservation area of South-to-North Water Diversion Project |
| Songhua River-Liaohe River basin | Eastern Inner Mongolia coal base | Shuangyashan water source conservation area of Anbang River |
| Huaihe River basin | Western Shandong coal base | Taierzhuang water diversion water source conservation area of Hanzhuang Canal |
| | | Jining water diversion water source conservation area of Liangji Canal |
| | | Xuzhou water diversion water source conservation area of a canal |
| | | Zibo water source conservation area of Zihe River |
| | | Water diversion water source conservation area of the lowest section of Nansi Lak |
| | | Water diversion water source conservation area of the highest section of Nansi Lake |
| | Huainan–Huaibei coal base | Dengfeng water source conservation area of Yinghe River |

According to Table 4.25:

(1) In 2006, the four coal bases of Shendong, Huanglong, central Hebei, and eastern Inner Mongolia had a very large constraint; the five coal bases of northern Shaanxi, Henan, eastern Shanxi, Huainan–Huaibei, and Yunnan–Guizhou suffered the large constraint of water resources; the four coal bases of eastern Ningxia, central Shanxi, northern Shanxi, and western Shandong had less constraint. If ranked by the level of constraint (from higher ones to lower ones), the coal bases had the following sequence: Shendong > central Hebei > eastern Inner Mongolia > Huanglong > Huainan–Huaibei > Henan > eastern Shanxi > northern Shaanxi > Yunnan–Guizhou > eastern Ningxia > central Shanxi > northern Shanxi > western Shandong.
(2) In 2020, both Shendong and Huanglong coal bases will suffer very large constraint of water resources; the northern Shaanxi, Henan, and the eastern Inner Mongolia coal bases will receive large constraint; the eastern Ningxia, central Shanxi, northern Shanxi, central Hebei, eastern Shanxi, western

**Table 4.23** Evaluation indexes for the constraints of regional water resources for coal bases and their assignment

| Constraint | Constraint evaluation index | Basis of constraint evaluation | Functional area | Assignment standard | Influence weight (%) |
|---|---|---|---|---|---|
| Water resource protection | Water deficiency ratio | – | – | 1 point for each 1% | 33.3 |
| Groundwater protection | Functional area of groundwater | Percent of the area of each type of functional areas in total area of all the functional areas | Conservation area | 0.5 point for each 1% | 33.3 |
| | | | Reservation area | 0.3 point for each 1% | |
| | | | Development and utilization area | 0.2 point for each 1% | |
| Water environment protection | (Surface) Functional area of water | Percent of the quantity of each type of functional areas in amount of all the functional areas | Conservation area | 0.4 point for each 1% | 33.3 |
| | | | Reservation area | 0.3 point for each 1% | |
| | | | Buffer area | 0.2 point for each 1% | |
| | | | Development and utilization area | 0.1 point for each 1% | |

**Table 4.24** Levels of the constraints of water resources for coal bases

| Constraint level | Value range | Characteristic |
|---|---|---|
| Very large constraint | 22–26 | (1) Extremely poor suitability of the coal base layout and the regional water resources; a big water resource gap<br>(2) A significant impact on groundwater protection<br>(3) A very large constraint by water environment functional areas;<br>(4) Avery large constraint of regional water resources for coal bases |
| Large constraint | 18–22 | (1) Poor suitability of the coal base layout and the regional water resources; a big water resource gap, which can be improved with actions<br>(2) A significant impact on groundwater protection<br>(3) A large constraint by water environment functional areas<br>(4) A moderate constraint of regional water resources for coal bases |
| Small constraint | 14–18 | (1) Good suitability of the coal base layout and the regional water resources (usually, the regional water resources can meet the water demands)<br>(2) A small impact on groundwater protection<br>(3) A small constraint by water environment functional area<br>(4) A small constraint of regional water resources for coal bases (in the general conditions) |

**Table 4.25** Overall scores of the constraints of regional water resources on the thirteen big coal bases in China

| Coal base name | Water resource shortage in 2006 (%) | Water resource shortage in 2020 (%) | Functional area of groundwater (%) | | | | Functional area of surface water (%) | | | | | 2006 | | 2020 | |
|---|---|---|---|---|---|---|---|---|---|---|---|---|---|---|---|
| | | | Conservation area | Development and utilization area | Reservation area | Total | Conservation area | Reservation area | Buffer area | Development and utilization area | Total | Score | Level of constraint | Score | Level of constraint |
| Shendong coal base | 21.10 | 20.90 | 5.04 | 35.35 | 1.23 | 41.62 | 0.00 | 3.75 | 2.50 | 7.5 | 13.75 | 25.24 | Very large | 25.17 | Very large |
| Ningdong coal base | 21.00 | 20.90 | 0.48 | 7.05 | 25.05 | 32.58 | 0.00 | 0.00 | 0.00 | 0.00 | 0.00 | 17.68 | Small | 17.65 | Small |
| Northern Shaanxi coal base | 4.00 | 7.00 | 12.1 | 19.75 | 0.00 | 31.85 | 10.00 | 0.00 | 5.00 | 5.00 | 20.00 | 18.43 | Large | 19.42 | Large |
| Central Shanxi coal base | 4.00 | 7.00 | 4.18 | 3.05 | 21.9 | 29.13 | 5.00 | 3.75 | 2.50 | 6.25 | 17.50 | 16.71 | Small | 17.70 | Small |
| Northern Shanxi coal base | 4.00 | 7.00 | 2.08 | 6.30 | 23.1 | 31.48 | 0.00 | 0.00 | 10.00 | 5.00 | 15.00 | 16.66 | Small | 17.65 | Small |
| Huanglong coal base | 14.40 | 13.0 | 8.18 | 29.55 | 0.00 | 37.73 | 0.00 | 7.50 | 5.00 | 5.00 | 17.50 | 22.98 | Very large | 22.52 | Very large |
| Henan coal base | 7.00 | 6.10 | 9.76 | 22.80 | 1.68 | 34.24 | 8.00 | 0.00 | 0.00 | 8.00 | 16.00 | 18.89 | Large | 18.59 | Large |
| Central Hebei coal base | 29.10 | 7.00 | 8.3 | 15.25 | 8.4 | 31.95 | 3.32 | 0.51 | 4.00 | 7.00 | 14.83 | 25.04 | Very large | 17.75 | Small |
| Eastern Shanxi coal base | 11.10 | 2.30 | 4.04 | 0.00 | 23.94 | 27.98 | 0.00 | 7.50 | 5.00 | 5.00 | 17.50 | 18.67 | Large | 15.77 | Small |
| Eastern Inner Mongolia coal base | 12.60 | 1.80 | 4.70 | 37.25 | 0.60 | 42.55 | 13.32 | 0.00 | 0.00 | 6.67 | 19.99 | 24.80 | Very large | 21.23 | Large |
| Western Shandong coal base | 9.45 | 4.30 | 16.6 | 7.10 | 0.84 | 24.54 | 7.52 | 0.00 | 1.24 | 7.50 | 16.26 | 16.58 | Small | 14.88 | Small |

(continued)

**Table 4.25** (continued)

| Coal base name | Water resource shortage in 2006 (%) | Water resource shortage in 2020 (%) | Functional area of groundwater (%) | | | | Functional area of surface water (%) | | | | | 2006 | | 2020 | |
|---|---|---|---|---|---|---|---|---|---|---|---|---|---|---|---|
| | | | Conservation area | Development and utilization area | Reservation area | Total | Conservation area | Reservation area | Buffer area | Development and utilization area | Total | Score | Level of constraint | Score | Level of constraint |
| Huainan–Huaibei coal base | 13.80 | 4.80 | 12.04 | 19.90 | 0.00 | 31.94 | 1.12 | 0.00 | 2.24 | 8.60 | 11.96 | 19.04 | Large | 16.07 | Small |
| Yunnan–Guizhou coal base | 3.60 | 1.10 | 11.8 | 15.90 | 2.76 | 30.46 | 0.00 | 17.10 | 0.00 | 4.29 | 21.39 | 18.30 | Large | 17.47 | Small |

Shandong, Huainan–Huaibei, and Yunnan–Guizhou coal bases will bear small constraint. If ranked by the level of constraint (from higher ones to lower ones), the coal bases had the following sequence: Shendong > Huanglong > eastern Inner Mongolia > northern Shaanxi > Henan > central Hebei > central Shanxi > eastern Ningxia > northern Shanxi > Yunnan–Guizhou > Huainan–Huaibei > eastern Shanxi > western Shandong.

(3) From 2006 to 2020, the central Hebei, eastern Inner Mongolia, Huainan–Huaibei, eastern Shanxi, western Shandong, Yunnan–Guizhou, Huanglong, Henan, Shendong, and eastern Ningxia coal bases present an increasingly smaller water resource constraint while the central Shanxi, northern Shaanxi, and northern Shanxi coal bases demonstrate an increasingly larger water resource constraint.

2. Constraint type

There are different levels of water resource constraints for the development of coal bases. However, it is seen that the water resource constraints for different coal bases have different characteristics. By characteristic, water resource constraints can be classified into water quantity constraint, water environment constraint, and overall constraint.

(1) Water quantity constraints: The coal bases with these constraints are characterized by serious supply–demand imbalance of regional water resources, highly insufficient supply of regional water resources, relatively smaller water environment constraint, and main limitation as water resources. The eastern Ningxia, eastern Shanxi, central Shanxi, northern Shanxi, Xinjiang, and Yunnan–Guizhou coal bases are all of this sort.
(2) Water environment constraints: The coal bases with these constraints are characterized by many regional water functional areas, serious damage of coal mining to groundwater, relatively more abundant water resources, and main limitation as water environment protection. The eastern Inner Mongolia, western Shandong, and the Huainan–Huaibei coal bases are all of this sort.
(3) Overall constraints: The coal bases with these constraints are characterized by certain gaps of regional water resources, many regional functional areas of water, and two limitations as water quantity and water environment. The Shendong, Huanglong, Henan, northern Shaanxi, and central Hebei coal bases are of this sort.

According to research achievements about water quantity constraints and water environment constraints, levels, and types of the water resource constraints for the coal bases are shown in Table 4.26.

**Table 4.26** Characteristics of water resource constraints of the coal bases

| Coal base name | Constraint level | | Constraint type |
|---|---|---|---|
| | 2006 | 2020 | |
| Shendong coal base | Very large constraint | Very large constraint | Overall constraint |
| Eastern Ningxia coal base | Small constraint | Small constraint | Water quantity constraint |
| Northern Shaanxi coal base | Large constraint | Large constraint | Overall constraint |
| Central Shanxi coal base | Small constraint | Small constraint | Water quantity constraint |
| Northern Shanxi coal base | Small constraint | Small constraint | Water quantity constraint |
| Huanglong coal base | Very large constraint | Very large constraint | Overall constraint |
| Henan coal base | Large constraint | Large constraint | Overall constraint |
| Central Hebei coal base | Very large constraint | Small constraint | Overall constraint |
| Eastern Shanxi coal base | Large constraint | Small constraint | Water quantity constraint |
| Eastern Inner Mongolia coal base | Very large constraint | Large constraint | Water environment constraint |
| Western Shandong coal base | Small constraint | Small constraint | Water environment constraint |
| Huainan–Huaibei coal base | Large constraint | Small constraint | Water environment constraint |
| Yunnan–Guizhou coal base | Large constraint | Small constraint | Water quantity constraint |

## 4.1.3 Countermeasures

### 4.1.3.1 Countermeasures for Reducing Water Resource Constraints for Coal Bases

The water resource constraints for different coal bases have different characteristics, so they should be reduced by different countermeasures which mainly include the following ones.

1. Inter-basin water diversion

Among the thirteen big coal bases in China, the Shendong, eastern Ningxia, northern Shaanxi, central Shanxi, northern Shanxi, Huanglong, and Henan coal bases all have water resource constraints and some of them have overdraft of groundwater at present and a small potential of surface water development and utilization. In the next 30 years, construction of the water and soil conservation project and the development and utilization of groundwater on the Loess Plateau will affect the changes of the runoff yield and runoff concentration relation in a unfavorable direction. Thus, the supply–demand imbalance of regional water resources will be very serious within a foreseeable period. There are two countermeasures taken for this imbalance: strengthening the rational allocation of regional water resources and the utilization of unconventional water, which is necessary for the Shendong coal base and the eastern Ningxia coal base (particularly the former); taking advantage of the national inter-basin water diversion projects (such as east and middle routes of the South-to-North Water Diversion Project) as an effective action to improve the water resource shortage of coal bases.

A Circular of the Report on the Plan for Distribution of Available Water of Yellow River jointly prepared by former National Planning Commission (now National Development and Reform Commission) and former Ministry of Water Resources and Electrical Power was forwarded by the General Office of the State Council on September 11, 1987. That official document (G.B.F. [1987] No. 61), which is the so-called Water Diversion Plan in 1987 for Yellow River was issued to all provinces (municipalities and autonomous regions) as the distribution plan for the available water of Yellow River before the South-to-North Water Diversion Project was initiated. According to the statistical data from 1980 to 2000, the Yellow River basin reaches below Sanmenxia had an annual water supply of 10.8 billion $m^3$ to other places out of the basin, including 1.936 billion $m^3$ to Henan, 8.820 billion $m^3$ to Shandong, and 0.076 billion $m^3$ to Tianjin. Table 4.27 shows the supplies and demands of water resources in the hydrological regions of the Yellow River basin where the coal bases are located after the east and middle routes of South-to-North Water Diversion Project come into practice in 2020 and 2030, respectively.

Thus, east and middle routes of South-to-North Water Diversion Project will improve water resource supply capacities of the above regions greatly and reduce water deficits of the regions significantly to meet water demands of the coal bases effectively. According to the *Comprehensive Planning of Water Resources of the Yellow River Basin*, the South-to-North Water Diversion Project is taken into effect, the 'energy bases' of the Yellow River basin, including the energy and chemical industry base in eastern Ningxia, the Ordos energy and chemical industry base, the Yulin energy and chemical industry base in northern Shaanxi and the Liliu energy base in Shanxi, will have serious water resource shortage because of insufficient water diversion from the Yellow River as a result of water allocation of the Yellow River for related provinces (municipalities and autonomous regions). After the South-to-North Water Diversion Project comes into product, the water demands of these 'energy bases' will be basically met with a surface water supply increased to 5.06 billion $m^3$ and a decrement of water resource shortage of 0.22 billion $m^3$ because of increase of the water diversion from the Yellow River following service of the west line of South-to-North Water Diversion Project.

2. Water right transfer

Increasing the water resource supply of an industry in a region by means of water right transfer is a much-concerned water resource countermeasure in recent years. In the year of 2005, Ordos had an annual unit use benefit of 2.88 yuan in agricultural water, and 125.04 yuan in industrial water. The latter is 43.4 times of the

**Table 4.27** Supplies and demands of water resources in the hydrological regions of the Yellow River basin where the coal bases are located after the east and middle routes of South-to-North Water Diversion Project come into practice in 2020 and 2030, respectively (unit: $10^8$ $m^3$)

| Coal base name | Yellow River basin Second-class hydrological region | Now to service of the east route and the middle route | | | Service of the east route and the middle route to service of the west line | | | After service of the west route | | |
|---|---|---|---|---|---|---|---|---|---|---|
| | | Water demand | Water supply | Water deficit | Water demand | Water supply | Water deficit | Water demand | Water supply | Water deficit |
| Shendong coal base | Lanzhou—Hekou Town | 205.04 | 152.58 | 52.46 | 197.26 | 157.80 | 39.46 | 205.63 | 194.75 | 10.87 |
| Eastern Ningxia coal base | | | | | | | | | | |
| Northern Shaanxi coal base | Hekou Town—Longmen | 18.21 | 15.91 | 2.30 | 26.20 | 23.48 | 2.72 | 32.37 | 32.07 | 0.29 |
| Central Shanxi coal base | | | | | | | | | | |
| Northern Shanxi coal base | | | | | | | | | | |
| Huanglong coal base | Longmen—Sanmenxia | 130.54 | 121.00 | 9.54 | 150.94 | 134.62 | 16.32 | 158.29 | 153.35 | 4.97 |
| Henan coal base | Sanmenxia—Huayuankou | 28.27 | 28.22 | 0.05 | 37.72 | 37.23 | 0.48 | 40.98 | 39.59 | 1.39 |
| Total | | 382.06 | 317.71 | 64.35 | 412.12 | 353.13 | 58.98 | 437.27 | 419.76 | 17.52 |

former. In the same year, Ordos initiated the Phase-I water right transfer of Yellow River. Three years later, the project resulted in huge economic and social benefits: An agricultural water right of 0.13 billion $m^3$ was successfully transferred to an industrial water right. In 2009, a plan about the transfer of a water right of 0.10 billion $m^3$ was approved. Thus, water right transfer is an effective means to reduce the water resource constraints for coal bases.

3. Water resource allocation

Water resource allocation is a scientific distribution of limited and different water resources to the water users in a specific basin or region in an effective, fair and sustainable way with help of engineering measures and non-engineering measures. The rule of water resource allocation between different industries is giving priority to domestic water while giving consideration to production water and ecological water. Water resource allocation can realize a balance among regional demands for domestic water, production water, and ecological water, and it can make the society, the economy and the environment develop in a sustainable and coordinated way.

Table 4.28 shows the water resource allocation among different industries in the second-class hydrological regions of the Huaihe River basin where the coal bases are located.

According to Table 4.28, the water supply volume allocated to the industry in the regions where both Huainan–Huaibei and western Shandong coal bases are located increases year by year from the base year to 2030, with a continuous annual increase of total regional water supply capacities. Especially in the western Shandong coal base, the industrial water takes up a greatest proportion in the water allocation of industries.

4. Water-preserved mining technology

The water-preserved mining is the first choice for protecting regional water resources in coal mining. Water-preserved mining is expected to protect groundwater resources and use mine water as a resource in coal mining. Its main content tasks are to conduct the comprehensive exploration for coal resources, water resources and ecological environment, ascertain the spatial distribution features of water resources, mark off different subareas, and distinguish the areas where the coal mining doesn't affect environment (no surface subsidence or groundwater level decline because of coal mining), coal mining subsidence area and transition area by indoor simulation experiments on the basis of related theories and methods of sciences such as coal field geology, hydrological geology, engineering geology, environmental geology, coal mining science, and rock mass mechanics. Moreover, it is used to plan and arrange the mining areas (coal fields), and guide coal enterprises to take different methods and corresponding engineering measures in coal mining, so as to minimize the impact of coal mining on both groundwater and the ecological environment. Before planning for the coal mining in coal bases, it is necessary to prepare detailed water-preserved mining plans to protect regional water

**Table 4.28** Water resource allocation among different industries in the second-class hydrological regions of the Huaihe River basin with some coal bases (unit: $10^8$)

| Coal base | Hydrological region | Target year | Water resource allocation among different water users | | | | | | | | Total |
|---|---|---|---|---|---|---|---|---|---|---|---|
| | | | Domestic water | | Production water | | Agricultural water | | Ecological water | | |
| | | | Quantity | Percent in the amount allocated (%) | Quantity | Percent in the amount allocated (%) | Quantity | Percent in the amount allocated (%) | Quantity | Percent in the amount allocated (%) | |
| Huainan–Huaibei coal base | Middle reaches of Huaihe River | Base year | 24.9 | 11 | 38.8 | 18 | 152.1 | 70 | 1.7 | 1.0 | 217.4 |
| | | 2020 | 41.6 | 16 | 49.7 | 19 | 165.0 | 64 | 3.1 | 1.0 | 259.5 |
| | | 2030 | 50.6 | 18 | 52.0 | 19 | 173.9 | 62 | 3.6 | 1.0 | 280.1 |
| Western Shandong coal base | Yishusi River basin | Base year | 17.1 | 10 | 23.7 | 13 | 136.9 | 77 | 0.5 | 0 | 178.2 |
| | | 2020 | 28.6 | 15 | 31.3 | 16 | 129.3 | 68 | 1.2 | 1.0 | 190.4 |
| | | 2030 | 33.8 | 17 | 34.9 | 18 | 127.4 | 64 | 1.6 | 1.0 | 197.7 |
| | Shandong Peninsula | Base year | 11.8 | 14 | 17.2 | 20 | 56.4 | 66 | 0.4 | 0 | 85.8 |
| | | 2020 | 21.7 | 22 | 22.7 | 23 | 55.1 | 55 | 0.7 | 1.0 | 100.3 |
| | | 2030 | 27.6 | 25 | 24.5 | 22 | 59.1 | 53 | 1.0 | 1.0 | 112.2 |

environments and groundwater systems effectively. Because of its great significance for reducing the constraints of water resource supply capacity and water environment of China for the big coal bases in China, this technology is applicable to the development of all the big coal bases in China.

#### 4.1.3.2 Water Resource Management Actions for Meeting the Water Demand of Coal Bases

(1) In the regions with water resource constraints, speed up the establishment of coal development plan-related water resource argumentation system and improve the water resource argumentation systems of energy development projects.

The water resource argumentation system is an important measure for improving the source control of water resource development and utilization. It is a complicated systemic work and a specific measure for determining water outputs and development scales. Its ultimate aim is to coordinate economic development scales and water resource carrying capacities. The water resource argumentation system is useful for better-unified management and sustainable utilization of water resources, water use as planned, water saving, etc. Thus, for the coal bases in regions with very large or large water resource constraints, it is necessary to speed up the establishment of their water resource argumentation systems, and further improve the water resource argumentation systems of energy developments. By this way, it can make sure the degree of development and utilization of water resources in the basins where the coal bases are located doesn't exceed the water resource carrying capacities of the basins, and profit-pursuing predatory development and competitive use of water resources can be absolutely avoided.

(2) Improve the water use quota standard systems for the coal industries in different provinces and autonomous regions and make the strict water quota management for the coal industry and coal enterprises.

The water use quota is a scientific basis for measuring the levels of water use by water users, tapping the potential of water saving and evaluating the effect of water saving. It is practically significant for the establishment of a water-saving society, reduction of water resource shortage, sustainable utilization of water resources, and sustainable development of the national economy. At present, the water use quota standard system for the coal industry in China is still imperfect and some provinces and municipalities have no water quota standards for their coal industries. For better supervision over water exploitation, utilization, and consumption, it is necessary to improve the water use quota standard systems for the coal industries in different provinces and autonomous regions as soon as possible in accordance with the regional water resource conditions, make the water quota management for both the coal industry and coal enterprises stricter, and strengthen the supervision over development and utilization of water resources.

(3) Speed up division of the functional areas of groundwater in main coal bases and work out the rules for groundwater protection in the coal production of different functional areas of groundwater.

Speed up the division of the groundwater functional areas in the regions where main coal bases are located in accordance with the system of functional area of groundwater in China. Work out the rules for groundwater protection in the coal production of different groundwater functional areas, and propose opinions about water resource protection in the coal mining of these functional areas in accordance with their functions. In both exploitation-forbidden areas and conservation areas, coal mining should be limited strictly or avoided.

(4) Work out the technical standards for the zoning and classification of utilization of the mine water in China, strengthen supervision over the utilization of mine water, and prepare technical standards for the utilization rates of mine water in accordance with different levels of water shortage of different regions.

Improving higher treatment rate and reuse rate of mine water are two effective methods to solve supply–demand imbalance and pollution of the water resources in mining areas. Mine water treatment with economical, rational, simple, and efficient methods can increase the water resources of mining areas and reduce environmental pollution caused by mine water drainage. The mining areas in China have quite different mine water quality: Some mine water is up to standard and can be utilized after simple treatment. However, some mine water is highly mineralized and can be utilized only after complicated treatment. It is therefore necessary to formulate the technical standards for the zoning and classification of utilization of the mine water in China in accordance with characteristics of mine water quality of the mining areas, strengthen supervision over the utilization of mine water, and prepare technical standards for the utilization rates of mine water in accordance with different levels of water shortage of different regions. By this way, it can promote the change of mine water in China into resources.

(5) Improve the technical standards for water drainage by coal enterprises and make the strict supervision over water drainage by coal enterprises. The wastewater is drained by coal mines with lower pH value, higher content of sulfate ions, higher hardness, higher content of total dissolved solids, higher iron ion content, more toxicologic components, and over-limit contents of Hg, Pb, Cr, F, and Mn. Direct drainage of the wastewater of coal mines without treatment results in pollution and quality degradation of regional groundwater and surface water. Thus, it is necessary to improve the technical standards for water drainage by coal enterprises and make strict supervision over the water drainage by coal enterprises so that the regional surface water and groundwater as well as regional water environment and ecology are not contaminated by the water drained in the coal production of coal enterprises.
(6) Strengthen water market construction in the regions with serious water shortage and improve the water safety in coal production by means of market allocation.

Water market is an institutional environment for the effective management of water resources with help of an economic mechanism. Water right market construction can improve the efficiency of water resources allocation among different water users or regions, and realize the optimal distribution of water resources in the socioeconomic system. For each region with serious water shortage where coal base (s) is located, the related government should establish a tradable water right system under effective control, speed up the construction of water market, and improve the water safety in coal production by means of market allocation, so as to realize coordinated development of resources, environment and society.

## 4.2 Constraints of Water Resources for the Development of Thermal Power

Due to the acceleration of its industrialization and urbanization, China now has a large increase in its power demand. Studies show that from 2015 to 2020, the generation in China will have an annual increase rate of 5.7% and, by the year of 2020, the total power consumption in its economic and social development and installed capacity of the power industry will reach 7.66 trillion kWh and 1.85 billion kW, respectively. In China, the main source of the power industry is coal, and the main form of power generation is thermal power generation. In the year of 2012, thermal power took up 78.7% of the total generation, and both its installed capacity and generation had a tendency of increase. Thus, studies about the constraints of water resources for the development of thermal power are significant for the normal development of thermal power.

### *4.2.1 Relation Between the Development of Thermal Power and Water Resources*

The production of thermal power plants is a process of energy conversion, where, the chemical energy of fuels (such as coal, petroleum or natural gas) is converted into thermal energy by means of combustion. The thermal energy is then transferred to the water in boilers to change water into steam with certain pressures and temperatures. After that, steam is sent to steam turbines to be expanded to convert the thermal energy into mechanical energy to drive steam turbine rotors to rotate. The rotating steam turbine rotors further drive generator rotors to rotate quickly to convert the mechanical energy into electric energy which is then transmitted to grids. According to the process of energy conversion above, the steam generated by water directly or after absorption of thermal energy is the working medium of thermal systems because of its important role in energy transmission. In addition, water is used for cooling the steam discharged by the turbines and the bearings of rotating equipment in the production of thermal power plants. Furthermore, water is necessary for cleaning in the flue gas desulfuration system and the coal transportation trestle spray system based on the wet process or semi-dry method. Thus, water is almost necessary for each link of the production in thermal power plants because both the working medium and the cooling medium are water or vaporous water.

#### 4.2.1.1 Water Demand Characteristics of Thermal Power

According to current operation level of the big thermal power plants in China, a thermal power plant with an installed capacity of 1000 MW can consume the fresh water supply at 35–40 $m^3$/s when a once-through water supply system is used, or 0.6–1 $m^3$/s when a circulating water supply system is used. To sum up, water is mainly used for the unit cooling, boiler water supplementation, removal of dust, ash and slag, domestic use, and fire control in thermal power plants. In recent years, more and more coal-fired power plants have begun to use fume gas desulfuration equipment because of the increasingly serious acid rain, which is a new link of water use.

The production of thermal power plants needs sufficient water within acceptable quality. Quality of the water used in the production of thermal power plant differs with application of the water. However, the general rule is that the sediments and scales in water supply systems or wear and corrosion of metallic parts should be avoided as much as possible. It is necessary to remove aquatic plants, foreign matters and coarse and hard sand in the condenser cooling water when a once-through water supply system is used. Some actions should be taken against aquatic organisms when seawater is used so as to avoid an excessively high content of SS when water is used for other applications. The boiler is required to make up for water in a high quality, and provided with high-quality, stable raw water as much as possible.

#### 4.2.1.2 Impacts of the Development of Thermal Power on Water Resources

In China, the industrial water is mainly consumed by thermal power, textile industry, petroleum chemical, papermaking, and metallurgy. According to statistical data of the year of 1999, the total water intaking quantity of these five industries was 77.5 billion $m^3$, accounting for 66.6% of total water intaking quantity of all industries. In other words, these five industries consumed two-thirds of total industrial water of the whole country. Among these five industries, thermal power is the one with the largest water intaking quantity and water consumption. Related statistical data show that the existing thermal power plants with a once-through cooling system in China have an average water intaking quantity of 40 $m^3$/s at a

1000 MW installed capacity. In the year of 2000, the water intaking quantity in all forms by thermal power plants came to 84.05 billion $m^3$, including 54.04 billion $m^3$ (including brackish water at the estuaries of rivers) of fresh water for once-through cooling systems.

Due to vulnerable water ecology because of serious water shortage, northern China with a large power load and many thermal power plants suffered very large constraints of water resources for the development of thermal power. The installed capacity of thermal power of the Yellow River basin was 26,260,000 kW in the year of 2002 and will be 147,310,000 and 176,310,000 kW, respectively, in 2020 and 2030. The installed capacity in 2030 will be 150,050,000 kW more than that in 2000. According to the forecasts in the *Comprehensive Planning of Water Resources of the Yellow River Basin*, the water demand of the thermal power industry of the Yellow River basin in the years of 2020 and 2030 will get to 1.41 and 1.55 billion $m^3$, respectively, 91% more than that in 2006, i.e., an increase of 2.2% per year; the water demand quota in 2030 will come to 221,000 $m^3$/kW less than that in the base year (the water intaking quotas in 2020 and 2030 are 96,000 and 88,000 $m^3$/kW, respectively). In 2020 and 2030, water deficit of the Yellow River basin will reach 7.894 and 10.420 billion kW, respectively. Thus, a support effect of the water resources in China does not comply with quick development of the thermal power industry in China, and the water resource is a largest constraint for development of the thermal power industry.

## 4.2.2 Constraints

### 4.2.2.1 Current Situation and Planning of the Development of Thermal Power

Table 4.29 shows the installed capacities of thermal power in different regions in China in different target years based on statistical data and forecast data from the China Electricity Council.

From 2008 to 2015, the national installed capacities of thermal power in most regions in China, except for east China, had been increasing at 3.3% each year on average; from 2015 to 2020, the installed capacities of thermal power in all regions in China will increase at 5.6% each year on average. On the whole, the installed capacity of thermal power in China has been increasing. In the Eleventh Five-Year Plan period, it increased quickest in northeast China, north China, and northwest China at a rate higher than the national average. After the Eleventh Five-Year Plan period, it increased quickest in northwest China at a rate higher than the national average.

**Table 4.29** Installed capacities of thermal power in different regions in China in different target years (unit: 10,000 kW)

| Region | Province | 2006 | 2007 | 2008 | 2015 | 2020 | Annual average increase rate between 2008 and 2015 (%) | Annual average increase rate between 2015 and 2020 (%) |
|---|---|---|---|---|---|---|---|---|
| Whole country | – | 48,382 | 55,607 | 60,286 | 89,200 | 116,900 | 3.3 | 5.6 |
| North China | Beijing | 398 | 390 | 476 | 28,300 | 36,000 | 7.2 | 4.9 |
| | Tianjin | 654 | 692 | 749 | | | | |
| | Hebei | 2609 | 2902 | 2987 | | | | |
| | Shanxi | 2666 | 3095 | 3525 | | | | |
| | Inner Mongolia | 2892 | 3987 | 4574 | | | | |
| | Total | 9219 | 11,066 | 12,311 | | | | |
| Northeast China | Liaoning Province | 1672 | 1972 | 1990 | 8800 | 10,400 | 5.8 | 3.4 |
| | Jilin Province | 704 | 758 | 835 | | | | |
| Northeast China | Heilongjiang | 1246 | 1408 | 1657 | 8800 | 10,400 | 5.8 | 3.4 |
| | Total | 3622 | 4138 | 4482 | | | | |
| East China | Shanghai | 1453 | 1415 | 1678 | 16,700 | 18,500 | −1.7 | 2.1 |
| | Jiangsu | 5178 | 5334 | 5068 | | | | |
| | Zhejiang | 3539 | 3949 | 4099 | | | | |
| | Anhui | 1413 | 1776 | 2482 | | | | |
| | Fujian | 1300 | 1391 | 1543 | | | | |
| | Shandong | 4940 | 5414 | 5593 | | | | |
| | Total | 17,823 | 19,279 | 20,463 | | | | |
| Central China | Henan | 3260 | 3854 | 4268 | 10,800 | 12,500 | 2.5 | 3.0 |
| | Hubei | 1162 | 1304 | 1421 | | | | |
| | Hunan | 1071 | 1336 | 1443 | | | | |

(continued)

**Table 4.29** (continued)

| Region | Province | 2006 | 2007 | 2008 | 2015 | 2020 | Annual average increase rate between 2008 and 2015 (%) | Annual average increase rate between 2015 and 2020 (%) |
|---|---|---|---|---|---|---|---|---|
| | Jiangxi | 657 | 927 | 934 | | | | |
| | Total | 6150 | 7421 | 8066 | | | | |
| South China | Guangdong | 4062 | 4471 | 4573 | 11,700 | 14,000 | 0.9 | 3.7 |
| | Guangxi Zhuang Autonomous Region | 543 | 931 | 1027 | | | | |
| | Hainan | 198 | 240 | 237 | | | | |
| | Chongqing | 559 | 637 | 666 | | | | |
| | Sichuan | 955 | 1200 | 1277 | | | | |
| | Guizhou | 1435 | 1596 | 1717 | | | | |
| | Yunnan | 856 | 1063 | 1003 | | | | |
| | Tibet Autonomous Region | 1 | 1 | 8 | | | | |
| | Total | 8609 | 10,139 | 10,508 | | | | |
| Northwest China | Shaanxi | 972 | 1229 | 1785 | 12,900 | 25,500 | 9.3 | 14.6 |
| | Gansu | 645 | 784 | 898 | | | | |
| | Qinghai | 152 | 190 | 200 | | | | |
| | Ningxia | 600 | 703 | 754 | | | | |
| | Xinjiang Autonomous Region | 594 | 656 | 820 | | | | |
| | Total | 2963 | 3562 | 4457 | | | | |

#### 4.2.2.2 Suitability of the Development of Thermal Power and Water Resources

1. Forecast of water intaking quantity of the thermal power industry

The draft quota for the thermal power industry in different regions in China are forecast based on Norm of Water Intake (Table 4.30) stated in Part 1: Fossil Fired Power Production (GB/T 18916.1—2010) as shown in Table 4.31.

From 2015 to 2020, east China, central China, and south China will always have a largest water-intaking quantity in the thermal power industry in China: Their total water-intaking quantities in the year of 2015 will be 35.1, 22.7, and 24.6%, respectively, in total quantities of the thermal power industry in China. The

**Table 4.30** Water intaking quotas for unit installed capacity (unit: $m^3$/s GW)

| Method of unit cooling | Unit installed capacity: <300 MW | Unit installed capacity: 300 MW | Unit installed capacity: 600 MW or above |
|---|---|---|---|
| Circulating cooling | $\leq 0.88$ | $\leq 0.77$ | $\leq 0.77$ |
| Once-through cooling | $\leq 0.19$ | $\leq 0.13$ | $\leq 0.11$ |
| Air cooling | $\leq 0.23$ | $\leq 0.15$ | $\leq 0.13$ |

**Table 4.31** Forecast of water intaking quotas of the thermal power industry in different regions in China in different target years

| Region | 2015 | | 2020 | | Increase rate of water intaking (%) |
|---|---|---|---|---|---|
| | Installed capacity (10,000 kW) | Water-intaking quota (10,000 $m^3$) | Installed capacity (10,000 kW) | Water-intaking quota (10,000 $m^3$) | |
| Whole country | 89,200 | 5942.81 | 116,900 | 7127.51 | 19.9 |
| North China | 28,300 | 596.00 | 36,000 | 758.16 | 27.2 |
| Northeast China | 8800 | 185.33 | 10,400 | 219.02 | 18.2 |
| East China | 16,700 | 2083.16 | 18,500 | 2307.69 | 10.8 |
| Central China | 10,800 | 1347.19 | 12,500 | 1559.25 | 15.7 |
| South China | 11,700 | 1459.46 | 14,000 | 1746.36 | 19.7 |
| Northwest China | 12,900 | 271.67 | 25,500 | 537.03 | 97.7 |

*Notes* 1. Water-intaking quantities of the units in Northwest China, north China and north China are calculated based on air cooling; water-intaking quantities of the units in central China, east China and south China are calculated based on circulating cooling

2. The unit installed capacities in different regions in China are all supposed as 600 MW or above

3. Operating time of units in the thermal power industry in China is supposed as 4500 h

water-intaking quantities in northwest China and north China will increase quickest: their water-intaking quantities in the year of 2020 will be 97.7 and 27.2%, respectively, higher than their values in the year of 2015.

2. Suitability of development of the thermal power industry in China and water resources

Even with actions including more effective water-saving actions, control over the excessive increase of water demands, further supporting and moderate development of new water sources for the existing facilities, rational allocation of water resources and guarantee of ecological water, ten first-class hydrological regions in China will still have supply–demand imbalance of water resources more or less from 2020 to 2030. For the purpose of suitability of the thermal power industries in different regions in China and the corresponding regional water resources, this study is intended to spatially match the regions with the corresponding first-class hydrological regions in the thermal power industry, as shown in Table 4.32.

Different regions in China have supply–demand imbalance of water resources. From the Twelfth Five-Year Plan period to the year of 2020, different regions in China, particularly north China and northwest China with the quickest water intaking increase in their thermal power industries, present the poor suitability of their development of thermal power and the corresponding regional water resources.

#### 4.2.2.3 Evaluation on Constraints of Water Resources for the Thermal Power Industry

Before the year of 2020, development of the thermal power industry in China, particularly in north China and northwest China, will show the poor suitability with the corresponding regional water resources because of the constraint of water resource shortage. In 2020, north China and northwest China (water deficiency ratio: 7.24%) will have serious water resource shortage and their thermal power industries will be constrained dramatically by water resources; north China (water deficiency ratio: 2.19%), east China, central China, and south China (water deficiency ratio: 1.58%) will have slight water resource shortage and their thermal power industries will suffer less constraints of water resources. In the Twelfth Five-Year Plan period, regional water resources will have larger constraints for thermal power industries, particularly in north China and northwest China.

### *4.2.3 Countermeasures*

(1) Give full consideration to water resource conditions and improve their spatial suitability for national and regional thermal power layouts.

**Table 4.32** Supply–demand balance of water resources in different regions in China in different target years

| Region | | | Water-intaking quantity of the thermal power industry ($10^8$ $m^3$) | | Basin |
|---|---|---|---|---|---|
| | 2020 | | | | |
| 2015 | 2020 | | Water deficit ($10^8$ $m^3$) | Water deficiency ratio (%) | |
| North China and northwest China | 0.09 | 0.13 | Yellow River basin | 65 | 12.46 |
| | | | Haihe River basin | 36 | 7.32 |
| | | | Rivers in northwest China | 12 | 1.94 |
| | | | Average | | 7.24 |
| Northeast China | 0.02 | 0.02 | Songhua River basin | 12 | 2.03 |
| | | | Liaohe River basin | 6 | 2.34 |
| | | | Average | | 2.19 |
| East China, central China and south China | 0.49 | 0.56 | Yangtze River basin | 13 | 0.57 |
| | | | Huaihe River basin | 31 | 4.19 |
| | | | Pearl River basin | 6 | 0.67 |
| | | | Rivers in southeast China | 2 | 0.60 |
| | | | Rivers in southwest China | 2 | 1.88 |
| | | | Average | | 1.58 |

In the Twelfth Five-Year Plan period and the Thirteenth Five-Year Plan period, regional water resource conditions should be fully considered for the thermal power layout in China to improve their spatial suitability. Coastal regions should make the best of their location advantages and use more seawater in thermal power production. The thermal power enterprises should be constructed as less as possible in regions with serious water resource shortage or serious overdraft of groundwater quantity.

(2) Do well the water resource argumentation management of thermal power projects and determine the cooling technologies and water supply sources for the thermal power enterprises in different regions scientifically.

Do well the water resource argumentation management in the regions with thermal power projects and do not issue water resource permits for the thermal power projects with water consumptions not compliant with other water resource carrying capacities. For each thermal power project with a water resource permit, strengthen the management, and supervision for its water saving and planned water use.

Cooling technologies and water supply sources of the thermal power enterprises in different regions should be determined scientifically in accordance with water resource characteristics and technological characteristics of the regions. For example, the regions with large water resource constraints should use the air cooling technology and unconventional water sources.

(3) Improve water use quota standard systems of the thermal power industries in different provinces and autonomous regions and do well water quota management of thermal power industries and thermal power enterprises.

At present, the water use quota standard system for the thermal power industry in China is still imperfect; some provinces and municipalities have no water use quota standards for their thermal power industries; the existing water use quota standards do not comply with the current water resource management. Thus, it is necessary to improve water use quota standard systems of the thermal power industries in different provinces and autonomous regions, conduct the strict water quota management of thermal power industries and thermal power enterprises, and strengthen the supervision over water resource utilization of the thermal power industry.

(4) Improve the regional water resource cost system and the water supply price system and promote water saving of the thermal power industry through economic measures.

In the regions with extensive thermal power, properly increase water resource costs and public water supply prices to increase the supply price gaps between conventional water resources and unconventional water resources such as reclaimed water; use a progressive price system for over-quota water consumptions, and arouse the enthusiasm of thermal power enterprises for new resource increasing and consumption reduction through economic measures.

(5) Strengthen external supervision over the water-intaking quantity and drainage of water by thermal power enterprises, and improve the internal water use and saving management systems of thermal power enterprises.

Strengthen the control over total water-intaking quotas by thermal power enterprises, the management of underground water-intaking permits and drainage outlets/drainage quantity and the supervision over the water qualities and quality at the water intaking sources and drainage of enterprises. Establish internal water use and saving management systems of enterprises, increase the utilization rates and reuse rates of water resources, and give full play to the output benefit of unit water resource.

# Chapter 5
# Countermeasures Against Climate Change Constraints for Energy Development

**Fan Yang and Jie Tan**

## 5.1 Situation of Countermeasures Against Climate Changes

### *5.1.1 International Situation*

#### 5.1.1.1 Global Climate Changes

Intergovernmental Panel on Climate Change (IPCC) is an intergovernmental agency founded in the year of 1988 by World Meteorological Organization (WMO) and United Nations Environment Programme (UNEP) to evaluate the current situation of scientific knowledge about climate changes, the impacts of climate changes on societies and economies, and the possible countermeasures for adaption to and mitigation of climate changes. Main research achievements of it are special evaluation reports summing up the current knowledge about climate changes and providing overall scientific, technical, social, and economic information about climate changes and their causes, possible impacts, and corresponding countermeasures. IPCC issued four formal Climate Changes Evaluation Reports in all in the years of 1990, 1995, 2001, and 2007, respectively, and the fifth one titled General Report in November 2014. All of these evaluation reports are internationally recognized research reports.

The fourth evaluation report of IPCC, i.e., Climate Change 2007, reached the following main conclusions:

---

F. Yang (✉)
China Electricity Council, Beijing, People's Republic of China
e-mail: yangfan@cec.org.cn

J. Tan
China Coal Processing & Utilization Association, Beijing,
People's Republic of China

 
Z. Wang (ed.), *Constraints and Solutions for Energy and Electricity Development*,
https://doi.org/10.1007/978-981-10-8304-4_5

(1) According to the global temperature observation data, eleven of the past twelve years (1995–2006) were among the twelve years; the linear warming tendency in the past one hundred years was 0.74 °C (0.56–0.92 °C), while that in the past fifty years was 0.13 °C (0.10–0.16 °C) each ten years (nearly doubling the former); in the past one hundred years, warming rate of the North Pole nearly doubled the global average warming rate and warming rate of lands was quicker than seas; since the year of 1961, the average temperature rise in seas in the world has reached a depth of 3000 m at least.
(2) According to the global temperature observation data, the global annual average rise rate of sea level was 1.8 mm (1.3–2.3 mm) from 1961 to 2003, and about 3.1 mm (2.4–3.8 mm) from 1993 to 2003; since 1993, the estimated contribution rates of sea level due to thermal expansion of seas, glaciers (including ice caps), and polar ice sheets have accounted for 57, 28, and 15%, respectively, of the estimated total contribution rate of all factors; in the uncertain regions, sum of the contribution rates complies with the data of sea level rise directly observed.
(3) According to observation data of the ice- and snow-covered areas, since the year of 1978, sea ice area of the North Pole has decreased at an average rate of 2.7% (2.1–3.3%) each ten years (up to 7.4% (5.0–9.8%) in summer) and the average areas of glaciers and snow in mountainous areas in the northern and southern hemispheres have been decreasing; since the year of 1900, in the northern hemisphere, the seasonally frozen ground has decreased by about 7% (up to 15% in spring); since the 1980s, upper parts of almost all the perennial frozen grounds in the North Pole have a temperature rise of 3 °C.
According to the observation data about current rise in the global average air temperature, the global average sea temperature and sea level, and extensive melting of snow and ice (Fig. 5.1), warming of the climate system is undoubted.
(4) According to the long-term observation data, it is highly believed that net impact of the human activities since the year of 1750 has a reason of warming; since middle of the twentieth century, most of the global average temperature rises observed were probably results of the greenhouse gas concentration increase due to human activities; in the past thirty years, the warming due to human activities has probably generated recognizable impacts on the changes observed in many natural systems and biological systems. See Fig. 5.2 for examples of the degrees of impacts on related systems and industries with the increasingly more temperature changes.

The fifth evaluation report of IPCC strengthened the scientific conclusion that climate changes are caused by human activities (data source: China's Policies and Actions for Addressing Climate Change). Some of the research conclusions are as below:

(1) The speeds of climate changes will probably be quicker than before, presenting a worse situation for the human beings in future. Intrinsic mechanism of climate

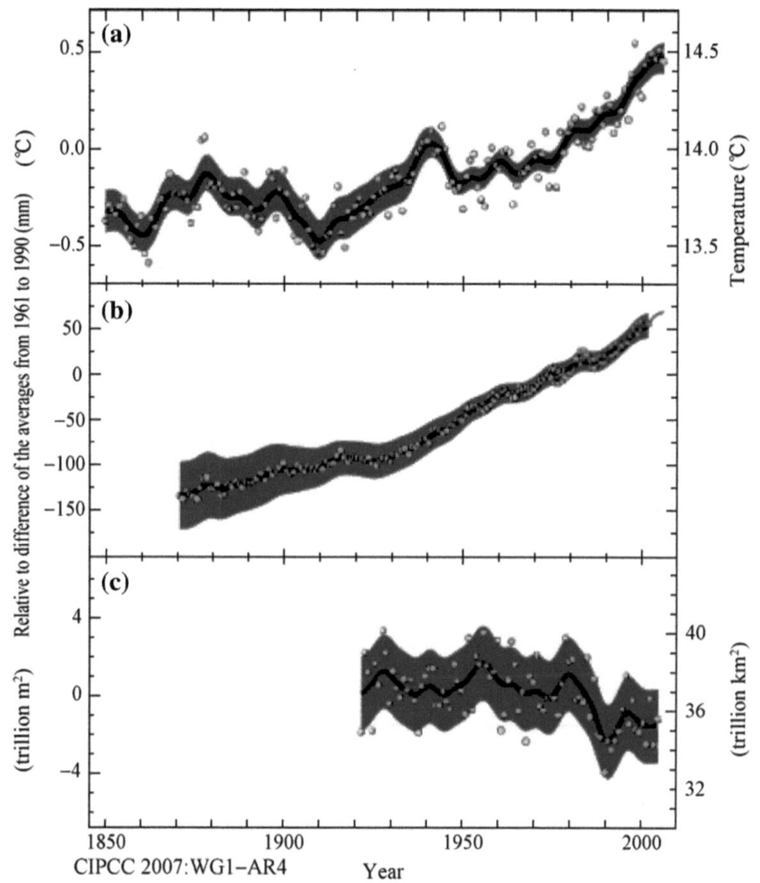


**Fig. 5.1** Changes of temperature, sea level, and the northern hemisphere's snow. **a** Global average temperature; **b** global average sea level; **c** snow area of the northern hemisphere. *Data source* Fourth evaluation report of IPCC for 2007

changes and their impacts is still not completely clear for people, resulting in high uncertainty in forecasts. However, the report above contains detailed evidence and it is highly certain (probability: 95%) that warming is caused by human activities (the probability was about 50% in 1995, 66% in 2001, and 90% in 2007).

(2) An increase in concentration of the greenhouse gases in air was observed. In the year of 2011, the $CO_2$ concentration in air was 391 ppm, 40% higher than the level in the year of 1750 before industrialization; 95 0.1 billion t of carbon (1 t of carbon is equivalent to 3.67 t of $CO_2$)) was emitted during the combustion of fossil fuels. Since the beginning of industrialization, the pH value of seawater has dropped by 0.1 or, in other words, the hydrogen ion concentration in seawater has increased by 26%. See Figs. 5.3 and 5.4.

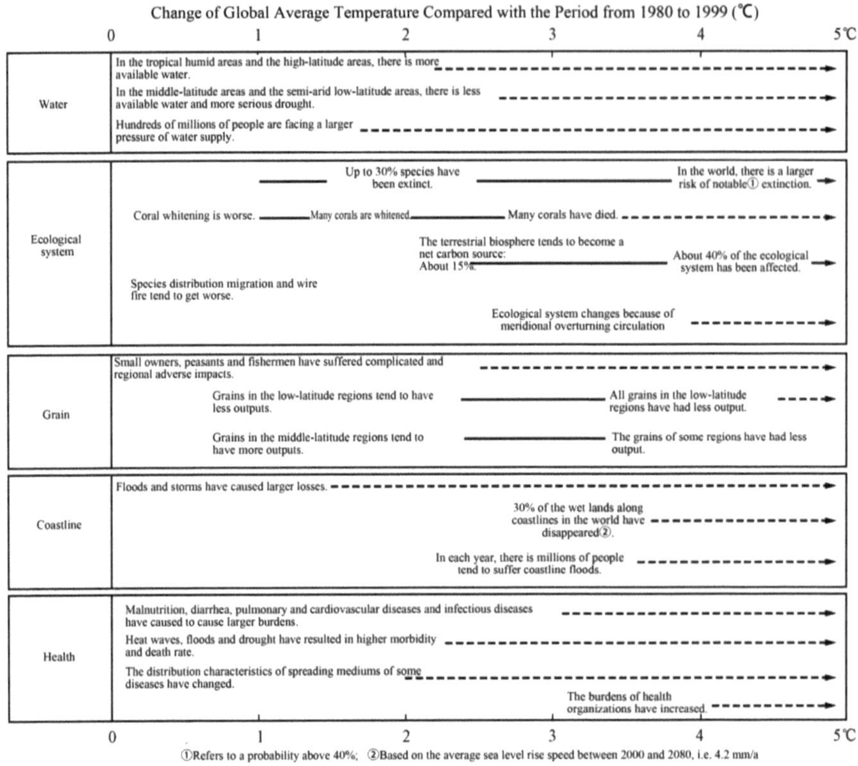


**Fig. 5.2** Examples of impacts on the global average temperature. *Data source* Fourth evaluation report of IPCC for 2007

**Fig. 5.3** Changes of $CO_2$ concentration in air. *Data source* Fifth evaluation report of IPCC for 2013—report of the first working team

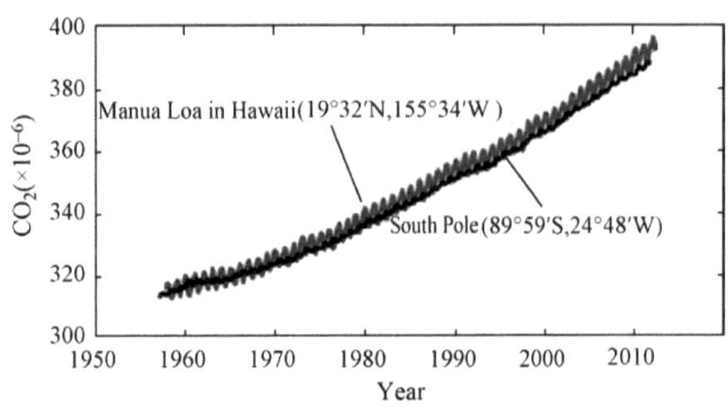


(3) Obvious global warming was observed. In the past thirty years since the year of 1850 with the earliest detailed meteorological records, the maximum temperature record was broken each ten years; the thirty years between 1983 and 2012 were probably the warmest years in the northern hemisphere since the year of

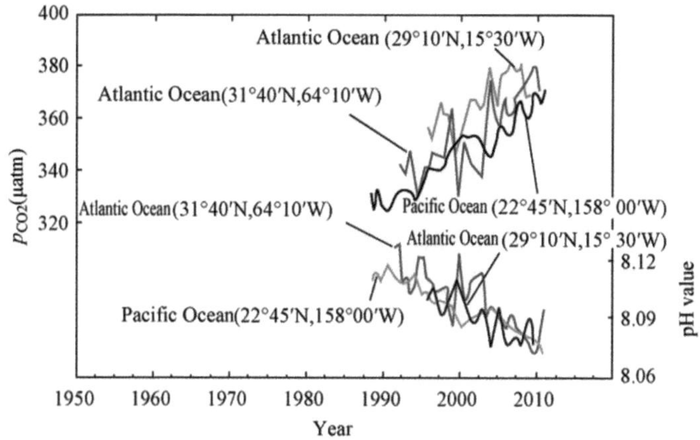


**Fig. 5.4** Changes of $CO_2$ pressure and pH value of sea surfaces. *Note* 1 atm = 1.013250 × $10^5$ Pa. *Data source* Fifth evaluation report of IPCC for 2013—report of the first working team

1400; from 1880 to 2012, the average temperature of the global sea surfaces and land surfaces had been increasing linearly and has increased by 0.85 °C; the annual average temperature between 2003 and 2012 was 0.78 °C higher than that between 1850 and 1900. Compared with lands, seawater has a larger area and a larger specific heat, so seas absorb the most heat newly increased. 90% of the heat accumulation from 1971 to 2010 was absorbed by seawater, of which 60% was absorbed by seawater below 700 m depth; average temperature of the shallow seawater with a depth below 75 m increased by 0.11 °C each ten years in that period. See Figs. 5.5 and 5.6.

#### 5.1.1.2 International Measures Against Climate Changes

The world climate conference was held in Geneva in the year of 1979. On the conference, climate changes were concerned by the international community. Since then, the international community has taken many countermeasures against climate changes.

IPCC was founded in 1988 by WMO and UNEP. It is expected to evaluate the scientific information about climate changes and environmental, social, and economic consequences of climate changes and work out practical countermeasures.

United Nations Framework Convention on Climate Change (UNFCCC) passed in 1992 was the world's first international convention for controlling the emission of greenhouse gases (GHGs) comprehensively in a bid to reduce the adverse impacts of global warming on human societies and economies. It has become the basic framework for the international community to address global climate changes. It came into force on March 21, 1994.

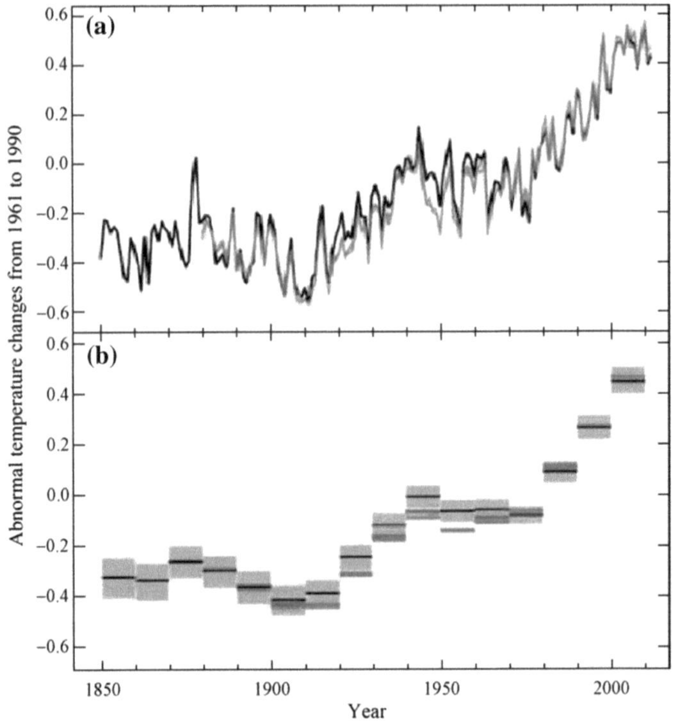


**Fig. 5.5** Changes of average temperature of the global sea surfaces and land surfaces between 1850 and 2012. **a** Annual average; **b** 10 annual average. *Note* The lines mean that the data originated from different data sets; the shadows refer to uncertainties. *Data source* Fifth evaluation report of IPCC for 2013—report of the first working team

**Fig. 5.6** Average heat changes of shallow seawater. *Note* The lines mean that the data originated from different data sets; the shadows refer to uncertainties. *Data source* Fifth evaluation report of IPCC for 2013—report of the first working team

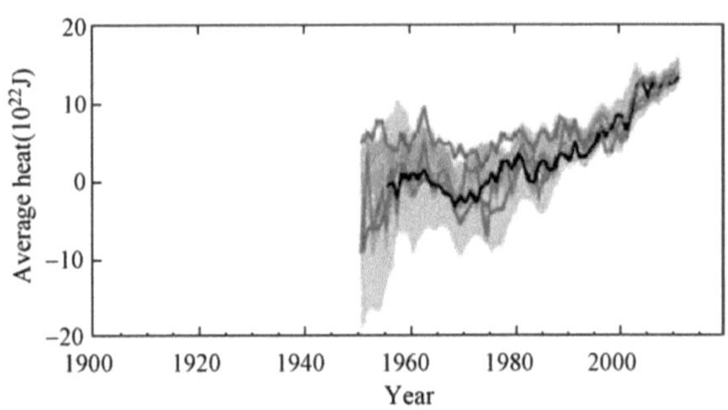


For the first time in the word, the Kyoto Protocol passed in 1997 limits the emission of greenhouse gases through a law and the emissions of greenhouse gases by some countries to alleviate the global warming. It took effect on February 16, 2005. According to it, by the year of 2010, greenhouse gas emissions of the contracting developed countries should be 5.2% less than in 1990, while the contracting developing countries have no obligation for such emission reduction.

In December 2007, the thirteenth conference of contracting countries was held in Bali, Indonesia. On the conference, the 'post-Kyoto' issues were mainly discussed and the Bali Roadmap passed on the 15th day of the month.

In December 2008, the fourteenth conference of contracting countries was held in Poznan, Poland, to set an agenda for the Copenhagen conference.

On July 8, 2008, the G8 leaders on the summit reached a consensus about the long-term objective for the emission reduction of greenhouse gases. On the summit, the G8 leaders seek a long-term objective for the emission reduction of global greenhouse gases by half at least by the year of 2050 through cooperation with other contracting countries of UNFCCC and determine it in the discussion with these contracting countries in related negotiations of UNFCCC.

In December 2009, a climate conference was held in Copenhagen. On the conference, Copenhagen Accord without a legal effect was reached under the leading role of the USA, China, India, Brazil, and South Africa. However, it was determined in Copenhagen Accord to found a green climate foundation and participating countries committed to provide funds to support developing countries to address climate changes. According to Copenhagen Accord, each participating country is expected to make an announcement before February 1, 2010, about its emission reduction objective before the year of 2020. In late 2010, the United Nations Climate Change Conference was held in Mexico to determine the responsibilities and objectives of the participating countries.

In the year of 2010, the sixteenth conference of contracting countries held in Cancun, Mexico. On the conference, the 193 contracting countries agreed to pass UNFCCC and Kyoto Protocol as countermeasures against climate changes.

On November 28, 2011, the seventeenth conference of contracting countries was held in Durban, South Africa. On the conference, it was decided to establish an Ad Hoc Working Group on the Durban Platform for Enhanced Action (AWG-DP), begin its work in the first half of 2012, work out a legal tool or achievement suitable for all contracting countries of UNFCCC before the year of 2015 and, since the year of 2013, carry out the Second Commitment Period of Kyoto Protocol, and initiate the Green Climate Fund.

On December 8, 2012, the eighteenth conference of contracting countries was held in Doha. In the end of the conference, the organizer forced attending countries to reach some decisions, consensuses, and achievements. On the conference, developing countries, particularly China, played an active role in promoting the passing of an amendment of Kyoto Protocol to guarantee implementation of the Second Commitment Period of Kyoto Protocol in the year of 2013; decisions about contents such as a long-term climate fund, achievements of the long-term cooperation working team of the UNFCCC, Durban Platform for Enhanced Action, loss and damage compensation mechanism were reached. However, the following issues remained unchanged in the conference: Firstly, Canada, Japan, New Zealand, and Russia had made it clear that they would not participate in the Second Commitment Period of Kyoto Protocol; the USA did not approve its state's participation in Kyoto

Protocol in 2001; Canada quitted Kyoto Protocol in 2011. Secondly, as for the climate fund most concerned by developing countries, it was reiterated in a decision in the conference that developed countries need to provide funds to developing countries to address climate changes and, before 2020, put Green Climate Fund into service and provide 100 billion USD each year, but developed countries said they were in an economic crisis and refused to provide specific schedules for fund provision. As a result, the conference had to put off the fund issue to the year of 2013. Thirdly, in the conference, the contracting countries of UNFCCC focused on negotiation about the Second Commitment Period of Kyoto Protocol and the long-term cooperation working team, they did not make substantial progress in the negotiation about Durban Platform for Enhanced Action.

Global efforts for addressing climate changes are an irresistible trend. In future, all countries in the world will strengthen their cooperation in alleviating and adapting to climate changes, reducing the emission of greenhouse gases, etc.

## 5.1.2 *Tendencies in China*

### 5.1.2.1 Climate Changes in China

Researches show that China faces a tendency of warming as the whole world does. According to the latest observation data issued by China Meteorological Administration, in the past one hundred years (1908–2007), the average land surface temperature in China had risen by 1.1 °C; since 1986, there have been 21 warm winters; 2007 was the warmest year since 1951 with the earliest systematic meteorological observation. In the past fifty years, the precipitation distribution in China has changed a lot: West China and South China now have more precipitation, while North China and most places in Northeast China now have less precipitation. Extreme climate events such as high temperatures, drought, and heavy precipitation tend to happen more frequently and powerfully. In summer, there are periods of high temperatures and heat waves; some regions, particularly North China, have worse drought; South China has more heavy precipitation; West China has more snow disasters. In the past thirty years, sea surface temperature and sea level of coastal areas in China have risen by 0.9 °C and 90 mm, respectively.

### 5.1.2.2 Existing Actions of China for Climate Changes

The Chinese government has fully recognized the significance and urgency to address climate changes. It has worked out and has been carrying out national plans against climate changes and taken many policies and actions against climate changes in accordance with requirements of the concept of scientific development and overall consideration of the relations between economic development and ecological construction, domestic factors and international factors, and present interests and long-term interests.

China joined UNFCCC and Kyoto Protocol in 1993 and 2002, respectively.

In 2004, the State Council approved Planning Outline for Medium- and Long-term Development of Energy Sources (2004–2020) (Draft). In the same year, National Development and Reform Commission issued the national first Medium- and Long-Term Special Plan of Energy Conservation of China.

In February 2005, National People's Congress approved Renewable Energy Law of the People's Republic of China. In August 2005, the State Council issued The Emphases of Building a Resources-saving Society of These Two Years. In December 2005, the State Council issued Decision for Issuing and Implementing Document Titled Interim Regulation on Promotion of Industrial Restructuring. In August 2006, the State Council issued Decision on Strengthening Energy Saving. These policy documents have provided a policy guarantee and a legal guarantee for higher capability of China to address climate changes.

On December 26, 2006, China's National Assessment Report on Climate Change was issued formally. It includes the history and tendencies of climate changes, the impacts of and adaption to climate changes, and the economic and social evaluation about the alleviation of climate changes. It systematically summarized the China's scientific research achievements about climate changes, evaluated the observed climate changes in the past one hundred years in China under the background of global climate changes and their impacts comprehensively, provided observation data about the tendencies of climate changes in the twenty-first century, analyzed and evaluated the possible impacts of climate changes and related international conventions on the ecology, environment, economy, and society in China, and proposed China's position, rules, advice, and related policies for addressing global climate changes.

In June 2007, the Chinese government issued China's National Programme to Address Climate Changes, setting the following overall objectives for China to address climate changes by the year of 2010: apparent effects of the policies and actions about control over the emission of greenhouse gases, better adaption to climate changes, a higher level of researches about climate changes, new progress in the scientific researches about climate changes, significant improvement of people's awareness of climate changes, and better systems and mechanisms for addressing climate changes.

On June 12, 2007, the State Council decided to set up a national leading group for climate change addressing, energy conservation, and emission reduction. The leading group is expected to serve as a national deliberation and coordination body for climate change addressing, energy conservation, and emission reduction to strengthen China's control and work of greenhouse gases. Main work of the leading group is as below: Study and work out important national strategies, policies, and countermeasures for addressing climate changes, make unified arrangement for climate change addressing, energy conservation, and emission reduction, study and discuss related international cooperation, negotiation plans, policies, and advice, coordinate and solve major issues in climate change addressing, and organize the implementation of related energy conservation and emission reduction-related policies of the State Council.

In October 2008, the State Council Information Office issued China's Policies and Actions for Addressing Climate Change to introduce China's actual conditions, impacts, policies, and action objectives of or for climate changes. On August 27, 2009, the 12th session of the Standing Committee of the 11th National People's Congress approved Decision of the Standing Committee of the National People's Congress on Addressing Climate Changes Positively.

On November 25, 2009, the Chinese government announced its objective to make its $CO_2$ emission of per unit of GDP in the year of 2020 40–45% lower than the value in 2005. According to the announcement, the objective would be included in the medium- and long-term plans for China's economic and social development as a constraint index; corresponding methods would be determined for the related statistics, monitoring, and evaluation in China; it was proposed to develop renewable energy sources vigorously and promote work such as nuclear power construction actively to make the consumption of non-fossil energy sources in China about 15% of the consumption of primary energy sources by the year of 2020.

On March 14, 2011, the 4th session of the 11th National People' s Congress approved The Twelfth Five-Year Plan for National Economic and Social Development of the People's Republic of China. The official document proposed objectives to make the consumption of non-fossil energy sources 11.4% of the consumption of primary energy sources and reduce the $CO_2$ emission of per unit of GDP by 17% and such specific active actions for addressing global climate changes as controlling the emission of greenhouse gases, promoting adaption to climate changes, and participating in international cooperation positively.

In 2012, National Development and Reform Commission, National People's Congress's Environmental and Resources Protection Committee and Legal Committee, Legislative Affairs Office of the State Council, and other related authorities jointly set up a leading group for drafting laws about climate changes to promote the drafting of related laws to form a legislative framework preliminarily. Shanxi and Qinghai issued Measures of Shanxi for Addressing Climate Changes and Measures of Qinghai for Addressing Climate Changes, respectively; Sichuan and Jiangsu are carrying out their legislative work for addressing climate changes; Shenzhen's Municipal People's Congress approved Regulations on the Carbon Emission Management of Shenzhen Special Economic Zone.

In July 2013, the State Council adjusted entities and individuals of the leading group for national leading group for climate change addressing: Premier Li Keqiang was appointed as leader of the leading group, and some departments were added.

In 2013, the national pilot of carbon emission permit trading was initiated formally. Related work has begun in Shenzhen, Beijing, and Shanghai, and the carbon market is now under orderly construction.

On November 18, 2013, nine ministries and commissions such as National Development and Reform Commission jointly issued National Strategy for Climate Change Addressing.

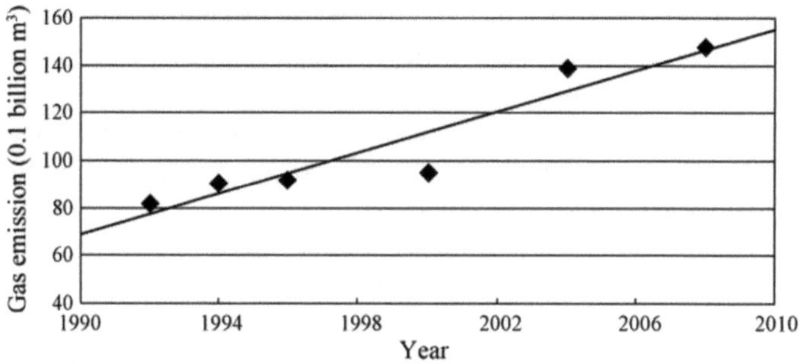


**Fig. 5.7** Mine gas emissions in China from 1990 to 2010

## 5.2 Current Emission and Emission Reduction of Greenhouse Gases in Main Energy Activities in China

### *5.2.1 Coal Development*

#### 5.2.1.1 Emission of Greenhouse Gases

The sources of emission of greenhouse gases in coal development mainly include the emission of mine gas, spontaneous combustion of the coal in mining areas, spontaneous combustion of coal gangue piles, emission of gases generated by the coal used in production. The emission of mine gas is the most source of greenhouse gas emission in coal development.

In coal mining, the emission of mine gas is inevitable. Methane, a main component of the mine gas, is a main member of greenhouse gases. Mine gas emission includes the emission in coal mining and emission of the residual gases before coal use (including coal storage and transportation). The emission in coal mining is the main part. Statistical data show that 1 t of raw coal emits about 5 $m^3$ of gas; 1 t of methane is equivalent to 21 t $CO_2$ if viewed from the greenhouse effect; 0.1 billion $m^3$ of methane is equivalent to 150 ten thousand t of $CO_2$ if viewed from the greenhouse effect and calculated with a concentration of 22.4 L/mol in the standard gas state. See Fig. 5.7 for the mine gas emissions in China from 1990 to 2010.

#### 5.2.1.2 Main Actions for Emission Reduction

The mine gas emitted in coal development is mainly reduced by means of overall utilization in the following main ways:

(1) Civil use: A civil use system for mine gas is usually made up of extraction pump(s), gas tank(s), a regulator station, and gas pipeline(s) and has a mine gas concentration of 30% at least.
(2) Power generation: The main processes include gas turbine power generation, steam turbine power generation, gas engine power generation, heating, power, and cooling cogeneration with mine gas.
(3) Chemical industry: The high-concentration pure mine gas exploited or extracted can generate a good economic benefit if used as a raw gas for chemical product production. A high-concentration mine gas can produce chemical products such as carbon black, formaldehyde, methanol, and chemical fertilizers; in the chemical industry, mine gas will be mainly used in the production of basic chemical raw materials such as chemical fertilizers and methanol; with a nature similar to natural gas, mine gas is suitable for the production of synthesis ammonia with low energy consumption, simple process flow, small investment, and good economic benefit.
(4) Industrial fuel: Mine gas can be used as a clean industrial furnace fuel to alleviate pollution and improve the quality of industrial products.
(5) Utilization technologies of low-concentration mine gas: At present, mine gas realizes overall utilization after purification. The main technologies include pressure swing absorption, low-concentration mine gas porous medium premixed combustion, low-concentration mine gas power generation, and mine ventilation air utilization.

#### 5.2.1.3 Emission Reduction

According to the forecast of related authorities for the coal output of China, the coal output of China in 2015 will be more than 3.9 billion t; the mine gas extracted in coal mining will have an extraction quantity of about 13 billion $m^3$ and a utilization rate of 6 billion $m^3$, suggesting a emission reduction of $CO_2$ of about 9000 ten thousand t.

See Table 5.1 for the coal output and the mine gas's extraction quantity and utilization rate in China in the twelfth Five-Year Plan period.

#### 5.2.1.4 Existing Problems and Countermeasures

1. Existing problems

The present extraction and utilization of mine gas in China are a unity of contradiction: On the one hand, the extraction of mine gas is the most important precondition of safe coal production and mine gas, in regardless of its volume and concentration, must be extracted. In spite of its high extraction quantity and rate, mine gas has a non-uniform extraction quantity and a varying extraction

**Table 5.1** Coal output and mine gas's extraction quantity and utilization rate in China in the Twelfth Five-Year Plan period

| Year | Coal output (0.1 billion t) | Extraction quantity of mine gas (pure) (0.1 billion $m^3$) | Utilization rate of mine gas (0.1 billion $m^3$) | Utilization rate of mine gas (%) |
|---|---|---|---|---|
| 2011 | 35.2 | 85 | 30 | 35.3 |
| 2012 | 36.5 | 95 | 40 | 42 |
| 2015 (forecast) | 39 | 130 | 60 | 46 |

*Data source* China National Coal Association

concentration. On the other hand, for civil use and equipment performance compliance, the mine gas used should have stable sources and a concentration of 30% at least.

2. Countermeasures

For good extraction and overall utilization of mine gas, mine gas extraction should be the foundation and safety production of mines should be the precondition. It is suggested to take the following countermeasures:

(1) Change ideas about mine gas extraction: Mine gas extraction should be regarded as a method to control and prevent mine gas explosion and an important part of coal development and utilization. Mine gas extraction should be managed in coal mining; the mine gas extracted should be operated as a coal product similar to coal and other inter-grown and associated minerals.
(2) Study the technologies of mine gas extraction and improve the level of mine gas extraction. For full and rational utilization of mine gas, new technologies, new processes, and new equipment of mine gas extraction must be studied and promoted to improve the level and concentration of mine gas extraction greatly in a short period to realize a stable extraction quantity to utilize mine gas well.
(3) Strengthen studies about the technologies of concentration, storage, and transportation of mine gas and eliminate the impacts of unfavorable factors including unstable quantity and concentration of mine gas extraction, high costs of storage, and long-distance transportation. For a reliable gas source supply service, the most basic action is to guarantee proper gas supplies and gas supply concentrations.
(4) Develop new methods of mine gas utilization, change the current situation that the mine gas extracted has simple applications (mainly for civil use), strengthen the studies and investments for the applications of mine gas in the chemical industry and power generation and as an industrial fuel, and vigorously develop the utilization of low-concentration mine gas, including porous medium combustion, low-concentration mine gas explosion-based power generation, use of mine ventilation air.

The utilization of mine gas in China has just begun, so mine gas now faces many difficulties in aspects such as technologies, equipment, and manpower; its utilization limits to high-concentration mine gas only; its utilization rate is low; there is a serious waste because most of the mine gas extracted is emitted. Thus, it is now urgent to develop new technologies and methods for rational use of mine gas. Promoting mine gas extraction by means of mine gas utilization stimulates the development of extraction technologies and is a new form of utilization of a clean energy source as well, protecting environment and realizing rational adjustment for the energy structure in China.

## 5.2.2 *Power Generation*

### 5.2.2.1 Main Actions for Emission Reduction

Emission reduction technologies for the greenhouse gases in the power industry include emission reduction through engineering measures, emission reduction through restructuring, emission reduction through management measures, and emission reduction through market mechanisms. Emission reduction through engineering measures refers to the process technologies, aiming at improving the energy efficiencies of units and reducing, capturing, and storing $CO_2$, such as the ultra-supercritical (supercritical) technology, the circulating fluidized bed technology, the integrated gasification combined cycle power generation technology, the cogeneration, oxygen-enriched combustion technology, and the carbon capture and storage technology. Emission reduction through restructuring aims at increasing proportion of carbon-free or low-carbon power generation technologies such as renewable energy sources and nuclear energy in the power structure, making these technologies replace thermal power with a high carbon emission and optimizing the power structure to reduce carbon emission. Emission reduction through management measures realizes carbon emission reduction by means of power dispatching sequence adjustment, generation right trading, auxiliary power reduction, etc. Emission reduction through market mechanisms aims at reducing $CO_2$ emission by participating in international cooperation through a clean development mechanism and obtaining verified carbon emission reductions through the funds and technologies transferred. See Table 5.2 for the related emission reduction technologies.

专栏解读

First IGCC demonstration project in China—Huaneng Tianjin IGCC Demonstration Power Plant

**Table 5.2** Main $CO_2$ emission reduction technologies

| Type of emission reduction technology | Emission reduction technology | Technical description | Technical progress |
|---|---|---|---|
| Emission reduction through engineering measures | Ultra-supercritical (supercritical) technology | Ultra-supercritical (supercritical) units have high thermal efficiencies and, in the same generation condition, consume less coal, thus reducing both energy consumptions and air pollutant emissions | Ultra-supercritical (supercritical) units are an internationally mature technology; the coal-fired units in China now have high capacities and parameters |
| | Integrated gasification combined cycle (IGCC) | IGCC power plants are significant for environmental protection and low-carbon development because they can realize nearly zero emission of pollutant emission and better capture of $CO_2$ | The IGCC power generation technology is now in commercial demonstration and is about to realize commercial application |
| | Carbon capture and storage (CCS) technology | In the CCS technology, $CO_2$ is separated from production activities in aspects such as industries or related energy sources, conveyed to the storage places for longtime isolation from air. With this technology, the carbon emission of unit generation can be reduced by 85–90%. Thus, it is a stable scheme for alleviating the greenhouse gas concentration in air | The technology of carbon capture and extensive utilization and storage is a new technology with a large emission reduction potential |
| Emission reduction through restructuring | Hydropower | Water resource development is significant for energy safety of the country, an optimal energy structure, less emission of pollutants and greenhouse gases, and coordinated regional development | Hydropower is a technically mature, economically feasible, and extensively developable renewable clean energy source |
| | Nuclear power | Nuclear energy is a clean, safe, and economical energy source and an alternative energy source allowing extensive development as well. With a main application in power generation, it has great potential in energy structure optimization, greenhouse gas emission reduction, and solving of energy problems | China has basically formed a complete nuclear fuel industry system |

(continued)

**Table 5.2** (continued)

| Type of emission reduction technology | Emission reduction technology | Technical description | Technical progress |
|---|---|---|---|
| | Wind power | Wind power is a clean renewable energy source. It can increase the energy supply of China, adjust the energy structure, and protect the ecological environment | The unit installed capacity of homemade has reached 1.5, 2 or 3 MW and the megawatt direct-driven variable-speed constant-frequency wind power units and megawatt doubly fed variable-speed constant-frequency wind power units both representing the tendency of technical development have realized mass production |
| | PV power generation | China has abundant solar energy resources and can become an alternative energy source in future | Extensive application of the PV power generation in China needs to be based on technical breakthrough, significant cost reduction, and technical and industrial preparations |
| Emission reduction through management measures | Energy-saving generation dispatching | In energy-saving generation dispatching, renewable generation resources are dispatched first according to the energy-saving, economic, and low-carbon rules and dispatch fossil generation resources in sequence according to a low-to-high rank of energy consumptions and pollutant emissions of the units to minimize energy consumptions, resource consumptions, and pollutant emissions | Quick and healthy development of the national economy and power system in China has laid a material base for energy-saving generation dispatching; energy-saving generation dispatching has certain experience and achievements |
| | Energy saving of power plants | The energy saving of thermal power plants is important for energy saving and consumption reduction. Power saving can reduce generation costs and alleviate environmental pollution by means of $CO_2$ emission reduction | – |

(continued)

**Table 5.2** (continued)

| Type of emission reduction technology | Emission reduction technology | Technical description | Technical progress |
|---|---|---|---|
| Emission reduction through market mechanisms | Clean development mechanism (CDM) | In the power industry, there are two types of potential CDM projects: power plant substitution, fuel substitution, and clean coal technology; substitution of fossil energy sources by energy sources generating less or no emission, cost reduction of power transmission and distribution, and DSM | In the past several years, the CDM projects in the power industry in China have had good performance and focused on wind power and hydropower |
| | Generation right trading | In generation right trading, the units with higher generating costs sell partial or all of their scheduled contract electricity to the units with lower generating costs to realize power generation, thus realizing an optimal power structure, and consumption and emission reduction | – |

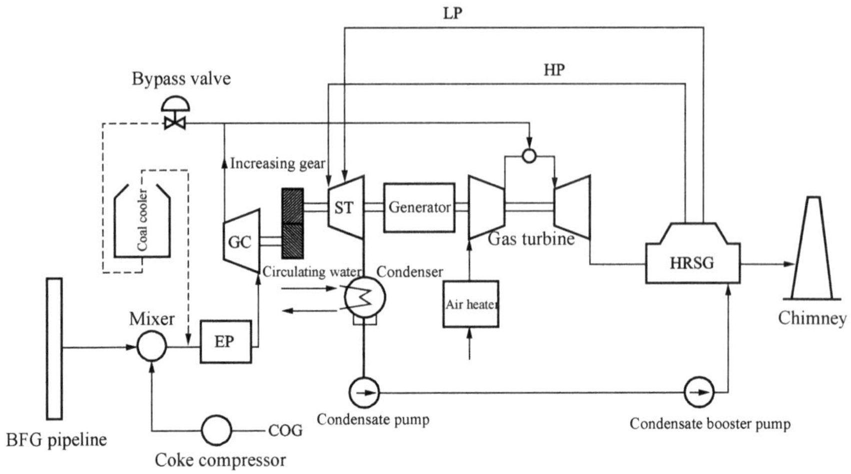


**Fig. 5.8** Process flow of IGCC. LP—low-pressure cylinder; HP—high-pressure cylinder; ST—steam turbine; GC—gas turbine; EP—compressor; COG—coke oven gas; BFG—blast furnace gas; HRSG—heat recovery steam generation

> Huaneng Tianjin IGCC Demonstration Power Plant was the first independently developed, designed, manufactured, and constructed IGCC demonstration project in China. The installed capacity of its phase I was 250 MW. As the most environment-friendly demonstration coal-fired power plant in China, the project has a generating efficiency up to 48%, a desulfurization efficiency above 99%, a low nitrogen oxide emission, and a capability to recover highly pure sulfur, making its pollutant emission indexes reach the level of gas turbine units. See Fig. 5.8 for its process flow.

Huaneng Tianjin IGCC Demonstration Power Plant is a substantial progress of the Green Coal-fired Power program with proprietary intellectual property rights and standing for leading level of the world's clean coal technologies. It has started a new era of the clean coal power generation technologies in China.

专栏解读

First $CO_2$ capture demonstration project in China —Carbon Capture Test Demonstration Unit of China Huaneng Group

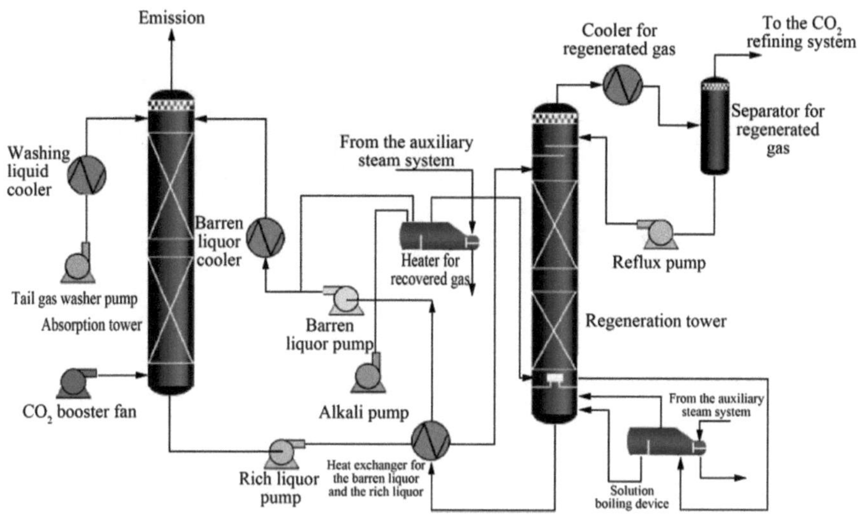


**Fig. 5.9** Process flow of $CO_2$ capture

The $CO_2$ Capture Test Demonstration Plant of Huaneng Beijing Thermal Power Co., Ltd. was the first project based on cooperation between China Huaneng Group and Australia and the first coal-fired power plant $CO_2$ capture demonstration project in China. The project was formally completed and put into service in July 2008. After construction, debugging, and commissioning, its $CO_2$ capture unit has a normal flue gas treatment capacity of 2372 m/h and a $CO_2$ recovery rate of 0.5 t/h and can operate stably at 60% to 120% of its rated capacity. It has an annual continuous operation time of 6000 h; the $CO_2$ separated and captured by it has a concentration above 95%. See Fig. 5.9 for its process flow.

Service of the project has provided effective technical research platform and technical support for China to master the core technologies of $CO_2$ capture and storage, become a leader in the future field of coal-fired power, and improve related independent technical innovation capability.

#### 5.2.2.2 Emission Reduction

1. Emission reduction effect

Compared with the situation in 2010, in 2013, the $CO_2$ emission intensity of per unit of GDP in China decreased by 10.68% and realized 62.8% of the objective that ‘the $CO_2$ emission of per unit of GDP in 2015 will decrease by 17%’ set in The

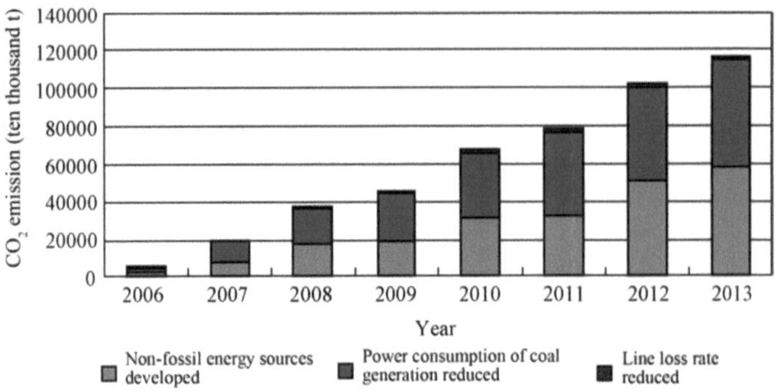


**Fig. 5.10** $CO_2$ emission reductions of different years (base year: 2005)

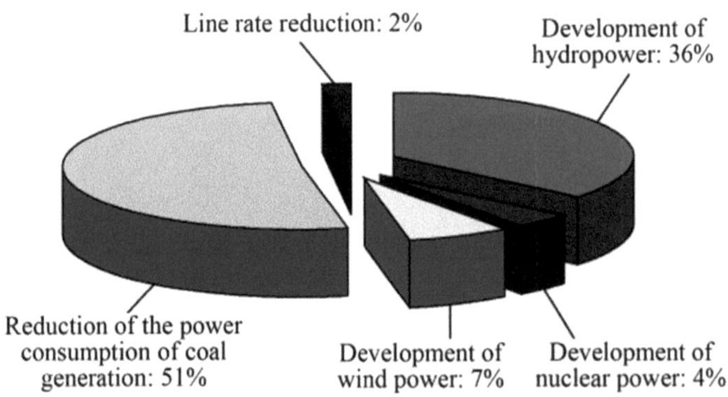


**Fig. 5.11** Contribution rates of the actions above between 2006 and 2013 to total $CO_2$ emission reduction (base year: 2005)

Twelfth Five-Year Plan for National Economic and Social Development of the People's Republic of China, which was basically the same as the schedule (60%).

From 2006 and 2013 (base year: 2005), total $CO_2$ emission reduction of the power industry through development of non-fossil energy sources, reduction of the coal consumption of power supply, and reduction of the line loss rate was about 4.73 billion t. Reduction of the coal consumption of power supply and development of non-fossil energy sources contributed 51 and 47%, respectively, to emission reduction of the power industry. When the base year was 2005, the $CO_2$ emission reductions in related years were introduced in Fig. 5.10 and contribution rates of the actions above between 2006 and 2013 to total $CO_2$ emission reduction were introduced in Fig. 5.11.

2. Climate change addressing

(1) Development of the power industry in China is still a difficult task: The power industry in China has reached great achievements, but it still has a big

gap of per capita level with developed countries. China is now in the middle and later periods of its industrialization, and its power industry as the basic industry should have appropriately advanced development in a bid to meet the overall, coordinated, and sustainable development of its economy and society. Thus, scientific development is still the most important task for the power industry in China and it is an urgent and arduous task to control emission of greenhouse gases in the power industry.

(2) The electrification level needs to be improved. The electrification level includes proportion of power generation energy sources in the consumption of primary energy sources and proportion of the electric energy in the final energy consumption. Researches show that optimizing the coal consumption structure by increasing the primary coal consumption of power generation can improve the electrification level, thus enhancing the energy efficiency and reducing the energy intensity.

(3) $CO_2$ emission reduction has an increasingly larger pressure. On the one hand, total $CO_2$ emission in China will keep increasing; on the other hand, most of the power plants constructed after the year of 2000 are coal-fired power plants which have a long time of service in future but an increasingly smaller potential for efficiency improvement. Thus, $CO_2$ emission reduction in the power industry will have a larger and larger pressure in future.

(4) External factors have larger impacts on the power industry. The power industry in China has been curbed, more or less, by global financial crisis, periodic regulation of the domestic economy, and the transmission difficulty of coal-fired power. The operation pressures of power enterprises keep increasing due to continuous increase in energy prices and high generating costs of renewable energy sources. The power industry is making efforts to guarantee electricity supply by overcoming many external unfavorable factors, which cannot remain for a long time if the external unfavorable factors are not solved from the aspect of mechanisms.

(5) Energy constraints affect one another. In North China, Northeast China, and Northwest China all with abundant coal reserves but water resource shortage, air cooling is employed for thermal power units, but the coal consumption rises by 10–15 g of standard coal/kWh.

## 5.3 Benefits of Greenhouse Gas Emission Reduction in the Main Energy Sectors

### *5.3.1 Coal Development*

1. Development and utilization of mine gas

The development and utilization of mine gas can not only reduce the emission of greenhouse gases but change wastes into resources and save energy resources. In recent years, the coal industry has strengthened control over mine gas and, as a result, the extraction quantity and utilization rate of mine gas have kept increasing. According to statistical data of China National Coal Association, the annual extraction quantity of coal measure stratum gas (mine gas) in 2013 was 15.5 billion $m^3$, including 12.5 $m^3$ of mine gas and 3 billion $m^3$ of surface coal measure stratum gas, 67% more than that in 2010. See Fig. 5.12 for the increase data. Most of the mine gas extracted and utilized in China is high-concentration mine gas, but a little of the low-concentration gas accounting for 90% of total mine gas emission and, in particular, the mine ventilation air gas with a concentration below 1% is basically not utilized. Mine ventilation air gas has a low methane content, but the gas emission is still considerable (tens of billions of $m^3$) because of the large ventilation rate.

For less carbon emission in mining areas, it is necessary to improve the extraction quantity and utilization rate of high-concentration mine gas and develop and utilize low-concentration mine gas extensively. The existing domestic and foreign mature technologies about the utilization of low-concentration mine gas mainly include power generation with low-concentration mine gas and utilization of ventilation air gas.

No mattered viewed from its technically mature, adaptability, product demands, and economical efficiency, power generation with low-concentration mine gas will be the main tendency in future. There are now two technologies about the utilization of ventilation air gas: catalytic flow reversal reaction (CFRR) technology aiming at lowering the spontaneous combustion temperature of methane with help of a catalyst and making methane oxidized and characterized by low reaction temperature, high catalyst price, and complicated processing; thermal flow reversal reaction technology characterized by high reaction temperature and methane oxidation ratio, simple processing, and low cost. Some institutions of higher learning and enterprises in China have begun to study the thermal flow reversal reaction technology and reached some achievements.

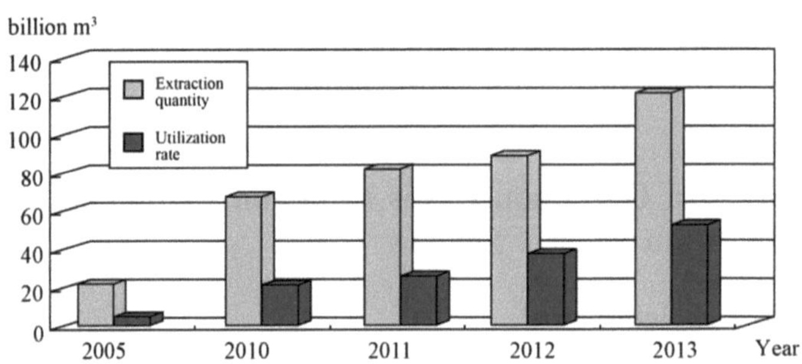


**Fig. 5.12** Extraction quantity and utilization rate of mine gas

2. Effective utilization of low-temperature heat sources in coal mines

The utilization of the low-temperature heat sources of coal mines is an effective method for the carbon emission reduction of mining areas. In coal production, there is usually much mine water drains from the water sumps at coal mine bottoms to surfaces. Thus, ventilation is necessary for the safety in mine production. Mine water, main return air of coal mines, and the pithead power plant condensate have basically constant temperatures all the year round and contain massive low-grade thermal energy. If this thermal energy can be utilized effectively, there will be significant reduction of energy waste and indirect reduction of $CO_2$ emission. This low-grade thermal energy is good heat source and heat sink for heat pump units and can be used for the heat preservation of coal mine shafts and the heating in industrial production site, office buildings, residential buildings and public facilities in winter, hot water provision for domestic use and washing and bathing by staffs all the year round, and cooling of office buildings, residential buildings, and public places in summer. In recent years, the heat pump technology has been developing and some coal mines in China have tried it. Table 5.3 refers to estimated reserves of the low-temperature thermal energy of coal mines in China. According to the estimates, low-temperature thermal energy of coal mines in China has a reserve of about 0.2 billion kW which is equivalent to 4390 ten thousand t standard coal/a. If the low-temperature thermal energy of coal mines is used completely, about 0.1 billion t of $CO_2$ emission will be reduced each year, so the potential of $CO_2$ emission reduction is large.

3. Energy consumption reduction of main mining equipment

Energy consumption reduction for main mining equipment in coal mines is another direct method for the carbon emission reduction of mining areas. Statistical data show that about 80% of total energy consumption of raw coal production is contributed by main fans, main drainage pumps, main hoists, and air compressors

**Table 5.3** Estimated reserves of the low-temperature thermal energy of coal mines in China in 2007

| Item | Available thermal energy (ten thousand kW) | Equivalent standard coal (ten thousand t) | $CO_2$ emission reducible (ten thousand t) | Total $CO_2$ emission reducible (ten thousand t) |
|---|---|---|---|---|
| Return air | 963 | 2293 | 5113 | 9789 |
| Mine water | 664 | 1384 | 3086 | |
| Pithead power plant condensate | 342 | 713 | 1590 | |

*Note* The national coal mine water drainage rate is supposed to be 5 billion $m^3$, and the average temperature difference is supposed to be 10 °C; total installed capacity of the pithead power plants is 184 ten thousand kW, and the annual generation is 8.7 billion kWh; standard coal $CO_2$ emission factor of each ton of standard coal is supposed to be 2.23

*Data source* Report on Industrial Analysis and Investment Consultation of the Coal Gangue in China in 2007 and 2008 on http://www.ocn.com.cn/

(the four types of main mining equipment) and operation of the main energy-consuming equipment (the four types of main mining equipment) of coal enterprises is very important for the energy consumption of coal enterprises.

According to statistics, about 40% of the energy-consuming equipments in all coal mines in China are backward and should be washed out; the original state-owned coal lines have an average efficiency of only 60% for their main fans and main drainage pumps and an operating efficiency below 55% for their systems, 15% lower than the foreign advanced level. The variable-frequency speed control technology can reduce the energy consumptions of mining equipment effectively. Some research institutes are studying key technologies and equipment such as the non-constant torque load, high-voltage, large-power and variable-frequency control over fans and pumps, local dynamic capacitance compensation for the power supply networks of coal mines, and energy-saving and centralized control over the belt conveyors for the coal transportation in coal mines, in a bid to develop variable-frequency controllers and dynamic reactive power compensation devices for coal mines, and energy-saving centralized control systems for the belt conveyors for the coal transportation in coal mines to realize energy saving in mine production.

4. Higher coal washing rate and higher thermal efficiency of boilers

Data show that the thermal efficiency of boilers reduces by 0.7–0.9% for each 1% of coal ash increased. In 2010, the national total coal consumption was 3.12 billion t, including 1.55 billion t for power generation and about 0.68 billion t for industrial boilers, and in other words, the coal consumption by boilers was 2.2 billion t and 705 of it was directly burned raw coal. If the raw coal is washed to remove the pure gangue (10–15%) before combustion, ash of the commercial coal will reduce by 7% and the thermal efficiency of boilers will increase by 5% at least. As a result, in each year, 0.1 billion t of coal will be saved and at least 0.22 billion t of $CO_2$ emission will be reduced. If the $CO_2$ emission during washing is detected (the indirect $CO_2$ emission during washing of the 2.2 billion t of raw coal was about 1470 ten thousand t), more than 0.2 billion t of $CO_2$ emission will be reduced.

Higher washing rate of coal can reduce the transportation load of coal gangue greatly and the energy consumption in long-distance transportation, thus reducing carbon emission. Furthermore, mining areas can take actions such as underground coal washing plants and simplified coal production systems to reduce carbon emission. Underground coal washing plants can avoid transportation of waste solids out of coal mines, thus reducing the energy consumption in transportation in mining areas; mine production systems can be simplified if condition permits, thus reducing $CO_2$ emission by means of simplified coal production systems and lower energy consumptions.

5. Potential of emission reduction

In the low-carbon development of mining areas, the most important thing is to control carbon sources. The mining areas in China mainly have the following two carbon emission sources: mine gas emission, which is more than 15 billion $m^3$ (equivalent to 0.225 billion t of $CO_2$) each year; indirect $CO_2$ emission due to energy consumption in coal production, which is about 0.1 billion t each year. Extensive development and utilization of low-concentration mine gas can reduce the energy consumptions of main mining equipment and further reduce $CO_2$ emission effectively. In addition, the low-temperature heat source of coal mines can be utilized effectively. China has a reserves of low-temperature heat source of coal mines of about 0.2 billion kW. In the case of complete utilization of the reserves, nearly 0.1 billion t of $CO_2$ emission can be reduced each year. Higher washing rate of coal can improve the thermal efficiency of boilers effectively, thus reducing the $CO_2$ emission. The $CO_2$ emission reduction is more than 0.12 billion t each year.

## 5.3.2 *Power Development*

### 5.3.2.1 Energy Saving and Consumption Reduction

1. Main actions for $CO_2$ emission reduction

(1) Optimize the structure of units. The current state policy for thermal power units is 'encourage large power plants and discourage small energy-inefficient power plants.' The conventional thermal power units with a unit installed capacity below 5 ten thousand kW will be closed down gradually; the conventional thermal power units with a service life of 20 years at least and a unit installed capacity below 10 ten thousand kW, the units with expired design service lives and a unit installed capacity below 20 ten thousand kW and the coal-fired units with a power supply standard coal consumption at least 10% higher than the provincial, municipal, or autonomous-region levels in the year of 2005 or at least 15% higher than the national average in the year will be closed down; conventional coal-fired thermal power units with a unit installed capacity below 30 ten thousand kW, the units with a power generation coal consumption higher than 300 g standard coal/kWh, and the air-cooled units with a power generation coal consumption higher than 305 g standard coal/kWh will be limited. In the twelfth Five-Year Plan period, even the small thermal power units with a unit installed capacity below 12.5 ten thousand kW or 20 ten thousand kW will probably be closed down, thus optimizing structure of the coal-fired units quickly.

The currently extensively used power generation technologies of direct coal combustion mainly include ultra-supercritical combustion, supercritical combustion, subcritical combustion, circulating fluidized bed combustion. For better generating efficiency, ultra-supercritical combustion and supercritical combustion are the main choices among all the power generation technologies of direct coal combustion.

(2) Improve the technical level of power generation. The main technologies of coal-based power generation in the world include conventional power generation through direct coal combustion and the new power generation through coal gasification.

In accordance with policies such as Twelfth Five-year Plan for Energy Development, Interim Provisions on Promoting Industrial Structure Adjustment, Several Opinions on Accelerating Decommissioning of Small Thermal Power Units, and Comprehensive Working Program on Energy Saving and Emission Elimination, the power industry will vigorously develop ultra-supercritical units with a unit installed capacity of at least 60 ten thousand kW, cogeneration units with a unit installed capacity of at least 30 ten thousand kW, heating, power, and cooling cogeneration units and big combined cycle units, use fluidized bed boilers for the units with a unit installed capacity of at least 20 ten thousand kW, utilize coal gangue or inferior coal for power generation, etc. In the regions mainly with industrial thermal loads, construct back pressure units mainly with thermal forces; in the regions with centralized heating loads or large potential, construct efficient and environment-friendly cogeneration units with a unit installed capacity of 30 ten thousand kW; in medium- and small-sized cities, construct heat, power, and gas cogeneration systems mainly based on the circulating fluidized bed technology, distributed cogeneration units, and heating and power and cooling cogeneration unit using clean energy sources as fuels, and change coal-fired small coal-fired boilers with decentralized heat supply into ones with centralized heat supply.

Researches show that, for $CO_2$ emission reduction, the coal gasification power generation technology is more technically and economically advantageous than the conventional coal-fired power generation technologies. Coal gasification power generation is the most important method to improve generating efficiency because it makes use of technologies such as gas turbine–steam turbine combined cycle and future high-temperature fuel cell—gas turbine combined cycle. According to the existing IGCC demonstration power plants in the world, the IGCC power plants with a unit installed capacity of 30 ten thousand kW have net efficiency up to the ultra-supercritical units with a unit installed capacity of 100 ten thousand kW.

(3) Improve suitability between the coal for power generation and power units. From 2004 to 2006, due to insufficient coal supply, inferior coal or other types of coal with large deviations from the design coal type had to be used in power generation, which affected the operating efficiencies of units and increased the unit consumption indexes of coal. It is estimated that in the next several years, coal supply and demand will be basically balanced and suitability between the coal for power generation and power units will continually be basically acceptable, helping to reduce the unit consumption indexes of coal consumptions by power generation and heat supply.

(4) Reduce auxiliary power ratios. The power saving of thermal power plants is important in energy saving and consumption reduction. Actions such as better production management, equipment selection, and technical upgrading can reduce the power consumptions of power plants and realize an economic benefit, i.e., generating cost reduction, and an environmental benefit, i.e., environmental pollution alleviation, because the power saved is the cleanest and useful for $CO_2$ emission reduction.
(5) Reduce line losses. After development of years, China now has an increasingly larger grid scale and more advanced grid technologies. The grids in China have 46.9 ten thousand km of transmission line circuits of 220 kV or above; inter-provincial and inter-regional 500 kV backbone grid frameworks have gradually formed; 750 kV power transmission demonstration projects have completed; 1000 kV extra high voltage and ±800 kV DC experiment and demonstration projects have been put into service. At present, the grids in China are the world largest and start grids are under positive research and development. In China, the resource allocation optimization on a larger scale has reduced line loss rates.

2. Energy saving, consumption reduction, and $CO_2$ emission reduction

In the power industry, energy saving, consumption reduction, and $CO_2$ emission reduction are mainly realized through the following ways: Reduce total coal consumption of the thermal power industry: In the case of the same generation, a smaller standard coal consumption will generate a smaller $CO_2$ emission; reduce line loss rates: In the case of the same power transmission capacity, a smaller power consumption in transmission will generate a smaller $CO_2$ emission because it equivalently saves the coal consumption of power generation.

This book analyzes potential and benefit of the greenhouse gas emission reduction in the power industry and focuses on the research ideas and methods. The index data assumed in this book are for reference only. According to change characteristics and laws of the indexes about coal consumption of power supply, auxiliary power ratio, and line loss rate, it is forecast that the national coal consumption of power supply, the power consumption rate of thermal power plants, and the line loss rate will drop to reduce to 330 g/kWh, 5.4, and 6.6%, respectively, by the year of 2015, and 320 g/kWh, 5.2, and 6.5%, respectively, by the year of

**Table 5.4** Changes of the carbon emission reduction indexes in coal-fired power generation

| Year | Coal consumption of power supply (g/kWh) | Power consumption rate of thermal power plant (%) | Coal consumption of power generation (g/kWh) | Line loss rate (%) |
|---|---|---|---|---|
| 2005 | 370 | 6.9 | 343 | 7.21 |
| 2010 | 337 | 5.6 | 318 | 6.70 |
| 2015 | 330 | 5.4 | 313 | 6.60 |
| 2020 | 320 | 5.2 | 306 | 6.50 |

**Table 5.5** Forecast generations in the power industry in China (trillion kWh)

| Item | 2011 | Standard of forecast | 2015 | 2020 |
|---|---|---|---|---|
| Generation | 3.9 | High | 6.08 | 7.67 |
| | | Middle | 5.59 | 7.21 |
| | | Low | 5.22 | 6.60 |

**Table 5.6** Forecast generation of thermal power and the forecast generation of coal-fired power in China (trillion kWh)

| Year | National generation | Generation of thermal power | Generation of coal-fired power |
|---|---|---|---|
| 2011 | 4.73 | 3.90 | 3.60 |
| 2015 | 6.08 | 4.57 | 4.43 |
| 2020 | 7.67 | 5.52 | 5.28 |

2020. Suppose these three indexes tend to reduce year by year, the changes will be as in Table 5.4.

(1) Estimate of the generation of thermal power: Table 5.5 shows the generations in the power industry in China in 2015 and 2020 forecast in accordance with speed and scale of the power development in China.
See Table 5.6 for the forecast generation of thermal power and the forecast generation of coal-fired power in China.
(2) Benefits of the carbon emission reduction of coal-fired power: According to the data of 2005 (base year) about coal consumption of power generation (343 g/kWh), line loss rate (7.21%), national generation (2.50 trillion kWh), generation of thermal power (2.04 trillion kWh), and national generation of coal-fired power (2.02 trillion kWh), the benefit of $CO_2$ emission reduction because of reduction of the coal consumption of power generation and reduction of the line loss rate in the coal-fired power generation in China are forecast (Table 5.7).

According to the forecast above, with reduction of total coal consumption and reduction of the line loss rate, the benefit of $CO_2$ emission reduction because of energy saving and consumption reduction in the power industry has been increasing. It is assumed that total $CO_2$ emission reduction of coal-fired power will be about 15.32 0.1 billion t in the twelfth Five-Year Plan period and about 38.82

**Table 5.7** Forecast of benefit of $CO_2$ emission reduction in coal-fired power (0.1 billion t)

| Year | Coal consumption reduction | Line loss reduction | Sub-total |
|---|---|---|---|
| 2010 | 2.24 | 0.16 | 2.40 |
| 2015 | 3.69 | 0.28 | 3.97 |
| 2020 | 5.43 | 0.42 | 5.85 |
| 2011–2015 (total) | 15.32 | 1.15 | 16.47 |
| 2011–2020 (total) | 38.82 | 2.96 | 41.78 |

0.1 billion t from 2011 to 2020. The action of reducing total coal consumption contributes more than 90% on average to $CO_2$ emission reduction, making it an important action for carbon emission reduction in the power industry in future. However, total coal consumption cannot reduce all the time due to the limitation by power generation technologies. In the long run, $CO_2$ emission reduction in the power industry has a bottleneck.

#### 5.3.2.2 Optimization of the Power Structure

1. Hydropower

Hydropower is a technically mature, economically feasible, and extensively developable renewable clean energy source. The development of hydropower resources is significant for energy safety of the state, energy structure optimization, reduction of both pollutants and emission of greenhouse gases, and coordinated regional development. The hydropower resources in China have advantages including abundance, low development level, and great development potential, but they have constraints including ecology, immigration, land submergence.

According to the result of the latest national hydropower resource recheck, installed capacities of the hydropower resources in China, including theoretical installed capacity, technically exploitable installed capacity, and economically exploitable installed capacity, are all the world largest. Of the hydropower resources in China, the theoretical installed capacity is 0.694 billion kW, with an annual theoretical generation of 6.08 trillion kWh; the technically exploitable installed capacity is 0.542 billion kW, with an annual technically exploitable generation of 2.47 trillion kWh; the economically exploitable installed capacity is 0.402 billion kW, with an economically exploitable generation of 1.75 trillion kWh. At present, the installed capacity of hydropower in China is 0.173 billion kW, suggesting a huge potential of development. Furthermore, thanks to the development of related hydropower technologies, the generating efficiency and utilization rate of hydropower will keep increasing.

Hydropower development is strategically significant because of the abundant hydropower resources and increasingly worse environmental situation in China. It can improve the energy supply capacity of China and optimize the energy structure of China. China is one of the few countries with energy production and consumption mainly based on coal. In China, the percent of coal in the primary energy source consumption structure is about 40% higher than the world average; the energy structure is simple; environmental benefit is small; and other quality substitute alternative energy sources are in an urgent need. Furthermore, hydropower development can alleviate the energy supply shortage in China, optimize the energy structure, and reduce the emission of greenhouse gases and environmental pollution. With progress of the industrialization process, the economy of China will continue to grow quickly and the energy demand of China will increase continually. The contracting countries of UNFCCC and Kyoto Protocol are main countries emitting greenhouse gases. In future, China will face an unprecedented pressure of emission reduction from the international community. Hydropower development is an active and effective action for positive addressing of global climate changes because of its role in reducing coal consumption, air pollutant emission, and $CO_2$ emission due to coal combustion.

In the early period of new China, the national installed capacity of hydropower was 36 ten thousand kW only. However, the value leaped to 0.196 billion kW in late 2009, becoming the world's first. At present, hydropower takes 22% of total installed capacity of the power industry in China and 16% of total generation of China, following hydropower in the whole power structure. Compared with other renewable energy sources in the whole power structure, hydropower has the largest percent, so it is strategically significant for energy balance and sustainable development of the energy industry as well.

2. Other clean energy sources

(1) Wind power: As a clean renewable energy source, wind power is useful for more energy supply, energy restructuring, and ecological environment protection in China.

China has abundant wind energy resources. According to the result of the latest survey about the wind energy resources in China, the wind energy resources at least 10 m above the surfaces of land areas have preliminarily verified total reserves of 43.5 billion kW, including about 3 billion kW which is technically exploitable. The wind energy resources at least above wind turbine hubs have a technically exploitable level of about 0.6 billion kW distributed in most regions in Northwest China, the northern part of North China, the northern part of Northeast China, the hinterland of Tibet Plateau, coastal regions, etc. In addition, according to observation data of the more than 900 meteorological stations in China, the wind energy reserves at least 10 m above offshore (seawater depth: 10 m) sea surfaces are about 0.75 billion kW.

In the twenty-first century, particularly after the issuing of Renewable Energy Law of the People's Republic of China in 2006, the wind power in China has realized leap-forward development: Total installed capacity of grid-connected wind power increased from 35 ten thousand kW in 2000 to 6142 ten thousand kW in 2012; technologies about and experience in the operation management of wind farms have accumulated, laying a foundation for the extensive development and utilization of wind power.

(2) Photovoltaic energy: China has abundant solar energy resources, which is a foundation for solar energy resources to become an alternative energy source in future. Driven by the international market, the solar PV industry in China has realized leap-forward development: For example, PV cell modules had productivity below 1 ten thousand kW in the year of 2000, but total installed capacity of grid-connected PV power generation reached 328 ten thousand kW in late 2012.

(3) Nuclear energy: Nuclear energy is a clean, safe, and economical energy source, an alternative energy source allowing extensive development and a major strategic choice for the safety of energy supply as well. With a main application in power generation, it has great potential in energy structure optimization, greenhouse gas emission reduction, and solving of energy problems. Development of nuclear energy in China began in 1985 when the construction of Qinshan Nuclear Power Plant began formally. In late 2012, total installed capacity of nuclear power was 1257 ten thousand kW, and the installed capacity of the units under construction was 3273 ten thousand kW. It is the strategic choice for the power industry in China to generate power with nuclear energy and increase the percent of nuclear power in the power structure.

3. Carbon emission reduction benefit of clean energy sources

The $CO_2$ emission reduction benefit of the clean energy sources used for power generation mainly lies in a substitution effect for the generation of coal-fired power: Compared with coal-fired power generation, the power generation with clean energy sources generates almost no greenhouse gases (the generation is the same). Thus, the larger the percent of clean energy sources in the power structure, the better the substitution effect will be.

Development tendency of the clean energy sources in China is analyzed and judged in accordance with related plans and research results. See Table 5.8.

Suppose the installed capacity and generation of clean energy sources will have a proportional increased from 2010 to 2020, the generation of clean energy sources will change as in Table 5.9.

In 2005 (base year), hydropower, nuclear power, and other renewable energy sources had a generation of 396.4 billion kWh, 53.1 billion kWh, and 4.3 kWh, respectively, and the coal consumption of power generation was 343 g/kWh. On this basis, the $CO_2$ emission reduction benefit because of substitution of coal-fired power generation by the net increase in generation of clean energy sources from 2011 to 2020 is forecast. See Tables 5.10 and 5.11.

According to the forecast above, the $CO_2$ emission reduction because of substitution of the generation of coal-fired power by the generation of clean energy sources has been increasing with increase in the installed capacity and generation of clean energy sources. Based on the research conditions, the power industry will have a total $CO_2$ emission reduction of about 3.334 billion t in the twelfth Five-Year Plan period and about 9.73 billion t in the period from 2011 to 2020 by means of energy source substitution. To $CO_2$ emission reduction, hydropower will make the largest contribution and nuclear power, wind power, and other clean energy sources will make quickly increased contributions, so there is huge potential of $CO_2$ emission reduction.

**Table 5.8** Installed capacities and generations in the power industry in China

| Year | | Total | Thermal power | Hydropower | Nuclear power | Renewable energy sources such as wind power and solar energy |
|---|---|---|---|---|---|---|
| 2005 | Installed capacity (0.1 billion kW) | 5.2 | 3.9 | 1.2 | 0.1 | 0.0 |
| | Generation (trillion kWh) | 2.50 | 2.04 | 0.40 | 0.05 | 0.00 |
| 2010 | Installed capacity (0.1 billion kW) | 9.52 | 7.00 | 1.96 | 0.10 | 0.32 |
| | Generation (trillion kWh) | 4.11 | 3.31 | 0.65 | 0.08 | 0.07 |
| 2015 (forecast) | Installed capacity (0.1 billion kW) | 14.35 | 9.7 | 2.9 | 0.45 | 1 |
| | Generation (trillion kWh) | 6.08 | 4.57 | 0.96 | 0.35 | 0.2 |
| 2020 (forecast) | Installed capacity (0.1 billion kW) | 18.52 | 11.8 | 3.5 | 0.8 | 1.8 |
| | Generation (trillion kWh) | 7.67 | 5.52 | 1.16 | 0.62 | 0.37 |

**Table 5.9** Changes of generation of clean energy sources (trillion kWh)

| Year | Hydropower | Nuclear power | Renewable energy sources such as wind power and solar energy |
|---|---|---|---|
| 2010 | 0.65 | 0.08 | 0.07 |
| 2015 | 0.96 | 0.35 | 0.20 |
| 2020 | 1.16 | 0.62 | 0.37 |

#### 5.3.2.3 Strengthening of DSM

Demand-side management (DSM) is a mode for power users to carry out power saving and load management. DSM projects can benefit the society, power

**Table 5.10** Generation substituted by clean energy sources from 2011 to 2020 (trillion kWh)

| Year | Hydropower | Nuclear power | Renewable energy sources such as wind power and solar energy | Total generation substituted |
|---|---|---|---|---|
| 2010 | 0.25 | 0.03 | 0.07 | 0.35 |
| 2015 | 0.56 | 0.30 | 0.20 | 1.06 |
| 2020 | 0.76 | 0.57 | 0.37 | 1.70 |
| 2011–2015 (total) | 2.15 | 0.81 | 0.66 | 3.62 |
| 2011–2020 (total) | 5.55 | 3.05 | 2.11 | 10.72 |

**Table 5.11** $CO_2$ emission reduction benefit of clean energy sources from 2011 to 2020 (0.1 billion t)

| Year | Emission reduction of hydropower | Emission reduction of nuclear power | Renewable energy sources such as wind power and solar energy emission reduction | Total emission reduction |
|---|---|---|---|---|
| 2010 | 2.38 | 0.28 | 0.62 | 3.27 |
| 2015 | 5.17 | 2.75 | 1.80 | 9.72 |
| 2020 | 6.79 | 5.07 | 3.25 | 15.12 |
| 2011–2015 (total) | 19.86 | 7.44 | 6.14 | 33.44 |
| 2011–2020 (total) | 50.49 | 27.66 | 19.15 | 97.30 |

companies, and power users. For the society, DSM can reduce power demands and the consumptions and pollutant emissions of primary energy sources; for governments, DSM can reduce power consumptions, promote sustainable development of the economy and upgrading of electrical equipment, increase the demand for efficient equipment, and realize increase in GDP; for power users, DSM can reduce power consumptions, cut down the production and operation costs and electric charges of enterprises, and improve the energy efficiencies and product competitiveness of enterprises; for power suppliers, DSM can lower the pressure of peak load regulation in peak periods, improve the power supply reliability and service, and, particularly in the supply–demand imbalance of power, lower the pressure of power rationing greatly; for power producers, DSM can improve the utilization rate of power generation equipment and reduce the generating costs, the coal consumption of power generation, and the investments for power construction.

In the DSM in the power industry, the capabilities to develop and dispatch demand-side resources can be observed from capacity of the efficiency power plant (EPP). According to related research data, the $CO_2$ emission reduction benefit because of strengthening of DSM in the power industry in China is described in Table 5.12.

**Table 5.12** $CO_2$ emission reduction benefit because of strengthening of DSM in the power industry

| Year | EPP (ten thousand kW) | Power saved (0.1 billion kWh) | $CO_2$ emission reduction (ten thousand t) |
|---|---|---|---|
| 2010 | 1480 | 119.1 | 1004 |
| 2015 (forecast) | 4672 | 785.8 | 6402 |
| 2020 (forecast) | 14,200 | 2855.9 | 23,268 |

#### 5.3.2.4 Carbon Emission Reduction in the Power Industry

Total $CO_2$ emission reduction in the power industry in China from 2011 to 2020 is forecasted in Table 5.13 and Figs. 5.13 and 5.14.

According to the forecast above and based on the research conditions, the power industry will have a total $CO_2$ emission reduction of about 5 billion t in the twelfth Five-Year Plan period and about 13.9 billion t by the year of 2020 by means of energy saving and consumption reduction in coal-fired power, line loss reduction, and development of clean energy sources. To $CO_2$ emission reduction, hydropower will make the largest contribution and nuclear power and other clean energy sources will make quickly increased contributions. In consideration of the technical difficulties in energy saving and consumption reduction in the power industry, power restructuring will be the main method for future $CO_2$ emission reduction in the power industry in China.

When no actions for $CO_2$ emission reduction are considered, the power industry will have a huge increase in $CO_2$ emission in future, i.e., 3.41 billion t in 2015 and

**Table 5.13** Forecast of total $CO_2$ emission reduction (0.1 billion t)

| Year | Energy saving and consumption reduction | | Power restructuring | | | Total emission reduction |
|---|---|---|---|---|---|---|
| | Reduction of the coal consumption | Reduction of the line loss rate | Substitution by hydropower | Substitution by nuclear power | Substitution by other renewable energy sources | |
| 2010 | 2.24 | 0.16 | 2.38 | 0.28 | 0.62 | 5.68 |
| 2015 | 3.69 | 0.28 | 5.17 | 2.75 | 1.80 | 13.69 |
| 2020 | 5.43 | 0.42 | 6.79 | 5.07 | 3.25 | 20.96 |
| 2011–2015 (total) | 15.32 | 1.15 | 19.86 | 7.44 | 6.14 | 49.91 |
| 2011–2020 (total) | 38.82 | 2.96 | 50.49 | 27.66 | 19.15 | 139.08 |

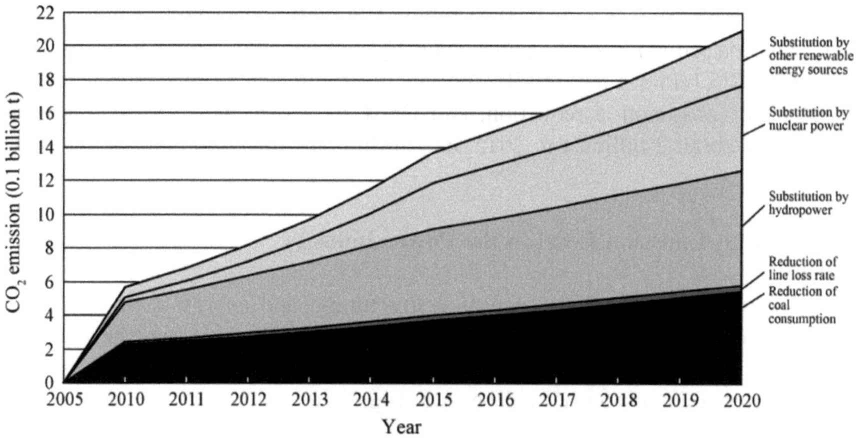


**Fig. 5.13** Forecast of $CO_2$ emission reduction in the power industry in China from 2011 to 2020 (base year: 2005)

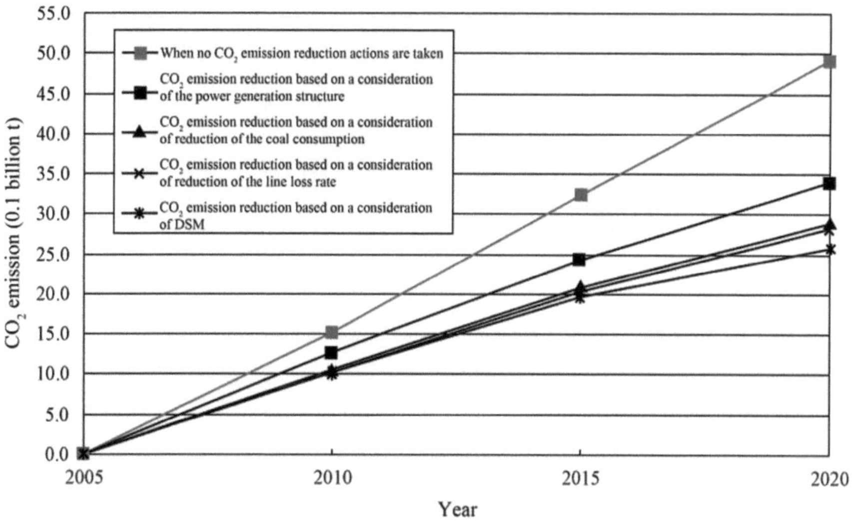


**Fig. 5.14** Effects of different emission reduction actions for total $CO_2$ emission reduction of the power industry (base year: 2005)

4.93 billion t in 2020, compared with 2005; when power restricting and substitution of partial coal-fired power by clean energy sources are considered, $CO_2$ emission reduction of the power industry will be about 0.97 billion t in 2015 and about 1.51 billion t in 2020; when reduction of total coal consumption is considered, $CO_2$ emission reduction of the power industry will be about 0.37 billion t in 2015 and about 0.54 billion t in 2020; when reduction of the line loss rate is considered, $CO_2$ emission reduction of the power industry will be about

0.03 billion t in 2015 and about 0.04 billion t in 2020; when DSM is considered, $CO_2$ emission reduction of the power industry will be about 0.06 billion t in 2015 and about 0.23 billion t in 2020. To sum up, if actions such as restructuring, energy saving, and consumption reduction, and DSM are taken, the $CO_2$ emission will increase by about 2 billion t in 2015 and 2.6 billion t than 2005.

#### 5.3.2.5 $CO_2$ Emission Level in the Power Industry

If emission reduction actions such as restructuring, and energy saving and consumption reduction are considered, the $CO_2$ emission in power generation in the power industry will be as in Fig. 5.15 and the $CO_2$ emission of unit generation (the generation of thermal power) will be as in Fig. 5.16.

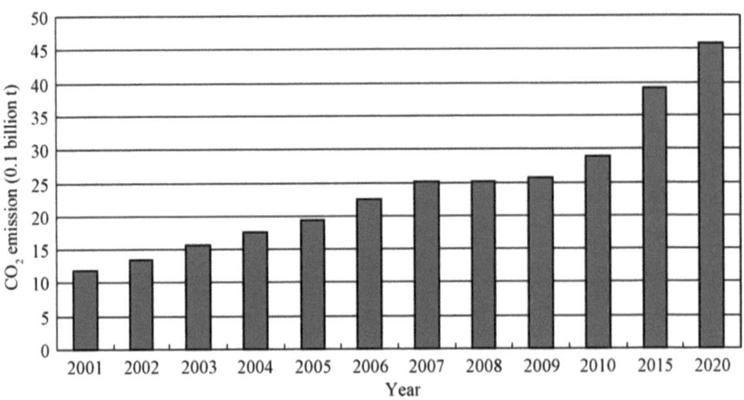


**Fig. 5.15** $CO_2$ emission in power generation in the power industry in China

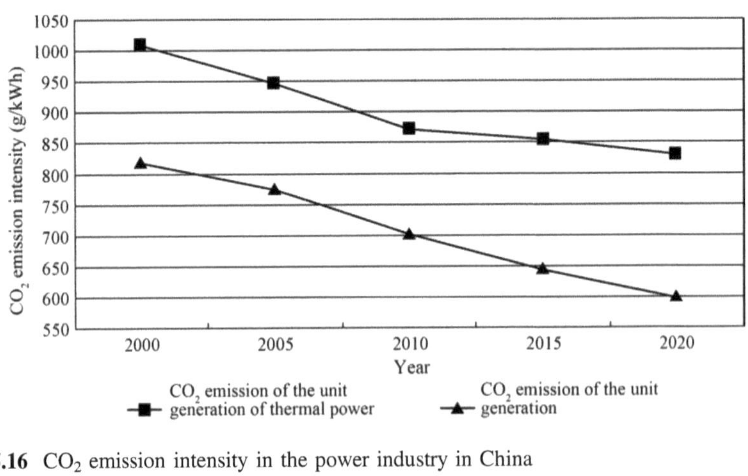


**Fig. 5.16** $CO_2$ emission intensity in the power industry in China

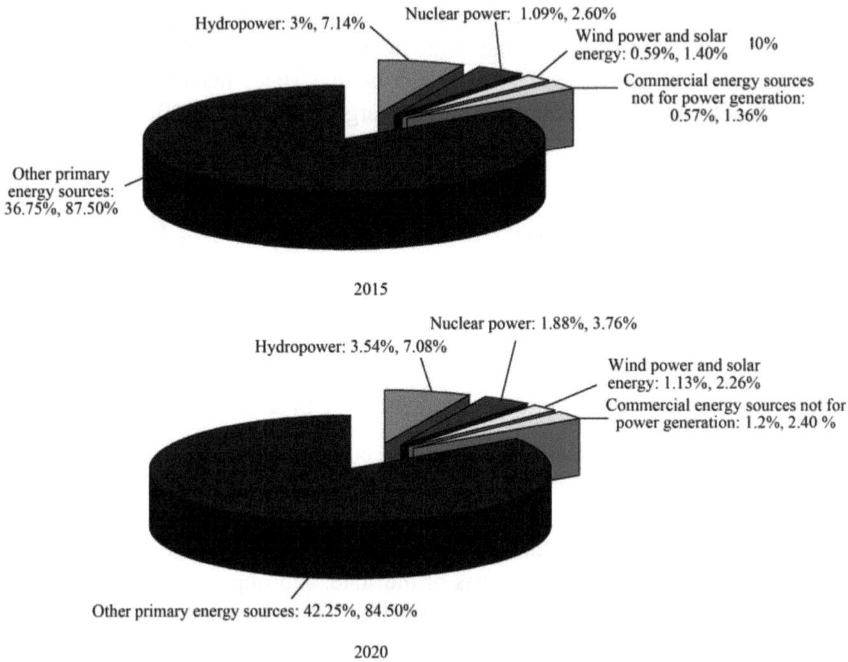


**Fig. 5.17** Percents of non-fossil energy sources in primary energy sources in China in 2015 and 2020

According to Figs. 5.15 and 5.16, $CO_2$ emission of the power industry in China will have the following characteristics:

(1) Larger total $CO_2$ emission but smaller increase rate: Net increase in $CO_2$ emission of the power industry was about 1 billion t in the eleventh Five-Year Plan period and will be about 1 billion t in the twelfth Five-Year Plan period and about 0.7 billion t in the thirteenth Five-Year Plan period.
(2) Decline of $CO_2$ emission intensity of the power industry: For years 2005, 2010, 2015, and 2020, the forecast $CO_2$ emissions of unit generation are about 774, 702, 643, and 598 g/kWh, respectively, and the forecast $CO_2$ emissions of the unit generation of thermal power are about 946, 872, 856, and 831 g/kWh, respectively.
(3) From 2011 to 2020, the $CO_2$ emission reductions of unit generation will be larger than the $CO_2$ emission reductions of the unit generation of thermal power, suggesting that power restructuring is an effective method contributing much to $CO_2$ emission reduction in the power industry.

As in Fig. 5.17, percents of the non-fossil energy sources in primary energy sources in years 2015 and 2020 are forecast with an equivalent caloricity method in this book.

According to Fig. 5.17, percents of the non-fossil energy sources in primary energy sources in years 2015 and 2020 will be 11.5 and 15.5%, respectively, basically realizing the objective proposed by the state; namely, the non-fossil energy sources should be 15% in primary energy sources in 2020.

## 5.4 Ideas of Main Energy Sectors for Addressing Climate Changes

### *5.4.1 Coal Industry*

1. Establish a guiding ideology of comprehensive treatment.

According to requirements of the state, the coal industry must reduce pollutant emissions vigorously, strengthen the comprehensive treatment of wastewater, waste gas ($CH_4$), waste solids (coal gangue), focus on solving the serious high-carbon problem influencing people following coal output increase, increase fund investments, try to gain preferential policies of the state, and improve environment legal system and the ecological environment recovery ratio by means of technological progress. Carry out restructuring for the coal industry to realize healthy development for the coal economy.

2. Strength management by low-carbon objectives.

In the twelfth Five-Year Plan period which is the key period for putting into practice the spirits of the 17th National Congress of CPC in a bid to build a well-off society in an all-around way in China, healthy development of the coal industry is particularly important, so it is necessary to strengthen management of the strategic objectives useful for ecological environment protection. By 2005, in addition to the increase in the coal output, total emission of the main pollutants of coal enterprises should be reduced by 10%; the tendency of worsening of the ecological environment should be basically be curbed; the mines where conditions permit should be built into new mines integrating coal, power, and gas.

3. Realize scientific planning and management.

Characteristics of the coal industry in China differ with the specific region, geological conditions, coal development mode, and ownership. Thus, it is necessary to carry out scientific zoning and detailed guidance and work out detailed implementation plans in accordance with the specific conditions. In the process of planning, highlight strategic emphases and use clean production technologies completely, in a bid to construct environment-friendly mining areas with good economic benefits, low resource consumptions and less serious environmental pollution.

### 5.4.2 Power Industry

1. Strengthen power restructuring.

Power restructuring in accordance with stage characteristics of the power industry in China and characteristics of primary energy sources is the most effective action for $CO_2$ emission reduction. On the precondition of ecological environment protection, develop hydropower orderly and vigorously because increase in total hydropower will reduce $CO_2$ emission effectively, in regardless of the impossibility of increase in the percent of hydropower because of its gradually decreased developable capacity; positively coordinate the relation between power generation with new energy sources mainly including wind power and solar energy and grids and energy power generation support; construct nuclear power projects actively; keep improving the generating efficiency and clean utilization of coal-fired power. Continue to implement the 'encourage large power plants and discourage small energy-inefficient power plants' policy, and increase the percent of heating, power, and cooling cogeneration. Keep optimizing the power structure and strengthen energy conservation and emission reduction for the existing power plants and technical upgrading for grid losses, in a bid to keep optimizing the reserves.

2. Speed up construction of the market mechanisms useful for the emission reduction of greenhouse gases.

Promote energy-saving generation dispatching actively; work out, improve, and implement the supporting actions in Measures for Energy-Saving Generation Dispatching; promote $CO_2$ emission trading mechanisms in accordance with the interests and characteristics of China; take part in international competition initiatively; and have the initiative and control $CO_2$ emission with the most economical actions.

3. Speed up the joint control over conventional pollution, combined pollution, and $CO_2$ emission.

Joint control usually has an overall benefit, and, in other words, an optimal overall benefit can be realized by means of saving and overall utilization of energy and water, and a cycle economy mode as well. Pollutant emission reduction, control over the energy consumption of per unit GDP, industrial water-saving, energy-saving objectives, and $CO_2$ emission intensity index in the twelfth Five-Year Plan period should be under multi-objective control and guaranteed by better comprehensive coordination and optimization. The key of joint control is to establish safe, clean, and economical full-life-cycle constraint evaluation index system and policy support system, design indexes and policies first according to the rules of stressing the essentials and circular progression, and coordinate from mechanisms.

4. Improve the law and regulation system and strengthen the supervision and administration according to law.

According to the concept of scientific development and the laws of market-oriented economy, keep strengthening the construction of related laws and regulations, speed up the issuing of an energy law and the revision of Electricity Law of the People's Republic of China, and legalize the ideas, indexes, systems, and industrial supervision duties about climate change addressing, energy conservation, and emission reduction. In addition, for energy conservation and emission reduction in the power industry, perform the functions of administration by governments and the duties of all parties involved, improve the law enforcement system, restore and improve the industrial monitoring and supervision system, and strengthen the supervision and administration according to law.

# Chapter 6
# Optimization of the Energy Development System

**Zhixuan Wang, Li Pan, Jingjie Zhang and Peitang Guo**

## 6.1 Ideas, Policies, and Objectives

### *6.1.1 Adhere to Correct Ideas About Energy Development*

Carry out the concept of scientific development firmly, change the way of energy development actively, and promote safe, clean, economical, and coordinated development of the energy industry. Take the domestic resource condition into consideration, follow the objective laws of energy development and economic development, and adhere to the strategy that coal takes the dominant position, electric power is the center, and oil, gas, and new energy sources develop comprehensively. Energy development should go with the tides, comply with the world's global, market-oriented, and low-carbon tendency of energy, and speed up the technical innovation of energy sources and the development of clean energy sources.

### *6.1.2 Adhere to Scientific Policies for Energy Development*

1. Insist appropriately advanced development of the energy industry continually.

Energy industry is the basic industry of a national economy. In the Twelfth Five-Year Plan period and the Thirteenth Five-Year Plan period, China will be in 'dual-speediness' development in the middle of industrialization, which is mainly represented by quick increase of energy development (the high- or medium-speed increase will be much slower than the increase in the two five-year periods above,

Z. Wang (✉) · L. Pan · J. Zhang · P. Guo
China Electricity Council, Beijing, People's Republic of China
e-mail: wangzhixuan@cec.org.cn

 
Z. Wang (ed.), *Constraints and Solutions for Energy and Electricity Development*,
https://doi.org/10.1007/978-981-10-8304-4_6

but will still be quickly viewed from the whole energy development course and the global energy development), which is determined by the laws of energy development and the essential characteristics of the development stage. This must be well understood. Thus, it is necessary to insist the policy of appropriately advanced development for the energy industry and keep enriching the meaning of this policy. Planning for total energy should be less strict than the approval of planned projects. With less strict planning, the earlier work of energy can be done in advance and projects can be optimized in the practical implementation of plans, in a bid to speed up restructuring, solve energy development problems which are still small, and improve the disorder in energy development. Incompliance between strict planning and development demands tends to worsen the disorderly construction of energy projects and cause more problems.

2. Insist on the energy development policy of conservation first continually.

The energy saved is the cleanest energy. Compared with the situation in advanced industrialized countries, the energy efficiency in China is still low, presenting potential of energy saving. However, the energy efficiency in China can be improved significantly by means of technological progress, economic restructuring, DSM, and energy-saving potential tapping in the energy industry. It is therefore necessary to continue to insist the policy of conservation first, speed up reform of the existing energy price policies, and promote energy saving through modern market-oriented economy systems.

Electric power is the best secondary energy source, and its saving is an important part of energy saving. The power industry should be regarded as a platform for the optimal allocation of energy resources and energy saving by the whole society. Electric power saving by the whole society should be promoted by means of coordinated development between grids and power sources (including distributed power sources), rational allocation of the units with different installed capacities, and development of smart grids.

3. Develop the domestic market and the international market simultaneously.

In future, coal should be mainly provided by domestic suppliers as before to comply with the increase of domestic demands. Make use of oil and gas resources on the international market vigorously while developing development of the oil and gas resources in China. Strengthen international multilateral cooperation to establish diversified petroleum supply systems. Create a bilateral and multilateral international environment useful for the cooperation in petroleum resources actively. Take positive political, economic, military, and diplomatic actions to make China develop better in overseas petroleum development. Speed up the development of natural gas through both domestic supply and foreign supply, and develop construction of both domestic main natural gas pipeline networks and natural gas pipeline networks for users.

### 6.1.3 *Set Scientific Constraint Indexes and Constraint Limits*

It is necessary in energy planning to set compulsory indexes and reference indexes for energy development. Compulsory indexes are constraint indexes such as coal consumption of power supply, water consumption of power generation, up-to-standard rate of $SO_2$ emission in the power industry, annual $SO_2$ emission in the power industry, up-to-standard rate of nitrogen oxide emission in the power industry, annual nitrogen oxide emission in the power industry (settable by region), and up-to-standard rate of dust emission in the power industry, and reference indexes are anticipated indexes such as overall utilization rate of coal ash, land reclamation rate of coal mines, utilization rate of coal gangue, and utilization rate of coal measure stratum gas.

## 6.2 Comprehensive Optimization

### 6.2.1 *Optimization by Means of Restructuring*

#### 6.2.1.1 Strengthen Coordinated Development of Coal and Power

1. Tendency of increasingly higher percent of the coal for power generation in total coal consumption.

In recent years, the consumption and increase of coal in China have mainly been happing in industries such as electric power, metallurgy, building materials, chemical industry, and agricultural production, and the coal consumptions in industries such as machinery and the coal consumption of living have been decreasing. Coal consumptions of the four industries including electric power, steel and iron, chemistry industry, and building materials have been the largest in China, accounting for 90% of the national coal consumption; the coal for power generation has accounted for more than 50% in of the national coal consumption: In the background that coal is the main part in the energy structure in China, this is an important sign of acceleration of the industrialization process and an inexorable law inevitably in this stage by all major industrialized countries in which coal is the main part of the energy structure.

The key is that China is now in the process of industrialization, while developed countries have realized economic transition. An energy-intensive and material-consuming export-oriented economy is an inevitable choice in the earlier period of modernization, increasing environmental and resource pressures of China. This industrial structure will have a longtime and far-reaching impact on the coal demand. Change of the economic growth mode is pressing, but it cannot be done at one stroke.

According to the domestic situation and the international division of labor, the heavy chemical industry in China will last for a long time.

From 1978 to 2013, the coal consumption of power generation increased by 17 times, from 0.109 billion to nearly 2 billion t. In 2013, the coal consumption of the power industry was about 2.05 billion t, taking 55.4% of the coal output of the year (Fig. 6.1). In the long run, percent of the coal for power generation will tend to increase with change of the economic growth mode in China. In the Twelfth Five-Year Plan period, coal has increased by 0.57 billion t, including 0.53 billion t as the coal for power generation. By 2020, coal will be 1.07 billion t more than that in 2010 and the coal for power generation will increase by 1.12 billion t; namely, all the coal increased will be consumed by power generation and the decreased 5000 ten thousand t of coal not used for the power industry will be consumed power generation too, resulting in a percent of the coal for power generation of 65.8% in the coal output. This tendency complies with the clean utilization requirement of coal.

China has been attaching great importance to adjustment and optimization of its power structure, but the generation of coal-fired power has been taking a percent of about 80%. In a long time, coal-fired power will remain as the most important mode of power generation in China and, with the process of modernization, percent of the coal consumption of power generation will increase while the coal consumptions in other industries will decrease gradually according to the related experience of developed countries. Thus, development of coal-fired power is the maximum power for development of the coal industry.

Electric power is the energy utilization method with the easiest centralized pollution control. Thanks to modern power technologies, the conventional pollutant emission of coal has been reduced to an economically acceptable low level completely meeting environmental quality requirement. Because of continuous promotion of the generating efficiency and continuous progress of cogeneration technologies, the emission of greenhouse gases in the power industry is now smaller than most other energy utilization methods such as industrial boilers. Therefore, compared with other coal-based energy utilization methods, electric

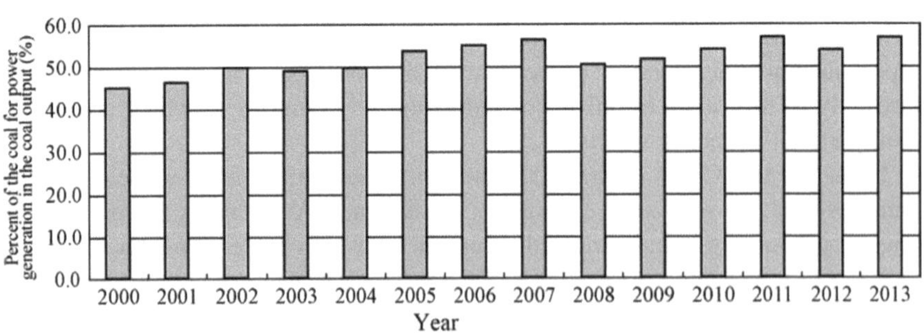


**Fig. 6.1** Percents of the coal for power generation in the coal outputs in China in different years. *Data source* China Electricity Council

power is the best choice for environmental protection. For the primary energy structure mainly including coal, increase of percent of the coal consumption of power generation is a main action for sustainable development of the energy industry. In most of industrialized countries in the world, their coal consumptions of power generation in their total coal consumptions are higher than 80% (Fig. 6.2). According to Fig. 6.2, percent of the coal consumption of power generation in China is lower than both the levels of developed countries and the world average.

The reasons include China does not have abundant petroleum or natural gas; coal is the main energy utilization method in other industries except for the power industry in China; coal is used as a raw material for chemical products, coke, etc.; there are hundreds of thousands of industrial boilers using coal in China.

2. Causes of the supply–demand imbalance of coal-fired power.

(1) In the first ten years in the twenty-first century, the heavy chemical industry in China had taken an increasingly larger percent in the national industrial added value and quick development of the energy-intensive industries had resulted in a supply shortage of coal, power, oil, and transportation. Excessively quick increase of coal, including the coal for power generation, and serious underestimate of the coal demand have resulted in a huge pressure for supply of the coal for power generation.

(2) Supply of the coal for power generation is limited. Both central and local governments tend to concern more about the safety of coal mines, strengthening the control over small coal pits and closing down the coal mines with lower capabilities and without necessary safety actions; many large- and medium-sized up-to-standard coal mines are still under construction. Thus, it is difficult to increase the coal output greatly soon. For example, after coal resource integration and coal mine merger and restructuring, both quantity and productivity of the coal mines in Shanxi have decreased: The original 2598 coal

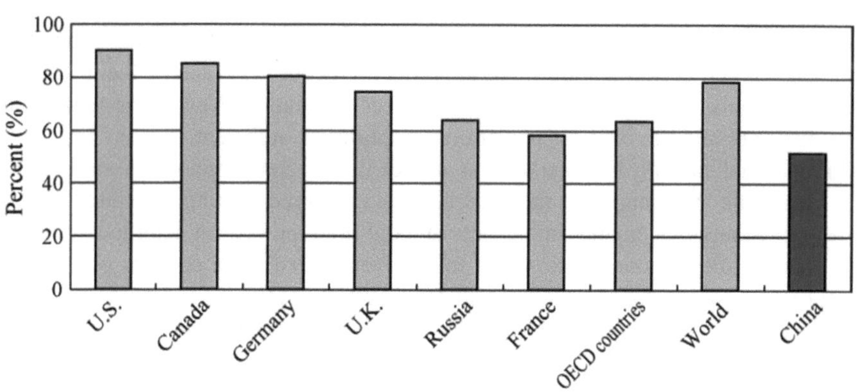


**Fig. 6.2** Percents of the coal consumptions of power generation in coal consumptions in main countries in the world

mines have decreased to 1053 ones, and all the coal mines with a productivity below 30 ten thousand t have been washed out.

(3) The geographical distribution of coal output and that of coal consumption are seriously imbalanced; the supply–demand imbalance of coal-fired power has been worsened by the limited transportation capability, particularly the limited railway transportation capability and local administrative barriers. The north-to-south coal diversion much depends on railway transportation.

1. The railway transportation capability increases quickly but still slows the coal demand increase, so the supply–demand imbalance of coal-fired power has been worsened by the transportation bottleneck.
2. The railway transportation efficiency needs to be improved; the railway transportation costs take larger percents in the coal costs (up to 50% in some regions).
3. Local administrative barriers interfere in both coal outputs and coal prices, affecting the suitability between regional coal supply and demands and resulting in supply shortage of the coal for power generation in some regions.
4. In coal circulation, some intermediate merchants hoard coal, affecting the normal supply of coal and resulting in supply shortage of the coal for power generation in some periods.

(4) An improper price mechanism is a deep-seated cause of the supply–demand imbalance of coal-fired power in China. For coal prices, China began to carry out a double-track pricing system and basically realized marketization in late 2005. In recent years, coal prices have been rising greatly, but the power prices in the related downstream industry have not been marketized; the 'coal-electricity price linkage' plan has not yet been implemented. As a result, coal-fired power enterprises have had much higher costs: Many of them suffered a large loss in 2008 and 2009. Consequently, these enterprises have become less active in supplying coal through market mechanisms. Furthermore, due to different opinions of coal enterprises and power enterprises on price of the coal for power generation, the contracts about coal for power generation signed and performed are small in number. Due to these factors, there will still be supply shortage of the coal for power generation even if the generating capacity and the coal for power consumption are sufficient.

(5) Seasonal factors: Factors such as higher power consumptions in summer by residents and municipal facilities and the coal-fired heating in winter tend to cause supply–demand imbalance of coal-fired power. In addition, extreme natural disasters and nationwide major events need more coal for power generation, affect supply of the coal for power generation, and curb the transportation capability, worsening the supply–demand imbalance of coal-fired power.

3. Optimum allocation of resources in a larger range by means of coordinated development of coal transportation and power transmission.

In China, coal resources and economic development have territorial imbalance: The logistic characteristics of 'north-to-south coal diversion,' 'west-to-east coal diversion' and 'coal output' and structural contradictions of the coal resources in China make the large-scale and long-distance transportation of energy sources in China inevitable. Thus, the relation between coal transportation and power transmission in the power structure in China is the key factor influencing China's distribution and efficient utilization of energy sources, environmental protection, and coordinated regional economy.

Most of the coal-fired power bases in west China and north China are basically 1000–3700 km from most of provinces and municipalities in east China and central China, and this distance is much larger than the technical transmission distance of 500 kV AC transmission lines (usually shorter than 900 km). Therefore, the extra high-voltage power transmission technology is now an important technical action for the long-distance and large-capacity energy transmission from main coal production bases to east China and central China and an important means for power transmission capability improvement.

Research data shows that the coal of a certain heat value has quite different prices in its places of origin and selling places: Longer economic distances of power transmission result in higher prices in intermediate links of coal transportation and higher economic efficiency of power transportation; compared with the coal with higher heat values, the coal with lower heat values is more economically efficient in power transmission than coal transportation; in the energy transmission from the coal-fired power bases in west China, northwest China, and southwest China to the load centers in east China, extra high-voltage AC power transportation and railway–sea coal transportation have basically the same energy transmission efficiency; in the energy transmission from the coal-fired power bases in Xinjiang to the load centers in east China and central China, extra high-voltage DC power transmission and railway coal transportation have basically the same energy transmission efficiency too.

In the Twelfth Five-Year Plan period, China should, based on the safety of electricity supply, carry out a national energy development strategy of extensive development and transmission of the energy sources of the coal-fired power bases to other regions in need in China, optimize the extra high-voltage power transmission technology as an important energy transmission method, speed up the construction of extra high-voltage power transmission projects, and promote the coordinated development between power sources and grids and between the grids of different voltage levels.

With breakthrough of the extra high-voltage power transmission technology, there will be a new method for coordinated development between coal and power. For example, develop the coal resources in Xinjiang rationally, find resource development intensity balance between inlands and remote regions, reduce the resource development intensities in inlands properly in a bid to reserve resources for future development, develop the resources in remote regions vigorously by increasing the local coal consumptions of power generation in a bid to provide clean energy sources for east China, central China, and the coastal regions in China. The state should actively construct transportation corridors for the transportation of coal from Xinjiang to other places in need in China and compile preferential policies to cut down the development and operation costs of coal enterprises. Furthermore, coal bases such as the Shendong coal base and the northern Shanxi coal base have a small quantity of pithead power plants and realize west-to-east coal diversion mainly through Shenhua Railway and Daqin Railway. For the new coal projects in places such as northern Shaanxi, Inner Mongolia, and eastern Ningxia, speed up the percentage of their coal consumptions of power generation through new efficient clean coal power generation technologies such as IGCC to realize local energy transformation.

#### 6.2.1.2 Compile Plans for Solving the Problems of Coal for Power Generation Effectively

1. Suitability between the thermal power structure and the coal supply system.

Strengthen the compilation of energy plans, and compile power plans and coal development plans in a coordinated and simultaneous way. Coordinate the power structure and grid development, plan for thermal power projects at the same time of the planning of the earlier work of coal supply projects and grid projects, and construct and put into service power source projects, coal resource development projects, and grid projects simultaneously, so as to guarantee fuel supply and power transmission.

In consideration of the layout and flow directions of the coal resources in China, give priority to the development of pithead power plants, construct coal-fired power bases, and carry out coal and power pool; optimize load center power plants, and construct advantageous port or crossroad power plants in the load centers in economically developed east China and central China with high power price affordability but shortage of primary energy sources, including heating and power cogeneration power plants, gas-fired power plants, and IGCC power plants; improve the capability to utilize external coal resources; use large-capacity, ultra-supercritical (supercritical), water-saving, and environment-friendly clean coal power generation technologies actively; carry out the policy of ‘encourage large power plants and discourage small energy-inefficient power plants’ continually, and upgrade old power plants to realize energy-saving and clean development for coal-fired power plants.

2. Adhere to the power-centered development strategy and the development mode of industrial cycle economies, encourage regions where conditions permit to realize coordinated development of coal and power, improve the coal-blending capability, and promote the development of regional or industrial cycle economies.

Construct power plants on coal-fired power bases, carry out coal and power pool, and promote regional cycle economies. The main purposes of coal and power pool are to occupy coal recourses, stabilize coal supply, affect the market prices of coal, and cut down generating costs. The coal resources occupied by means of coal and power pool should have relatively more stable prices when used for power generation and be controlled in a rational range to meet the basic demand of power generation (the remaining demands should be met by means of market purchases). The occupied coal resources can be used for affecting the market prices of coal to make the fuel costs rational, thus realizing flexible fuel cost adjustment and benefiting competition of the power market. Carry out rational utilization, processing, and outward transportation of coal simultaneously, improve the industrial chain including coal, power, aluminum, chemicals, roads, and ports, and form big industrial clusters for the sake of synergy and a scale advantage.

The low thermal efficiencies of coal-fired boilers in China are mainly because of inconformity between the coal natures and the boiler types as a result of use of different types of coal with quite different combustion characteristics in coal design and practical combustion. In the power industry, the main solution for this problem is boiler retrofitting for the sake of adaption to the coal natures. As a result, many new boilers have to be retrofitted soon after installation. Thus, the state invests a lot of money each year to retrofit the boilers with a clinkering tendency, overtemperature, and serious wear. Therefore, close cooperation with power enterprises is necessary in coal resource development for the sake of conformity of the mining, addressing, transportation, blending, and storage of coal. Coal blending is a process to proportionally mix the coal with different quality levels by means of screening and crushing, so as to make the coal fed into boilers basically the same as the coal used in design, make full use of coal resources, optimize the coal structure, and realize mutual complementation for different coal qualities to realize safe, stable, and economical operation for boilers and improve the coal combustion efficiencies of boilers.

Coal blending can turn low-quality inferior coal into a quality fuel, improve the characteristics of environmental protection, and promote pollution control; it can turn coal with easy clinkering into coal with difficult clinkering to decrease boiler faults and improve boiler lives; it can increase the utilization rate of inferior coal to make full use of local coal resources; it can guarantee coal with stable quality for power plants and improve the ignition stability and combustion efficiency of coal to realize stable low-load combustion for boiler coal; it can cut down the transportation costs of coal to realize rational utilization of coal and improve the circulation benefit of coal. Compared with the situations in USA and Europe, the coal for power generation in China has very poor quality. In USA and Europe, ash of the coal for power generation is stably below 10%; in China, it is unstably at 27% or so on average, resulting in unstable combustion of boiler coal and shorter lives of units, going against pollution control, and increasing the ineffective transportation capabilities of railways. See Fig. 6.3 for quality of the coal for power generation in USA from 1995 to 2007.

3. Spare no efforts to solve the problems of coal-fired power, particularly the problems about coal price and power price.

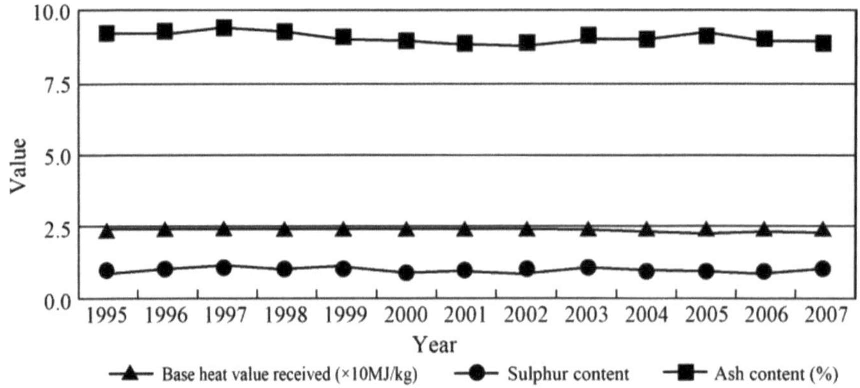


**Fig. 6.3** Quality of the coal for power generation in USA from 1995 to 2007. *Data source* U.S. Energy Information Administration

The domestic coal market and the international coal market are closely linked. A main characteristic of the international coal market is that there are less long-term contracts but more spot contracts which are now more important, thus making coal prices more uncertain. Domestic enterprises engaged in coal production and consumption participate in the market-oriented economy directly to face market competition. As a result, the purchasing prices and selling prices of coal will depend on the market completely and change more frequently. In order to rationally avoid the coal price risks due to the coal price reform in China and realize longtime stable and economical operation for the power industry, it is necessary to solve policy-related losses of the thermal power industry, actively seek the power price reform policies issued by related authorities of the state, and promote the formation of a rational power price mechanism first. The main actions are as below: Regulate the order and environment of coal development, standardize the charges of coal from its leaving from coal mines to arrival at power plants, and avoid irrational charges in intermediate links such as transportation. Establish a rational coal–power price relation gradually, and promote determination of power purchase prices through market competition; carry out the policy of progressive power prices gradually to drive the whole society to use and save power rationally; establish rational direct supply power prices for major users, and promote direct trading between major users and power producers; continually promote and improve the policy of differential power prices, the policy of peak-valley power prices, and the policy of desulfuration power price.

4. Encourage enterprise groups to construct big coal storage yards to stabilize coal prices and increase energy supply safety.

The market prices of coal have much to do with production situations of the downstream coal-using enterprises: In the coal supply shortage in each year, the coal prices are 50–60 yuan per t higher than usual. Thus, enterprise groups are encouraged to construct big coal storage yards for dynamic coal storage: Residual coal is stored on big coal storage yards in off seasons including spring and autumn and used in busy seasons including summer and winter to solve the problem of coal supply shortage in the peak periods of power consumption and such natural disasters as snow disasters and earthquakes and stabilize the coal prices. Coal storage bases can be constructed in coal-producing regions or entrepots such as ports to enable coal to avoid busy railway transportation and reach ports directly for circulation.

## 6.2.2 *Optimization by Means of Technological Progress*

The main actions of technological progress and optimization for the main constraints for energy development include the technologies related to energy efficiency, clean energy sources, and energy environment. According to the strategic road map for the energy technology development in China worked out in 2007 by Chinese Academy of Sciences, the key fields of energy technology development before 2020 will be as below: energy-saving and clean energy source technologies, new-generation zero-emission polygeneration-integrated gasification combined cycle, pressurized fluidized bed combustion–combined cycle (PFBC-CC), CCC key technology demonstration, efficient coal liquefaction; the technology of substitution of oil by coal-based alcohol ether and olefin (this technology should have project demonstration and extensive use); safe and clean nuclear energy technologies and technologies about other renewable energy sources except for hydropower vigorously (these technologies should be developed vigorously); technologies about the forward-looking deployment of energy sources except for the conventional fossil energy sources.

### 6.2.2.1 Improve the Efficiency and Energy-Saving Technologies of Energy Utilization

1. Clean and efficient development, utilization, liquefaction, and polygeneration of coal.

Attach importance to developing the technologies and supporting equipment for efficient coal exploitation and efficient power generation technologies and equipment such as heavy-duty gas turbines, IGCC, high-parameter ultra-supercritical units, and supercritical big circulating fluidized beds, and vigorously develop coal conversion technologies such as coal liquefaction, coal gasification and coal chemical, the polygeneration system technology based on coal gasification, the technologies and equipment for overall control and utilization of coal pollutants, the geological theories and resource evaluation about occurrence of the coal measure stratum gas, and advanced technologies about coal exploration and mining.

2. Smart power system-based extra high-voltage power transmission technology, superconductivity technology, storage technologies, and new energy source technologies.

Attach importance to developing the large-capacity long-distance DC power transmission technology, the extra high-voltage AC power transmission technology and corresponding equipment, the technologies about grid connection, transmission and distribution of interrupted power sources, the technologies about monitoring and control of electric energy quality, the safety technologies for big interconnected grids, the key technologies used in the West-to-East Power Diversion Project, the automation technologies about grid dispatching, and the information technologies and systems about efficient power distribution and supply, master equipment and technologies about the extra high-voltage power transmission in high-altitude extreme cold areas, and carry out preliminary studies about the superconductivity power technology, including the studies of power transmission and distribution equipment such as superconducting transmission equipment, superconducting switches, superconducting cables, and superconducting fault current limiters.

3. System energy-saving technologies and process energy-saving technologies.

The system energy-saving technologies aim at promoting the energy efficiencies of energy-intensive processes (such as steel and iron, and aluminum), and they are mainly developed for the design, integration, and systematic optimization of systems rather than individual equipment. The process energy-saving technologies include selective catalysts, advanced separation technologies, measurement and control systems, new materials, and advanced motor systems. In the long run, biological technologies and biomass materials will be particularly attractive.

4. Distributed energy sources.

Distributed energy sources can cut down the costs of transmission and distribution, and cogeneration can realize high efficiencies and reduce the emission of both $CO_2$ and other air pollutants. At present, it is studied to promote the system efficiencies of distributed power generation, improve equipment (advanced gas turbines, micro-gas turbines, natural gas engines, and fuel cells), and utilize renewable energy sources and dual-fuel power generation systems.

5. Energy storage.

Study the advanced big storage technologies and equipment with high efficiencies, large capacities, low costs, and low pollution, and put into service the existing energy storage technologies including pumped storage, compressed air energy storage, and chemical-cell energy storage. The technologies are in development include mechanical storage technologies (flywheels and compressed air), electrochemical storage technologies (advanced cells and reversible fuel cells, and hydrogen), and electrical storage technologies (supercapacitors and superconducting magnets).

#### 6.2.2.2 Technologies of Clean Energy Sources

1. Low-cost and extensive development and utilization of renewable energy sources.

Attach importance to developing large wind power generation equipment, construction technologies and equipment for the wind farms in offshore areas and on lands and for the regions with centralized wind energy resources in west China, the technologies about cost-effective solar PV cells and their utilization, the solar thermal power generation technologies, the solar building integration technologies and the technologies about development and utilization of biomass energy, geothermal energy, etc.

2. Nuclear power technologies.

Emphases in the studies about nuclear power technologies are: advanced pressurized water reactors and high temperature gas-cooled reactors; design of new efficient reactors; technologies about nonproliferation reactors and fuels; design and application of advanced small-power reactors; new advanced technologies about the storage of nuclear fuel and nuclear waste solids; fast breeder reactors; nuclear fusion technologies.

3. Hydrogen energy

Emphasis in the studies about hydrogen energy is the following new technologies about the production, storage, transportation, and final use of hydrogen energy: photocatalytic decomposition of water to hydrogen energy with solar energy; decomposition of water into hydrogen energy by means of thermochemical cycles; hydrogen energy preparation by means of biomass oxidation and microorganisms; hydrogen energy storage in metallic oxides, carbon nanotubes, or organic compounds; onboard proton exchange membrane fuel cells; solar energy–hydrogen energy systems; liquid hydrogen used as an aviation fuel.

#### 6.2.2.3 Increase the Economic Investment for Water-Saving Technologies

In the China Water-Saving Technology Policy jointly compiled by National Development and Reform Commission, Ministry of Science and Technology, Ministry of Water Resources, Ministry of Housing and Urban-Rural Development, and Ministry of Agriculture, some water-saving technologies are proposed. The unconventional water utilization technologies mainly using reclaimed water, seawater, and rainwater, the coal mining water retention technology, and the thermal power air-cooling technology are the water-saving technologies extensively used in China's coal industry and thermal power industry. A larger economic investment for water-saving technologies will improve water-saving benefits in the energy industry in China greatly and arouse enthusiasm of the energy industry for water saving. It is now a most effective and quickest action for alleviating the water resource constraints for the energy development in China.

1. Economic analysis for the utilization of unconventional water.

(1) Utilization of reclaimed water: After simple treatment, reclaimed water can be used for mining area greening and environmental sanitation watering, thus increasing the regional available water resources and solving water environment pollution.

Usually, the cost of reclaimed water mainly depends on the technology and scale of its treatment. A high treatment level of wastewater results in higher pollutant removal rate and reclaimed water quality, a larger range of recycling and more sewage recycling investment and operating cost. For example, the cost of centralized treatment and recycling of sewage is 0.6–3.0 yuan per $m^3$.

(2) Utilization of seawater: The thermal power plant in coastal areas can make full use of the location advantage of these areas to develop technologies such as seawater cooling and seawater desalination and scrubbing vigorously to increase the seawater supply to reduce the pressure of increasingly serious regional freshwater supply. At present, seawater utilization methods such as seawater cooling, seawater desulfuration, and toilet flushing with seawater are used. The unit average investment of direct seawater utilization in China is 2000–3500 yuan (h $m^3$). The seawater desalination cost which is complicated is an important constraint for the promotion and application of seawater desalination. The seawater desalination cost is mainly made up of an investment cost and an operation management cost and affected by factors including seawater desalination method and scale, raw water quality and temperature, geographical, geological and climate conditions, energy price, desalinated water quality requirement, material selection in design, operating rate, safety capacity, service life, investment source, interest rate, taxes, etc. When the cost of water diversion is not considered, total cost of seawater desalination on the international market is 0.26–2.1 USD per $m^3$, or 0.47 USD per $m^3$ on average, several times lower than the 10 USD per $m^3$ 40 years ago and about 50% lower than the 1.02 USD per $m^3$ in 1985. On the international market, the ex-factory price of desalinated water is usually 0.6–0.9 USD per $m^3$. Thanks to technological progress and market competition, costs of the main seawater desalination equipment have decreased by nearly half than ten years ago. Generally speaking, the seawater desalination cost is still a little higher than the running water prices in most coastal cities. In China, seawater desalination now has a total cost of 4.0–7.5 yuan per $m^3$, an operating cost of 3–4 yuan per $m^3$, and an investment cost of 7500–15,000 yuan ($m^3$ day), depending on the specific process used.

According to the comparison between seawater desalination long-distance water diversion or industrial reclaimed water in Table 6.1, the actual unit cost of seawater desalination is increasingly competitive. With continuous progress and large-scale and localized development of seawater desalination technologies in a period in future, the seawater desalination cost will decrease continually. It is estimated that investment for the desalination of each ton of seawater will decrease continually and reach about 6000 yuan per $m^3$ in the year of 2020.

(3) Utilization of rainwater: Rainwater utilization is significant for the mining areas in semiarid regions and arid regions because rainwater can be locally collected, and simply treated and used for the greening, scrubbing and production of mining areas and the cooling, scrubbing and production of thermal power plants. According to survey data of Beijing, the annual cost of collection, storage, and utilization of rainwater is 0.1–1.2 yuan per $m^3$ and the unit investment of it is 20–160 yuan per $m^3$.

Viewed from technologies and costs, the utilization of unconventional water sources has had mature technologies; cost of the secondary treatment of reclaimed water is 1–2 yuan per $m^3$, and the treatment cost of seawater has been decreasing; the urban areas have relatively higher economic affordability. Thus, unconventional water sources are now ready for extensive use and promotion in the coal industry and the thermal power industry.

2. Economic analysis for the coal mining water retention technology.

(1) Economic investment analysis: With the coal mining water retention technology, water inflows of the thirteen big coal bases in China will all reduce by 10–15% than the current year and, in detail, their water inflows will reduce by 39,841 ten thousand $m^3$ and 64,828 ten thousand $m^3$, respectively, in 2015 and 2020. According to the estimate based on the current technical level, in the coal mining water retention technology, an investment of 4 yuan is necessary for the

**Table 6.1** Cost comparison between seawater desalination and long-distance water diversion or industrial reclaimed water

| No. | Form of water source | Average cost (yuan/$m^3$) | Water quality |
|---|---|---|---|
| 1 | Long-distance water diversion | Water diversion from Luanhe River to Tianjin City: 2.3 (direct cost) | Used after treatment |
| | | South-to-North Water Diversion Project: >5 | Used after treatment |
| 2 | Power plants or industrial boilers 用水再生处理 | >9 | Used directly |
| 3 | Seawater desalination | Seawater: 4–7 (total cost) | Used directly |
| | | Brackish water: 2–4 (total cost) | Used directly |

water inflow reduction by 1 $m^3$; the investment for the technology will be 1.593 billion yuan in 2015 and 2.593 billion yuan in 2020. See Table 6.2.

(2) Calculation of the ecological service values: The coal mining water retention technology has a huge ecological service value because it protects surface water resources, groundwater resources, and water environment and maintains safety for the surface water and groundwater circulation system. In order to evaluate the ecological service values, ecological services can be classified into a supply service, a regulation service, a cultural service, and a support service. See Tables 6.3 and 6.4 for calculation methods for values of the ecological services.

According to the forecast, the water inflow reductions of the thirteen big coal bases in China in 2015 and 2020 because of the coal mining water retention technology will be 39,841 ten thousand $m^3$ and 64,828 ten thousand $m^3$, respectively, and the corresponding ecological service values increased will be 25.897 billion yuan and 42.138 billion yuan, respectively. According to a comparison between the technical investments in these two years, there will be a huge ecological benefit. Therefore, this technology is an effective measure for alleviating the water resource constraints for the coal industry in China because its extensive use can realize both economic growth and ecological protection.

3. Economic analysis for the air-cooling technology.

(1) Economic analysis for water saving: The water consumption of a 600 MW air-cooled unit is about 1200 t/h less than that of a 600 MW water-cooled unit

**Table 6.2** Estimate of investments for coal mining water retention in different coal bases

| Coal base name | Water inflow reduction because of coal mining water retention (ten thousand $m^3$) | | Investment for coal mining water retention (ten thousand yuan) | |
|---|---|---|---|---|
| | 2015 | 2020 | 2015 | 2020 |
| Shendong coal base | 6462 | 10,973 | 25,848 | 43,892 |
| Eastern Ningxia coal base | 912 | 1776 | 3650 | 7104 |
| Northern Shaanxi coal base | 777 | 1653 | 3108 | 6612 |
| Central Shanxi coal base | 2375 | 3705 | 9500 | 14,820 |
| Northern Shanxi coal base | 5985 | 9405 | 23,940 | 37,620 |
| Huanglong coal base | 1653 | 2708 | 6612 | 10,830 |
| Henan coal base | 2755 | 4133 | 11,020 | 16,530 |
| Central Hebei coal base | 1385 | 2117 | 5542 | 8467 |
| Eastern Shanxi coal base | 4560 | 7980 | 18,240 | 31,920 |
| Western Shandong coal base | 2755 | 3990 | 11,020 | 15,960 |
| Huainan–Huaibei coal base | 2185 | 3563 | 8740 | 14,250 |
| Yunnan–Guizhou coal base | 3572 | 5985 | 14,288 | 23,940 |
| Eastern Inner Mongolia coal base | 4465 | 6840 | 17,860 | 27,360 |
| Total | 39,841 | 64,828 | 159,368 | 259,303 |

**Table 6.3** Calculation methods for the ecological service values of each cubic meters of coal mining water retention

| Service function | | U/P (yuan/ $m^3$) | Basis of calculation |
|---|---|---|---|
| Supply service | Emergency water supply | 15.0 | Value of water supply service for particularly arid years: about 15 yuan/$m^3$ |
| Regulation service | Water quality regulation for pollution prevention | 7.0 | The treatment of polluted groundwater needs a high cost and a long time. According to French experience in recovery from groundwater pollution, the recovery of each cubic meters of polluted groundwater needs 5.7 yuan. Due to characteristics of the groundwater in China, such as overdraft and pollution, the value of regulation service will be correspondingly increased |
| Cultural service | Spring culture | 5.0 | Reference to Jinci Temple: income of cultural tourism—0.15 billion yuan (flow: 1 $m^3$/s, annual consumption: 3000 ten thousand $m^3$) |
| Support service | Support for groundwater circulation | 38.0 | (1) Support for groundwater circulation: Sum of the surface water cost, the recharge cost is the surface water cost price (reference to price of a recovered groundwater flow field)<br>(2) Support for surface subsidence prevention, 10 yuan/$m^3$ |
| Total | | 65.0 | |

(obtained based on the design water consumptions of multiple projects through a weighted average method). If calculated based on an annual operation time of 6000 h and a water charge of 0.15 yuan per t (including groundwater and surface water), the annual water charge of a 600 MW air-cooled unit will be 360 ten thousand yuan less than that of a 600 MW water-cooled unit; if urban reclaimed water is used as the power plant water source and needs deep treatment because it does not comply with the quality requirement for the water used by power plants and the water charge is 2.5–3.5 yuan per t, the annual water charge of a 600 MW air-cooled unit will be 1800 ten thousand yuan to 2520 ten thousand yuan less than that of a 600 MW water-cooled unit. Because of their less water consumptions, air-cooled units have much smaller investments for their water source projects than water-cooled units.

(2) Economic investment analysis: Air-cooled units are an important measure for solving the problems in development of the thermal power industries in the regions with abundant coal but water shortage in China because they can save water by more than 75% than the water-cooled units of the same type. For example, Datong Second Power Plant in Shanxi uses water-saving air-cooled

**Table 6.4** Calculation of the ecological service values of coal mining water retention in different coal bases

| Coal base name | Water inflow reduction because of coal mining water retention (ten thousand $m^3$) | | Ecological service value (ten thousand yuan) | |
|---|---|---|---|---|
| | 2015 | 2020 | 2015 | 2020 |
| Shendong coal base | 6462 | 10,973 | 420,030 | 713,245 |
| Eastern Ningxia coal base | 912 | 1776 | 59,280 | 115,440 |
| Northern Shaanxi coal base | 777 | 1653 | 50,505 | 107,445 |
| Central Shanxi coal base | 2375 | 3705 | 154,375 | 240,825 |
| Northern Shanxi coal base | 5985 | 9405 | 389,025 | 611,325 |
| Huanglong coal base | 1653 | 2708 | 107,445 | 176,020 |
| Henan coal base | 2755 | 4133 | 179,075 | 268,645 |
| Central Hebei coal base | 1385 | 2117 | 90,025 | 137,605 |
| Eastern Shanxi coal base | 4560 | 7980 | 296,400 | 518,700 |
| Western Shandong coal base | 2755 | 3990 | 179,075 | 259,350 |
| Huainan–Huaibei coal base | 2185 | 3563 | 142,025 | 231,595 |
| Yunnan–Guizhou coal base | 3572 | 5985 | 232,180 | 389,025 |
| Eastern Inner Mongolia coal base | 4465 | 6840 | 290,225 | 444,600 |
| Total | 39,841 | 64,828 | 2,589,665 | 4,213,820 |

units and, as a result, the water consumption of each 200 MW unit has reduced from 719 $m^3/h$ (annual average) to 252 $m^3/h$ (annual average), presenting a water-saving rate of 65% and a water-saving quantity of 467 $m^3/h$.

Compared with water-cooled units, air-cooled units will have larger coal consumptions of power supply, which is the cost of water saving by air-cooled units. According to rough statistics, the average coal consumption of power supply of air-cooled units is about 14 g/kWh larger than that of conventional water-cooled units.

Utilization of unconventional water resources and application of the coal mining water retention technology and the thermal power air-cooling technology need investments which differ with the specific region. In China, it is necessary to make the coal industry and the thermal power industry based on the characteristics of regional water resources and take the most economical and most effective water-saving actions in accordance with regional water conditions to alleviate the constraints of regional water resources for these two industries.

### 6.2.3 Optimization by Means of Management Actions

Strengthen DSM and carry out overall resource planning. In order to meet energy demands, the main methods in conventional planning are to strengthen exploration, build new power plants, and increase installed capacities. In the new situations of energy shortage and development mode emphasizing, China has begun to attach importance to demand side resources. DSM mainly refers to the management of the electricity market by both suppliers and purchasers to achieve the purpose of improving the reliability of power supply and reducing energy consumption and supply and demand costs. In order to realize DSM, it is necessary to consider the energy-saving resources obtained by means of higher energy use efficiencies of users and reform of the energy use methods of users as allocatable resources in planning, realize competition between demand side resources and supply side resources, and conduct overall resource planning for them.

In overall resource planning, all resources on both supply side and demand side of the power industry are integrated as a whole for power planning and used in an efficient, economical, and rational way to realize the minimum total cost for the planning, while the needed energy service level is maintained.

In recent years, China has attached increasingly larger importance to DSM: Official documents such as Circular of the State Council on the Recent Emphases in Building a Resource-efficient Society (G.F. [2005] No. 21), Circular of the State Council on Printing and Distributing an Overall Plan for Energy Saving and Emission Reduction (G.F. [2007] No. 15), Circular of the State Council on Printing and Distributing a National Plan for Climate Change Addressing (G.F. [2007] No. 17), Urgent Circular of the General Office of the State Council on Orderly Power Consumption by Means of Better DSM in the Power Industry (G.B.F.D. [2008] No. 13), Circular of the National Development and Reform Commission on Power Operation in 2008 (F.G.Y.X. [2009] No. 568) proposed specific requirements for the improvement of applicable laws and regulations, compilation of effective incentive polices, standardization of orderly power consumption and establishment of long-term mechanisms and the requirement of continuing to strengthen the DSM in the power industry and giving full play to its role in improvement of energy efficiency and strengthening of orderly power consumption to avoid the tendency of less strict DSM after supply–demand imbalance in the power industry is alleviated.

At present, DSM has no incentive actions in terms of finance, tax, price, etc., and, particularly, grid companies need to start DSM with a lot of materials, money, and manpower. However, in the power industry, DSM now has no stable fund sources and DSM goes against the operation objectives of grid companies. For example, by means of green actions and DSM strengthening, a grid company has saved its power consumption, but its profit reduces correspondingly, which has lowered enthusiasm of the grid company for strengthening DSM. In the Twelfth Five-Year Plan period, DSM should be strengthened in the following aspects:

(1) Consider DSM of power in unified management of the industry system, carry out overall planning for power demands, power source construction, grid construction, power purchase contract, and energy-saving service, realize optimal allocation for energy resources of the whole society through rational and effective utilization of the energy resources on both the supply side and the demand side, and embody power-saving effect or potential of the demand side to the planning and construction of the supply side. Meet the future power demands with the minimum costs by means of overall resource planning.
(2) Adjust the grid company profit assessment mechanism properly. In the operation performance assessment of grid companies, do not consider their volumes of power sales only but regard their effects of DSM of power as an assessment index and encourage them to implement DSM of power, so as to promote power saving by the whole society with help of the profit assessment mechanism and lay a foundation for the establishment of an energy efficiency management long-term mechanism.
(3) Establish a DSM fund. The efforts of grid companies for seeking potential power resources (energy-efficient power sources) and realizing the minimum energy losses and the maximum energy efficiencies are actions of positive externality and should be compensated rationally by the beneficiaries. In order to encourage grid companies to carry out DSM of power to improve their energy utilization efficiencies, it is suggested to promote the practice of the government of Yunnan of establishing a special DSM fund or, based on some money extracted from power prices, a special power demand side fund to implement policy and corresponding mechanisms of DSM of power.
(4) Speed up the studies and compilation of energy efficiency standards. At present, most of industrial energy-consuming equipment, general-purpose equipment, household appliances, and lighting appliances in China have no energy-saving standard systems. It is therefore suggested to speed up the compilation of standard systems to standardize the design, production, and use of energy-consuming equipment and other energy-consuming products. It is also suggested to standardize, guide, and promote all types of energy-saving services.
(5) When compiling power price policies, fully reflect the quality differences of electric energy on the market, increase the range of a peak-valley power price system for both power purchase prices and power-selling prices, and form peak-valley power prices differences rationally; in the power price catalogue, set interruptible power prices to realize rational power costs for the users carrying out peak load shifting consciously.
(6) Make the limitation on the service fields of grid companies properly loose and allow grid companies to carry out related services except for power purchase and sales (such as energy-saving service).

## *6.2.4 Supply and Demand System Optimization Based on Smart Grids*

### 6.2.4.1 Speed Up the Construction of Smart Grids

At present, the international market has no unified definition for smart grids, but it has the consistent core idea, i.e., to promote the intelligence of grids by means of modern information, communication, and control technologies to adapt to new energy source power, meet the demands for bilateral and interactive grid services, and realize safe, reliable, economical, efficient, and sustainable power supply. The development of smart grids is the key for energy safety, the development of renewable energy sources, and better DSM.

(1) Improve grids' operating efficiency and capability of optimum allocation of resources to alleviate supply–demand imbalance of the energy resources in China.

According to the conditions of China, quicker construction of smart grids with powerful backbone grid frameworks is an effective means for optimal allocation of resources and higher operating efficiencies of power systems. Smart grids using advanced power transmission technologies can realize large-scale, long-distance, and low-loss power transmission effectively, improve the transmission capabilities and operation control flexibility of grids significantly, and make the best of grids' effect of optimal allocation of resources. Improve the overall utilization efficiencies of power generation equipment, the transmission efficiencies of grids, and the power consumption efficiencies of users to alleviate supply–demand imbalance of the energy resources in China.

(2) Promote the development of clean energy sources to solve the problems of shortage and unbalanced structure of the energy resources in China. The power generated by renewable energy sources such as solar power, wind power, small-sized hydropower, and biomass is difficult to be connected to power systems because its quality does not meet the expectation of people. Such access of massive electric energy to power systems endangers the economical operation and safety of the conventional grids receiving power passively and other power sources. Based on integration of advanced technologies about information, automation, energy storage, operation control, and dispatching, smart grids can realize accurate forecasts and optimal dispatching for all resources including clean energy sources to avoid the adverse impact of access of massive clean energy source into power systems on the safe and stable operation of grids, effectively promote the capabilities of grids to receive clean energy sources, improve the sustainable development and absorption of clean energy sources, and solve the problems of shortage and unbalanced structure of the energy resources in China.

(3) Give play to the advantage of intelligent control of smart grids to realize DSM of energy resources. Due to the current interest pattern of power systems, DSM is difficult to optimize the power structure and balance the operation of power systems. However, making use of related technologies about information and intelligent control to have DSM included in the design and system construction of each electrical product and realize response and exchange between terminals and grids and distributed energy sources can completely solve the problems expected to be solved through DSM.

#### 6.2.4.2 Develop Pumped-Storage Power Plants Vigorously

With their construction and nationwide connection to grids, pumped-storage power plants not only play a role in some grids but also give play to a complementary effect in regional and inter-regional interconnected grids. In the case of large-scale connection to grids and long-distance power transmission, it is more necessary to construct pumped-storage power plants which are reserve capacities with powerful functions and quick regulation close to the load centers of receiving ends. Pumped-storage power plants are now effective regulating power sources for the safe and table operation of modern big grids.

Pumped-storage units have good regulation performance and high economic benefit and social benefit. They provide an effective means for the peak and frequency regulation of power systems because of their time of start, stop, and regulation; they can operate in both pumping phase modulation and power generation phase modulation and serve as a continuously adjustable reactive power source in operation, showing great capabilities of phase modulation and voltage regulation; they have an obvious advantage of energy storage: They can pump water in valley periods and generate power in peak periods, thus realizing forced reduction of generating capacity of thermal power and centralized connection of wind power to grids in valley periods; they can improve the operation conditions of thermal power units and nuclear power units and realize backup and black start in the emergencies of power systems. In the earlier period, pumped-storage power plants were mainly used for improving the poor regulation performance of nuclear power. However, with the connection of new energy sources such as wind power to grids, the energy storage characteristic and regulation advantage of pumped-storage power plants are found complementary with the operating characteristics of nuclear power and wind power. According to the current technical level, this cannot be realized by other power sources.

The pumped-storage power plants in China appeared later, but their construction has a higher starting point because of the experience in the construction and management of many conventional hydropower plants. In late 2012, the pumped-storage power plants which had been constructed and put into operation had a total installed capacity of 2065 ten thousand kW which is mainly distributed in the grids of economically developed regions in China, including 44% in the East

China Grid, 23% in the South China Grid, and some in the North China Grid. On the whole, pumped-storage power plants have a small scale: Their total installed capacity is about 1.8% of the national installed capacity of the power industry.

According to statistical data of China Renewable Energy Engineering Institute, the currently constructed pumped-storage power plants have a total installed capacity of 1195 ten thousand kW; the pumped-storage power plants in a feasibility study have a total installed capacity of 1608 ten thousand kW; the pumped-storage power plants in pre-feasibility study and planning have a total installed capacity of 3090 ten thousand kW. According to structure and layout of the power sources in China, percent of the pumped-storage power plants in a grid mainly including thermal power should be 6–10% of total installed capacity of the power units; for a grid with much hydropower with functions of peak and frequency regulation, the percent above should be 4–7%; for a grid with nuclear power units or a large wind power capacity, the percent above should be 40–50%. China has abundant wind energy resources but insufficient hydropower resources. The instability of wind power makes peak regulation of the grids in related regions more difficult. Thus, it is necessary to construct pumped-storage power plants in the regions with centralized wind power for the purpose of better operation of wind power. However, the pumped-storage power plants in the regions with abundant wind energy resources are small and highly insufficient to meet the demands of related grids.

In late 2012, the power industry in China had a total installed capacity of 1.147 billion kW. Pumped-storage power plants with a total installed capacity of 5700 ten thousand kW were expected to be constructed according to a percent of 5% supposed. The pumped-storage power plants completed those under construction and those in a feasibility study have a total installed capacity close to this value. The nuclear power plants and wind power plants that are now under construction will be put into operation and connected to grids soon. Thus, it is pressing to construct supporting pumped-storage power plants of corresponding scales.

According to Planning Outline for Medium and Long-term Development of Energy Sources, in the year of 2020, the pumped-storage power plants in China will have a total installed capacity of 5000 ten thousand kW and those in the operating regions of State Grid Corporation of China have a total installed capacity of nearly 4000 ten thousand kW. Research data shows that total installed capacity of the pumped-storage power plants in China will need to be 6000 ten thousand kW at least in a total installed capacity of the power industry of 1.8 billion kW in China in the year of 2020 and this value will need to be 10,641 ten thousand kW to 15,051 ten thousand kW in the year of 2030. See Table 6.5.

In the forecast above, the demand for pumped-storage power plants due to connection of a lot of wind turbines to power systems and the demand of some regions with abundant hydropower in southwest China for pumped-storage power plants are not considered. If these two demands are considered, the demands of China in 2020 and 2030 for pumped-storage power plant will be larger.

**Table 6.5** Forecast of demands for pumped-storage power plants in China (ten thousand kW)

| Power system | Target year | |
|---|---|---|
| | 2020 | 2030 |
| Total demand | 6000 | 10,641–15,051 |
| Including: North China Grid | 1620 | 2430–4300 |
| East China Grid | 2220 | About 4022 |
| Central China Grid | 870 | 1679–3639 |
| Northeast China Grid | 470 | 900–1480 |
| Guangdong Grid | 820 | About 1610 |

### 6.2.5 *Optimization of Interbasin Water Resource Management*

Imbalance between China's water resource distribution and energy structure is the largest constraint for the energy development in China. The situation of abundant energy but insufficient water resources in west China cannot be solved completely soon. In consideration of the increasingly pressing energy demand in the economic development of China, quicker construction of interbasin water diversion projects is an effective action for reducing the water resource constraints for the energy development in China completely.

China has been planning and implementing many interbasin water diversion projects. The water diversion projects planned and implemented for the coal industries and thermal power industries in regions with water resource shortage include east, middle, and west routes of South-to-North Water Diversion Project, Hanjiang-to-Weihe River Water Diversion Project, Xiaojiang River Water Diversion Project, etc. See Table 6.6 for basic information of these main water diversion projects.

Operation of the interbasin water diversion projects will increase supply capacities of water resources of the water-receiving regions greatly, alleviate the supply–demand imbalance of regional water resources, provide more industrial water, and obtain more opportunities for development of the coal industries and thermal power industries of these regions. See Table 6.7 for supply and demand of water resources in the hydrological regions of coal bases between the beginning time of operation of the east and middle routes of South-to-North Water Diversion Project and the beginning time of operation of the west route of South-to-North Water Diversion Project.

Water diversion projects can alleviate the supply–demand imbalance of water resources of the water-receiving regions, but their large economic investments and negative ecological and social impacts should not be ignored.

Thus, the development of coal industries and thermal power industries in different regions should be mainly based on increase of regional water supply capacities and vigorous and active promotion of clean and water-saving technologies. If the water resource constraints for regional energy development cannot be alleviated by means of actions of new resource increasing and consumption

**Table 6.6** Information of interbasin water diversion projects

| Project | Phase I of the west route of South-to-North Water Diversion Project | Hanjiang-to-Weihe River Water Diversion Project | East route of South-to-North Water Diversion Project | Middle route of South-to-North Water Diversion Project | Xiaojiang River Water Diversion Project |
|---|---|---|---|---|---|
| Scope of water supply | Qinghai, Gansu, Ningxia, Inner Mongolia, Shaanxi and Shanxi, and the Shiyang River Basin nearby | Central Shaanxi in the Weihe River Basin | Jiangsu, Anhui, Hebei, Tianjin, eastern part of the Huang-Huai-Hai Plain in Shandong, and Shandong Peninsula | Beijing, Tianjin, Hebei, and Henan | Central Shaanxi in the Weihe River Basin and the reaches of the main stream of Yellow River below Tongguan |
| Objective of water supply | Important cities, energy bases, Heishanxia ecological irrigation district, and Shiyang River Basin | Industrial water and domestic water in cities (main objective), agricultural irrigation water and ecological water | Domestic water, industrial water, and environmental water in cities | Domestic water and industrial water in key cities (main objective), agricultural water and other applications along the key cities | Supplement for the ecological water and domestic water in the cities in the Weihe River Basin and ecological water for the downstream riverway of Yellow River in the non-flood seasons |
| Available water for diversion (0.1 billion $m^3$) | 80 | 15.3 | 148.17 | 130 | 55.2 |
| Water supply cost (yuan/$m^3$) | 0.569 | 0.406 | 0.43 | 0.489 | 0.82 |

**Table 6.7** Supply and demand of water resources in the hydrological regions of coal bases from the beginning time of operation of the east and middle routes of South-to-North Water Diversion Project to the beginning time of operation of the west route of South-to-North Water Diversion Project (0.1 billion m3)

| Coal base name | Second-class hydrological region of the Yellow River Basin | Now to beginning time of operation of the east and middle routes | | | Beginning time of operation of the east and middle routes to the beginning time of operation of the west route | | | Beginning time of operation of the west route | | |
|---|---|---|---|---|---|---|---|---|---|---|
| | | Water demand | Water supply | Water deficit | Water demand | Water supply | Water deficit | Water demand | Water supply | Water deficit |
| Shendong coal base | Lanzhou–Hekou Town | 205.04 | 152.58 | 52.46 | 197.26 | 157.80 | 39.46 | 205.63 | 194.75 | 10.87 |
| Eastern Ningxia coal base | | | | | | | | | | |
| Northern Shaanxi coal base | Hekou Town–Longmen | 18.21 | 15.91 | 2.30 | 26.20 | 23.48 | 2.72 | 32.37 | 32.07 | 0.29 |
| Central Shanxi coal base | | | | | | | | | | |
| Northern Shanxi coal base | | | | | | | | | | |
| Huanglong coal base | Longmen–Sanmenxia | 130.54 | 121.00 | 9.54 | 150.94 | 134.62 | 16.32 | 158.29 | 153.35 | 4.97 |
| Henan coal base | Sanmenxia–Huayuankou | 28.27 | 28.22 | 0.05 | 37.72 | 37.23 | 0.48 | 40.98 | 39.59 | 1.39 |
| Total | | 382.06 | 317.71 | 64.35 | 412.12 | 353.13 | 58.98 | 437.27 | 419.76 | 17.52 |

*Note* In each of the project names in the table above, 'South-to-North Water Diversion Project' is omitted

reduction, the management of interbasin water resources based on the minimum impacts of water diversion projects will be the most effective and basic action for alleviating the constraints.

In a word, on the basis of the supposed total energy demand and energy structure, corresponding actions and system optimization for resource endowment, environmental protection, water resources, and climate change addressing can realize safe, clean, and economical development for the energy industry.

## 6.3 Suggestions for Policies and Actions

### *6.3.1 Energy Structure Optimization*

Continue to attach importance to the coal development in the western regions in China, and develop the coal resources in the western and northern regions in China vigorously. Particularly, Xinjiang Uygur Autonomous Region is the region with the most coal resources in China and important energy sustainability region and strategic energy reserve region in China as well. It is therefore necessary to strengthen development of the coal resources in Xinjiang and carry out a 'Xinjiang-to-east China coal diversion' policy.

Determine productivity of the coal industry in accordance with the water resource condition, limit development of the heavy chemical industry enterprises with large water consumptions strictly, and develop cycle economy, ecological economy, and low-carbon economy vigorously to promote coal development; carry out protective exploitation for groundwater and surface water. Combine coal development and water resource protection and water drainage and water supply to minimize the impact of coal mining on the water resources and the water environment. The coal chemical industry is a highly water-consuming industry. Due to shortage of the freshwater resources in China, it is necessary to consider the carrying capacities of regional water resources and water environment fully in the construction of new coal chemical bases and follow the rule of 'arrange close to water resources and determine productivity in accordance with water resources' in the site selection and scales of coal chemical projects.

In the Twelfth Five-Year Plan period and the Thirteenth Five-Year Plan period, the thermal power layout should be determined based on a full consideration of regional water resource conditions to improve their spatial suitability. Coastal regions can make the best of their location advantages and use more seawater in their thermal power production; regions with serious water resource shortage and regions with serious overdraft of groundwater should not construct thermal power projects or, if such construction is necessary, take stricter water-saving actions for them. Manage the water resource argumentation of the regions with thermal power projects strictly, and not issue water resource permits to the thermal power projects with imbalance between their water consumptions and the corresponding regional water resource carrying capacities. For the thermal power projects with water resource permits in different regions, determine their cooling technologies and water supply sources scientifically in accordance with water resource characteristics and technological characteristics of the corresponding regions and strengthen supervision over their water use as planned and water saving.

### 6.3.2 Determination of a Rational Energy Structure

Determine a rational energy structure in accordance with the conditions of China, related natural and economic laws, and overall balance among energy safety, economic development, and environmental protection. There is no universally applicable rational energy structure at all in the world but different rational energy structures applicable to different countries only. The energy development in China is mainly affected by factors such as resource endowment, environmental protection, water resources, and climate change addressing. Of the energy industry in China, its structure is more affected than its total energy demand, making it necessary and urgent to adjust the energy structure in the Twelfth Five-Year Plan period.

Spare no efforts to speed up the development of non-fossil energy sources to realize the state's objective for percent of non-fossil energy sources in primary energy sources, thus making a positive contribution to $CO_2$ emission reduction.

Optimize and develop coal-fired power continually. The energy structure mainly including coal is the foundation for energy safety and power safety of China. In the Twelfth Five-Year Plan period or even a longer period, percent of the coal consumption in total energy consumption in China will still be higher than 60%. Thus, it is necessary to improve percent of the coal consumption of power generation and switch the emphasis on coal transportation to both coal transportation and power transmission. As for organizational structure of the coal industry, continue to encourage the establishment of big coal enterprise groups.

In the Twelfth Five-Year Plan period, develop natural gas vigorously; strengthen the construction of natural gas reserve systems, and give play to role of the price levers to guarantee stable supply of natural gas; strengthen the development of unconventional natural gas resources such as coal measure stratum gas, shale gas, and marsh gas generated by municipal wastes; strengthen construction of the imported pipes of natural gas and liquefied natural gas, import more natural gas and liquefied natural gas, adjust the policies about natural gas utilization whenever necessary, encourage the substitution of oil by gas, and construct regulating power plants using natural gas or liquefied natural gas at proper time and in applicable places; improve the percent of heating, power and cooling cogeneration or heat and power cogeneration.

### 6.3.3 Energy Industry Promotion Through Technological Progress

Concentrate on researching and developing the technologies for the efficiency improvement of energy resources, strengthen the researches and development of key technologies about resource saving and recycling, and develop resource saving and overall utilization-related technologies with a promotion significance. Introduce, absorb, and innovate technologies, and establish a technical support system for resource saving by means of technological progress and innovation. Take actions such as technological progress fund establishment to encourage and guide enterprises to employ advanced technologies about the saving of energy, water, and materials to improve the general level of resource saving. Strengthen technical upgrading for the energy conservation, emission reduction, and grid loss reduction of the existing power plants at the same time of power structure optimization, thus keeping optimizing the reserves.

Speed up development and application of the clean utilization technologies of energy sources. Speed up comprehensive treatment of the land subsidence areas due to coal mining, attach importance to the overall utilization of resources, develop actions and technologies for carbon emission reduction in the coal industry, enhance the extraction rate and utilization rate of coal measure stratum gas continually, reform the mine layout and coal mining technologies, reduce pollutants greatly, strengthen the comprehensive treatment of wastewater, waste gas, and waste solids, improve environmental law system, and increase the ecological environment recovery ratio both through technological progress.

Concentrate on researching and developing big wind power generation equipment, construction technologies, and equipment for the wind farms in offshore areas and on lands and for the regions with centralized wind energy resources in west China, the technologies about cost-effective solar PV cells and their utilization, the solar thermal power generation technologies, the solar building integration technologies and the technologies about development and utilization of biomass energy, geothermal energy, etc.

Speed up the studies and compilation of energy efficiency standards, and standardize the design, production, and use of energy-consuming equipment and products. Meanwhile, standardize, guide, and promote all types of energy-saving services. Carry out unified scientific management for the energy industry. Encourage enterprises to strengthen the transformation of their technical achievements, introduce and promote new technologies and products, arrange and cooperate in the key pilot tests and demonstration projects in the development of key projects and technologies about technical upgrading, create enough engineering conditions for scientific researches and the application of their achievements, and make scientific research achievements engineered, commercialized, and industrialized.

### 6.3.4 Quicker Elimination of Outdated Capacities

Eliminate outdated and energy-intensive processes and capacities quicker. Keep closing down small coal mines, and quit from the mining areas with serious natural disasters. The power industry should continue to implement the 'encourage large power plants and discourage small energy-inefficient power plants' policy and close down outdated condensing small thermal power units quicker to optimize the coal-fired power structure.

### 6.3.5 Improvement of Applicable Laws, Regulations, Economic Policies, and Management Actions

Strengthen the construction of applicable laws and regulations. Speed up the issuing of an energy law and the revision of Electricity Law of the People's Republic of China, and legalize the ideas, indexes, systems, and industrial supervision duties about climate change addressing, energy conservation, and emission reduction. In addition, for energy conservation and emission reduction in the power industry, perform the functions of administration by governments and the duties of all parties involved, improve the law enforcement system, restore and improve the industrial monitoring and supervision system, and strengthen the supervision and administration according to law.

Establish a rational coal–power price relation gradually, promote heat price adjustment, and reduce the long-standing heat supply loss. Continue to improve related policies of power purchase prices in the power generation with new energy sources, set different power purchase prices for different power plants, scales, and regions, improve corresponding supervision mechanisms and actions according to law, and ensure implementation of laws, regulations, and policies about power purchase prices. Set rational prices for power transmission and distribution in accordance with the rules of rational investments and returns and guiding grid companies to carry out DSM actively to make the operation and development of grids guaranteed by mechanisms. Carry out power-selling price reform actively, combine different power price types, adjust them rationally, establish a power-selling price mechanism useful for fair sharing and study and form general service policies for the power industry to solve the cross-subsidization of power prices completely.

By means of policy adjustment, make coal, power, oil, and transportation supplement one another and develop in a coordinated way to realize coordinated development for coal and power. Make the development and utilization of renewable energy sources more commercialized and extensive. With help of related market mechanisms, governments should guide and encourage market objects to develop and utilize renewable energy sources while creating markets, compiling market rules, standardize market activities, etc.

Strengthen coal dressing and processing, construct supporting coal dressing facilities for the new coal mines with high-sulfur high-ash coal exploited, and wash the coal exploited. For the existing coal mines with high-sulfur high-ash coal exploited and the high-sulfur coal for power generation with stable supply channels, encourage the construction of desulfuration process facilities which can realize sulfur resource recycling.

Improve water use quota standard systems of the thermal power industries in different provinces and autonomous regions, make the water quota management for both the thermal power industry and coal enterprises strict, and improve the supervision over utilization of the water resources in the thermal power industry.

Speed up the construction of market mechanisms useful for the emission of greenhouse gases. At the same time of implementation of a CDM mechanism, establish a carbon emission trading mechanism in accordance with the conditions of China, take part in international competition actively, and promote the control of $CO_2$ emission with the most economical means.

Strengthen DSM to promote energy and power saving by the whole society. Establish and improve the supporting actions for Measures for Energy-Saving Generation Dispatching; strengthen the DSM of power comprehensively, strengthen system design, and revise and improve the laws, regulations, and policies about the DSM of power to ensure effective implementation of the DSM of power from mechanisms and policies. Revise applicable laws, regulations, and policies such as Measures for the Implementation of Demand Side Management whenever necessary, and make clear organization, fund source, and supervision mechanism of the DSM of power to legalize and standardize the energy efficiency and DSM in the power industry.

### 6.3.6 Strengthening of the Coordination Between Energy Development Planning and Other Related Planning

Enhance the coordination between energy development planning and other related planning. The coordination between coal and power is the main one among all energy contradictions, and new energy source development and grid development and the construction of supporting power sources form a new contradiction in transformation of the energy industry. To handle this main coordination and this new coordination, it is necessary to give overall consideration to the effective linkage between upstream and downstream, central interests and local interests, urban planning and energy planning, and overall planning and special planning and the coordination among energy development, fund raising, and manpower.

Carry out scientific zoning, detailed guidance, and compilation of detailed and feasible plans for the coal industry in accordance with different regions, geological conditions, development modes, and ownerships. Resource guarantee, productivity and market construction and transportation capability should be planned comprehensively to meet the increasing coal demand and environmental safety requirement.

Strengthen implementation of the plans of renewable energy sources, take supporting actions for the plans about grids and regulating power sources, solve the problems in new energy sources' connection to grids and absorption, and attach importance to the quicker approval of conventional big hydropower projects, the quicker construction of pumped-storage power plants, and issue power price policies for pumped-storage power plants. Speed up the construction of grid capacities for wind power receiving and the compilation of supporting actions; speed up the planning of solar power generation projects and the compilation of power price policies to ensure realization of the objectives of energy source planning.

# Afterword

It is worthy to note the research background of this book is based on the "Research on Main Constraints of Energy Development", one of the major studies for energy development planning during the "12th Five-Year Plan" that carried out in 2009. Both research basis and the resolved problems are related to the energy development during the 12th FYP, and some data and trend analyses are as well mainly reflect the situation of that period. However, this book was completed after 5 years in 2014. In the past 5 years, significant changes have been taken place in both international and domestic economic and energy situations. For example, China's coal price has fallen sharply from the peak, and the coal-fired power industry has significantly relieved from suffering losses. Smog has frequently occurred across a large area of China since 2013, which aroused great concern from the government and the general public. Important regulations and standards such as the *Environmental Protection Law of the People's Republic of China* and the *Ambient Air Quality Standards* have been revised, and a series of measures for smog control and structure adjustment have been promptly released. Meanwhile, since the "new normal" of economic development presented in 2013, China's economy is currently at an obvious turning point, transiting from high-speed to medium–high growth. Because of this book focusing on theoretical and regular study, the aforementioned changes do not have a considerable impact on the conclusions, which on the order hand even proves the reliability of the research viewpoints present in this book to a certain extent. Therefore, although the above-mentioned situations have been concerned in this book, the analysis is without in-depth discussion. The main opinions, analysis approach and main conclusions in this book should still be used as references for energy researchers.

 
Z. Wang (ed.), *Constraints and Solutions for Energy and Electricity Development*,
https://doi.org/10.1007/978-981-10-8304-4

# References


1. Daiel Yergin. The Quest - Energy, Security, and the Remaking of the Modern World [M]. Translated by Niu Yuben, Yan Zhimin. Beijing: Petroleum Industry Press, 2012 (P8).
2. Jeremy Rifkin. The Third Industrial Revolution - How Lateral Power is Transforming Energy, the Economy, and the World [M]. Translated by Sun Tiwei, Sun Yuning. Beijing: CITIC Press, 2012 (P9).
3. Wei Yiming et al. Progress in Modeling for Energy-Economy-Environment Complex System and Its Application [J]. Chinese Journal of Management, 2015, 2: 159–170 (P13).
4. Eliyahu M. Goldratt, Jeff Cox. The Goal-A Process of Ongoing Improvement [M]. North River Press, 2004 (P17).
5. Eliyahu M. Goldratt. Critical Chain [M]. Translated by Luo Jiaying. Beijing: Publishing House of Electronics Industry, 2006 (P17).
6. Huang Kunzhi. Introduction and Application of Theory of Constraints. Accessed via http://www.doc88.com/p-676139595777 (P17).
7. Huge Fortune Door. Understanding of TOC Through the Movie The Goal. Accessed via http://www.docin.com/p-441518842.html (P17).
8. Editorial Board of China Electric Power Encyclopedia. China Electric Power Encyclopedia (Synthetic) [B]. 3rd Edition. Beijing: China Electric Power Press, 2014 (P19).
9. Wang Zhixuan. The Key to Solving the Energy Problem—System Optimization [M]. China Power Enterprise Management, June, 2014 (P22).
10. Wu Qiaosheng, Cheng Jinhua. Study on Energy Constraints and China's Industrialization Development [M]. Beijing: Science Press, 2009 (P28).
11. Wang Zhixuan. The Key to Haze Control in China—Coal Power [N]. China Energy News, 2014 (P30).
12. Xia Yishan, Fu Quanzhang, et al. China's Perspective on International Energy Development Strategy [M]. Beijing: World Affairs Press, 2009 (P53).
13. Jin Bei, et al. Resources and Growth [M]. Beijing: Economy and Management Press, 2008 (P55).
14. Cui Minxuan. China Energy Development Report 2007 [R]. Beijing: China Water & Power Press, 2007 (P63).
15. Wang Anjian, Wang Gaoshang, et al. Energy and National Economic Development [M]. Beijing: Geological Publishing House, 2008 (P70).
16. Liu Chunhong. Analysis on the Current Situation of the Coal Industry in Shanxi Province and Its Strategies [J]. China Energy, 2003, 25 (12): pp. 36–38 (P82).
17. China Electricity Council. Annual Development Report of China's Power Industry 2013 [M]. Beijing: China Market Press, 2013 (P85).
18. China Electricity Council. China's Electric Power in the Thirty Years of Reform and Opening-up [M]. Beijing: China Electric Power Press, 2008 (P85).
19. China Electricity Council. Annual Development Report on China's Power Industry 2014 [M]. Beijing: China Market Press, 2014 (P85).




Z. Wang (ed.), *Constraints and Solutions for Energy and Electricity Development*,
https://doi.org/10.1007/978-981-10-8304-4

20. State Grid Energy Research Institute. Annual Report on the Analysis of Energy and Power [R]. Beijing: China Electric Press, 2013 (P90).
21. Shang Lei, Liu Zhongbo. A Study on the Causes of Crisis and Countermeasures in the Coal Industry [J]. Journal of Beijing University of Chemical Technology (Social Sciences Edition), 2008, 64 (4): pp. 24–27 (P99).
22. China National Coal Association. Research Report on the Development of China's Coal Industry 2013 [R]. Beijing: China Economic Publishing House, 2013 (P104).
23. Wu Qiang, Liu Wenchang, Liduo. Theories and Practice of Research in the Mine Environment [M]. Beijing: Geological Publishing House, 2005 (P104).
24. Qiu Ruihua. Discussion on the Development of Circular Economy in Coal Enterprises [J]. Journal of Huainan Vocational Technical College, 2009, 30 (9): pp. 48–50 (P113).
25. Li Xinghua, Duan Lei, et al. Design and Application of a Dilution Sampling System for Measuring Particulate Matter from Stationary Combustion Sources [J]. Journal of Environmental Sciences, 2008, 28 (3): pp. 458–463 (P117).
26. Wang Zhixuan. Zhu Fahua, Liu Simei, et al. Environmental Impacts of $SO_2$ and Control Measures in Thermal Power Plants [M]. Beijing: China Environmental Science Press, 2002 (P118).
27. Wang Zhixuan. Problems on the Emission Reduction of $SO_2$ in Thermal Power Plants [J]. China Power, 2008, 41 (2): pp. 44–47 (P121).
28. Wang Zhixuan et al. Current Situation of Control of Air Pollutants from Coal-fired Power Plants [M]. Beijing: China Electric Power Press, 2009 (P122).
29. Cheng Junfeng, Zeng Hancai, Han Jun. Emission and Control of Trace Heavy Metal in Boilers of Coal-fired Power Plants [J]. Thermal Power Generation, 2002, 31 (2): pp. 23–30 (P126).
30. Stein ED, Cohem Y, Winer AM. Environmental distribution and transformation of mercury Compounds [J]. Crit Rev Environ Sci Techno, 1996, (26): pp. 1–43 (P126).
31. Wang Zhixuan. Prospects of Energy-saving and Emission Reduction for the Power Sector During the "Twelfth Five-Year Plan" Period [J]. Power Decisions Reference, 2009, 53 (4): pp. 14–20 (P130).
32. Edited by Chinese Society of Hydroelectric Engineering. 60 Years of Hydropower in China (1949–2009) [M]. Beijing: China Electric Power Press, 2009 (P147).
33. Zhou Jianping, Qian Gangliang. The planning and development status of the thirteen hydropower bases [J]. Journal of water conservancy and hydropower construction, 2011, (124): pp. 1–7 (P164).
34. Zhou Jianping, Qian Gangliang. Planning and the Development Status of the 13 Large-sized Hydropower Bases in China [J]. Water Conservancy and Hydropower Construction, 2011, 1: pp. 1–7 (P169).
35. Zhang Guoliang. China's Water Supply and Demand in the 21$^{st}$ Century [M]. Beijing: China Water & Power Press, 1999 (P176).
36. Zhang Yue. China's Water Resources and Sustainable Development [M]. Nanning: Guangxi Science & Technology Press, 2000 (P176).
37. Li Guangxin et al. Guide Series on Ecological Environmental Protection [M]. Department of Environmental Protection of Shanxi Province (P178).
38. Lu Jianbin. Analysis on the Influence of Coal Mining on the Groundwater Resources in Jinzhong City [J]. Ground Water, 2009, 31 (3): pp. 58–59 (P195).
39. Wang Chunhong, Hao Binbin, Meng Yushuang. Balancing Coal Resources Development and the Regional Environment in Shendong Coal Base [J]. Journal of Hebei University of Engineering (Natural Science Edition), 2009, 26 (1): pp. 19–20 (P207).
40. Hou Guangcai, Zhang Maosheng. Research on the Exploration of Groundwater in Ordos Basin [M]. Beijing: Geological Publishing House, 2008 (P210).
41. Li Xiaoyun, Zhang Hua, Xi Bing, et al. Water Utilization and Water-saving Technology in Thermal Power Plants [M]. Beijing: China Water & Power Press, 2008 (P215).

42. Compiled by Control Division of Water Resources under Ministry of Water Resources, China Institute of Water Resources and Hydropower Research, the Second Editorial Office of China Standards Press. A Collection of Water Intake and Use Quota Standards and Laws & Regulations (Volume 1 and Volume 2) [M]. Beijing: China Standards Press, 2006 (P218).
43. Song Xuan, Geng Leihua, Du Xia, et al. Analysis on Water Intake & Consumption and Quotas in China's Thermal Plant Industry [J]. Journal of Water Resources and Water Engineering, 2008, 19 (6): pp. 64–66 (P218).
44. Jiang Zisheng, Han Mailiang. Water Resources Utilization and Countermeasures of Thermal Power Units [J]. Huadian Technology, 2008, 30 (6): pp. 1–5 (P220).
45. National Development and Reform Commission. China's Policies and Actions for Addressing Climate Change (2013). Accessed via http://qhs.ndrc.gov.cn/zcfg/201311/t20131107_565920.html (P227).
46. Liu Xiuli et al. Greenhouse Gas Emission Trends in China and the U.S. and Calculation of China's Energy-saving and Emission Reduction Potentials [J]. Research and Investigation, 2009, 9: pp. 14–16 (P235).
47. Lin Boqiang. Power Demands and Key Principles of Scientific Planning in the "Twelfth Five-Year Plan" Period [J]. Power Decisions Reference, 2009, 53 (4): pp. 3–11 (P310).
48. Wang Qingyi. Energy Science and Technology in the 21$^{st}$ Century [J]. Electric Power Technologic Economics, 2008, 2: pp. 11–17 (P328).
49. Xu Weiyi, Miao Xiangjun, Qiu Jiarong, He Weiping, He Ruilin, Guo Zongyong. On the Significance of and Measures for the Protection of Fish in Nujiang River [J]. Yunnan Agriculture, 2008, 3: pp. 24–26 (P328).
50. Editorial Board of China's Coal Economy Research Collections. Coal Economy Research Collections 2008 [J]. Beijing: Coal Industry Publishing House, 2008 (P328).

# Index




Z. Wang (ed.), *Constraints and Solutions for Energy and Electricity Development*,
https://doi.org/10.1007/978-981-10-8304-4

MIX
Papier aus verantwortungsvollen Quellen
Paper from responsible sources
FSC® C105338

If you have any concerns about our products,
you can contact us on
**ProductSafety@springernature.com**

In case Publisher is established outside the EU,
the EU authorized representative is:
**Springer Nature Customer Service Center GmbH**
**Europaplatz 3, 69115 Heidelberg, Germany**

Printed by Libri Plureos GmbH
in Hamburg, Germany